Graduate Texts in Contemporary Physics

Based on Lectures
at the University of Maryland
College Park

Graduate Texts in Contemporary Physics

R.N. Mohapatra: **Unification and Supersymmetry: The Frontiers of Quark-Lepton Physics**

R.E. Prange and S.M. Girvin (eds.): **The Quantum Hall Effect, 2nd ed.**

M. Kaku: **Introduction to Superstrings**

J.W. Lynn (ed.): **High Temperature Superconductivity**

H.V. Klapdor (ed.): **Neutrinos**

J.H. Hinken: **Superconductor Electronics: Fundamentals and Microwave Applications**

High Temperature Superconductivity

Edited by
Jeffrey W. Lynn

With Contributions by
Philip B. Allen, Fernand D. Bedard, Dietrich Belitz, Jack E. Crow, Richard A. Ferrell, Jeffrey W. Lynn, Nai-Phuan Ong, Anthony Santoro, Robert N. Shelton, and Ching-ping S. Wang

With 125 Illustrations

Springer-Verlag
New York Berlin Heidelberg
London Paris Tokyo Hong Kong

Jeffrey W. Lynn
Center for Superconductivity Research
Department of Physics
University of Maryland
College Park, MD 20742-4111, USA

Library of Congress Cataloging-in-Publication Data
High temperature superconductivity / Jeffrey W. Lynn, editor.
p. cm. — (Graduate texts in contemporary physics)
Includes bibliographical references.
ISBN 0-387-96770-2 (alk. paper)
1. High temperature superconductivity. I. Lynn, Jeffrey W.
II. Series.
QC611.98.H54H5433 1990
537.6′23—dc20 90-9436

Printed on acid-free paper.

Camera-ready text provided by the editor.
Printed and bound by R.R. Donnelley & Sons, Harrisonburg, Virginia.
Printed in the United States of America.

9 8 7 6 5 4 3 2 1

ISBN 0-387-96770-2 Springer-Verlag New York Berlin Heidelberg
ISBN 3-540-96770-2 Springer-Verlag Berlin Heidelberg New York

To

The Memory of Captain Robert C. Mayo, USN

and

My Gorgeous Mother-In-Law

Preface

One of the most exciting developments in modern physics has been the discovery of the new class of oxide materials which have high superconducting transition temperatures. Systems with T_C well above liquid nitrogen temperature are already a reality, and there is evidence for new materials with higher T_C's yet. Indeed, the science fiction allusion of a room-temperature superconducting system, which just a short time ago was considered ludicrous, appears today to be a distinct possible outcome of materials physics research.

These materials are interesting for two basic reasons. First, they present exciting new physics. A large variety of experimental techniques are being employed to investigate these materials, yet the physical mechanism responsible for the high T_C's has not been identified. Speculation, though, is rampant. The basic physics of these systems is being explored, and the level of effort is immense. The phenomenal popular interest, on the other hand, stems from the enormous potential of these materials for practical applications. A number of devices have already been fabricated, which make it clear that oxide superconductors could have a substantial economic impact. They are also intrinsically capable of supporting large critical currents, and if high-current devices can be successfully developed, then these new materials should have a diversity of applications. Nature, however, has been reluctant in revealing her secrets, and the road to economic success is a difficult one. On the optimistic side, though, it is clear that if room-temperature superconductivity ever becomes a reality, then superconductivity will have a profound influence on our everyday lives which could rival developments such as the transistor and the laser.

For both of the above reasons, a large number of scientists, engineers, and technicians are being thrust into the field of superconductivity. To address the need for an advanced course on superconductivity, we arranged a graduate-level course at the University of Maryland in the Fall of 87. The purpose of the lectures was to reach graduate students as well as the many researchers at the University and in the surrounding Washington area. It was quite difficult to find an appropriate text for such a course, since there are very few books on superconductivity which are presently in print, and most of these are out-of-date and/or inappropriate as a general text on oxide superconductivity. We therefore decided to undertake the production of an up-to-date introduction to superconductivity, written in the style of a textbook, and then a set of chapters on the new high T_C materials. A series of introductory lectures was given at the beginning of the course, and then experts in the field of superconductivity were invited to present an extended lecture, and follow up with a written chapter. The result is the present book.

The first chapter is quite basic. However, for the beginner I would recommend first reading an introduction to superconductivity such as is found in solid state texts like Kittel or Ashcroft and Mermin. Of course, it may be skipped by the more advanced reader. The next two chapters review the essential properties of "conventional" superconductors: Chapter 2 is devoted to the theory of type-II superconductors, since the new oxides are all strong type-II materials, and Chapter 3 discusses the Josephson effect both because of the interesting quantum physics aspect as well as the important applications that derive from this effect.

Chapters 4-9 are devoted to the properties of the oxide superconductors, and are a mix of experiment and theory. Chapter 10 gives an overview of applications. These have been very difficult chapters to write because of the enormous volume of literature already available, because of the incredible pace of work presently under way worldwide, and because of the pressing needs of our own research programs. With a tutorial text in mind, we found it rather impossible to cite all the relevant articles: The citations alone could easily have filled the approximately 400 page limit. There are quite a number of conference proceedings which are already available, and there will be many more to come, and more extensive references can be found there. Nevertheless, we would like to apologize in advance for any work which may have been overlooked during the preparation of the present chapters.

I would like to thank my coauthors for the excellent lectures they presented, and for their assistance and encouragement during the preparation of these chapters. I would also like to thank Victor Korenman for his critical reading of the first chapter, and Richard Prange for his verbal and compensatory encouragement.

Finally, I would like to thank Cay Horstmann, who was very helpful in his assistance with the word processing program which produced this book. All the chapters were produced using ChiWriter (Horstmann Software). I would also like to thank Julian Noble (Noble House Software) for his assistance with the program that produced the index (Indexchi).

College Park

Jeffrey W. Lynn

Contents

Contributors

Philip B. Allen
Department of Physics
State University of New York
Stony Brook, NY 11794

Fernand D. Bedard
Department of Defense
Fort Meade, MD 20755

Dietrich Belitz
Department of Physics, and
Materials Science Institute
University of Oregon
Eugene, OR 97403

Jack E. Crow
Physics Department
Temple University
Philadelphia, PA 19122

Richard A. Ferrell
Center for Superconductivity
Research
Department of Physics
Universtiy of Maryland
College Park, MD 20742

Jeffrey W. Lynn
Center for Superconductivity Research
Department of Physics
University of Maryland
College Park, MD 20742

Nai-Phuan Ong
Department of Physics
Princeton University
Princeton, NJ 08544

Anthony Santoro
Reactor Division
National Institute of Standards
and Technology
Gaithersburg, MD 20899

Robert N. Shelton
Department of Physics
University of California–Davis
Davis, CA 95616

Ching-ping S. Wang
Center for Superconductivity Research
Department of Physics
University of Maryland
College Park, MD 20742

CHAPTER 1

Survey of Superconductivity

Jeffrey W. Lynn

1.1. Introduction

The venerable subject of superconductivity was discovered over half a century ago by Kammerlingh Onnes [1], but until quite recently it was strictly a low temperature phenomenon. The discovery of the new oxide superconductors, with transition temperatures up to 125 K or higher, has generated tremendous excitement for two reasons. First, they open a new temperature realm for superconducting devices which should have widespread commercial applications, and these potential benefits have attracted phenomenal attention from the public. Second, the conventional electron-phonon interaction appears not to be the origin of the superconductivity in these materials, leaving the fundamental physics open to investigation. Indeed, it is becoming apparent that many of the properties of these new materials are unusual, and a proper understanding will require developing and extending concepts from many areas of condensed matter physics. Nevertheless, the superconducting state appears to be associated with a pairing of electrons, and hence the overall superconducting behavior of the new systems will be similar in many respects to the conventional systems. In fact most of the familiar phenomena which are a manifestation of the superconducting state—persistent currents, Josephson tunneling, vortex lattice—have been established in the new materials. Thus it will be quite valuable to review the basic phenomenology of superconductivity, which will serve as a foundation for the subjects in the remainder of the book. To this end the present Chapter presents a general overview of superconductivity. In addition to this general review, a more detailed discussion of type-II systems will be given in Chapter 2 since the new materials are all strong type-II superconductors, while Chapter 3 is devoted to the theory of the Josephson effect because it is an interesting macroscopic quantum phenomenon which is the basis for many applications. The remainder of the book addresses the properties of the oxide superconductors.

1.2. dc Electrical Resistance

For a normal metal the resistance to the flow of electrical current is caused by the scattering of electrons in the system. One source of scattering is from other electrons in the material, and another is from elementary excitations in the systems such as lattice vibrations (phonons). This electron scattering is intrinsic to the material, and gives rise to the intrinsic electrical resistivity ρ_I. Resistance can also be caused by defects in the lattice such as impurities, grain boundaries, and twin domains, which depend on the quality of the sample. The overall resistivity can usually be written as (Mattheissen's rule)

$$\rho(T) = \rho_I + \rho_{Defects} \quad . \tag{1.1}$$

The extrinsic term $\rho_{Defects}$ is typically insensitive to the temperature, while the intrinsic term ρ_i can be strongly temperature dependent. For example, at low temperatures the electron-phonon interaction yields $\rho_I \propto AT^5$.

The general behavior of a normal conductor and a superconductor is shown in Fig. 1.1. The superconducting state is associated with the precipitous drop of the resistance to an immeasurable small value at a specific critical temperature T_c. In the example shown we have chosen the resistance above T_c to be considerably higher for the superconductor than for the non-superconductor to illustrate an important trend in metals: Systems which are good ordinary conductors, such as the noble metals Cu, Ag, and Au, do not become superconducting, while many poor conductors are good superconductors. The reason for the characteristically high resistivity above T_c in conventional superconductors originates from a strong electron-phonon interaction which causes the large electron scattering rate, and it is this same interaction which is responsible for the electron pairing in the superconducting state. Hence a strong electron-phonon interaction produces a relatively high superconducting transition, whereas a weak interaction will be too feeble to produce an electron pairing. The oxide materials are also quite poor electrical conductors in the normal state, which indicates that a strong electron scattering mechanism operates in these systems as well.

The most familiar property of a superconductor is the lack of any resistance to the flow of electrical current. Classically we call this perfect conductivity. However, the resistanceless state is much more than just perfect conductivity, and in fact cannot be understood at all on the basis of classical physics; hence the name *super*conductor. It is not often that something exists in nature which is better than perfect, and it is enlightening to delineate the difference between a perfect conductor and a superconductor.

Consider then a classical gas of electrons. The resistivity (shown in Fig. 1.1 by the dashed curve) is given by the Drude result:

$$\rho_I(T) = \frac{m}{ne^2\tau} \tag{1.2}$$

where m is the electron mass, e is the charge, n is the density, and τ is the mean time between collisions. We can take this result over to a metal by noting that in a perfect crystal as the temperature approaches absolute zero there are progressively fewer excitations from which electrons can scatter, and we find that $\tau \to \infty$ as $T \to 0$. This gives $\rho_I(T=0) \equiv 0$, and classically the conductivity is perfect.

Now consider initiating a current in a closed loop of wire. For a perfect conductor we might at first expect the current to continue forever. However, the electrons circulating in the loop of wire are in an accelerating reference frame, and accelerating charges radiate energy. Including these classical radiation effects

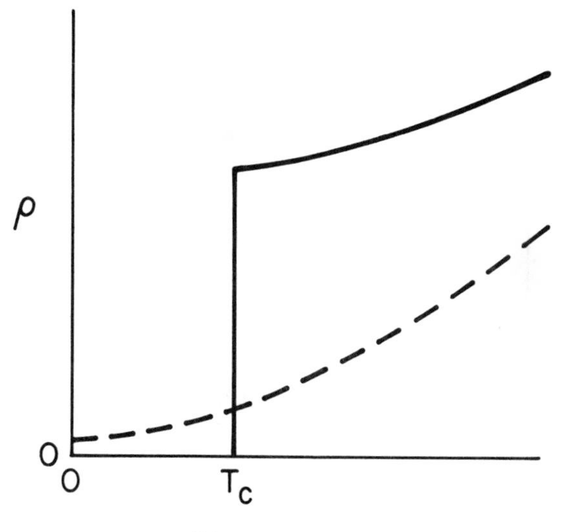

Fig. 1.1

yields a current which decays with time, and hence there is resistance. In a superconductor, however, there is no observable decay of the supercurrent (with an experimental half-life exceeding 10^5 years). The observation of supercurrents which do not decay in time is an explicit demonstration that superconductivity is a macroscopic quantum mechanical phenomenon.[1] Although in some cases classical considerations may seem to provide insight into the behavior of superconductors, it is important to keep in mind that a proper and complete understanding must rest in the quantum theory.

1.3. Perfect Diamagnetism

Soon after the discovery of superconductivity, Onnes observed that the electrical resistance reappears if a sufficiently large current is passed through the conductor, or if a magnetic field above a critical value is applied to the sample. The phase diagram for a typical elemental (type-I) superconductor is shown in Fig. 1.2. Note that the value of the critical field H_c approaches zero in a continuous fashion as a function of temperature, which contrasts with the apparent sudden drop of the electrical resistivity at T_c shown in Fig. 1.1. This apparently discontinuous drop in R can be

[1]This is really just a macroscopic analog to the Bohr atom, which of course is unstable classically, but stable quantum mechanically.

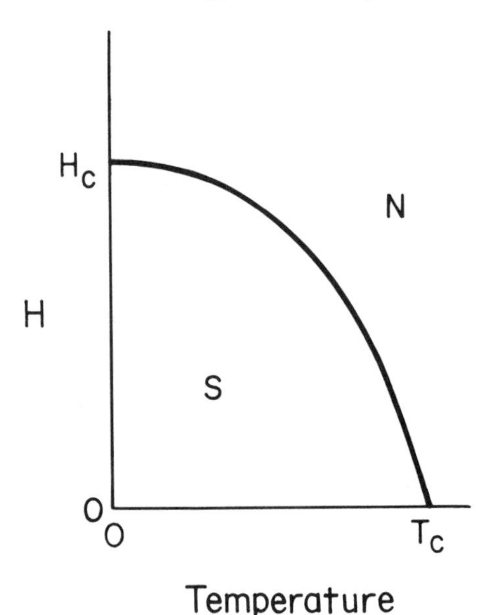

Fig. 1.2

misleading: All the properties of the superconducting state in fact develop gradually as shown in the Fig. 1.2, a behavior which is characteristic of a second order or continuous phase transition. The reason for the apparent discontinuity in the resistivity originates from the large intrinsic length scale associated with fluctuations in the superconducting order parameter, and we will discuss this further below.

For a simple (type-I) superconductor there is a direct relation between the critical current and the critical field, which can be appreciated by the following example. Consider a superconductor which is initially in a field-free region, and then apply a magnetic field to the system. Since the initial field is zero, **H** is a function of time, and we know from Maxwell's equations that a time-dependent magnetic field will generate an electric field. In an ordinary metal this gives rise to eddy currents, but for a superconductor persistent currents will be set up. The currents in turn generate a magnetic field of their own, which opposes the applied field. A simple analysis of this problem starts with Newton's law:

$$m^* \frac{d\mathbf{v}}{dt} = -e\mathbf{E} \quad , \tag{1.3}$$

where m^* is the effective mass of the electron, and $-e$ is the elementary charge. If we multiply both sides by $-en_s$, where n_s is the density of electrons, then we have

$$\frac{d\mathbf{j}}{dt} = \frac{n_s e^2}{m^*} \mathbf{E} \quad , \tag{1.4}$$

where $\mathbf{j} = -en_s\mathbf{v}$ is the induced current. Taking the curl of both sides gives

$$\frac{d}{dt}\left[\nabla\times\mathbf{j}\right] - \frac{n_s e^2}{m^*} \nabla\times\mathbf{E} = \frac{d}{dt}\left[\nabla\times\mathbf{j} + \frac{n_s e^2}{m^* c} \mathbf{B}\right] = 0 \tag{1.5}$$

where we have used Faraday's law

$$\nabla\times\mathbf{E} = -\frac{1}{c}\frac{\partial \mathbf{B}}{\partial t} \quad . \tag{1.6}$$

We may relate **B** and **j** by another of Maxwell's equations,

$$\nabla\times\mathbf{B} = \frac{4\pi}{c}\mathbf{j} + \frac{\partial \mathbf{D}}{\partial t} \quad , \tag{1.7}$$

which is particularly simple at low frequencies where the displacement current may be neglected. The double cross product may be rewritten with use of the vector identity $\nabla \times (\nabla \times \mathbf{B}) = \nabla(\nabla \cdot \mathbf{B}) - \nabla^2\mathbf{B}$. Since $\nabla \cdot \mathbf{B} \equiv 0$, we obtain finally

$$\frac{d}{dt}\left\{\nabla^2\mathbf{B} - \frac{4\pi n_s e^2}{m^* c^2}\mathbf{B}\right\} = 0 \quad . \tag{1.8}$$

Clearly any **B** field which is time independent will satisfy the above equation. Hence if $\mathbf{B} = 0$ initially, the field inside the material will remain zero when a field is applied.

The magnetization which is induced in the system can be obtained via

$$\mathbf{B} = \mathbf{H} + 4\pi\mathbf{M} = \mathbf{H} + 4\pi\chi\mathbf{H} \equiv 0 \quad , \tag{1.9}$$

from which we obtain the result that the susceptibility $\chi = -1/4\pi$. The response is termed diamagnetic since $\chi < 0$, and a superconductor is a perfect diamagnet since the induced magnetization completely cancels the applied field. The total exclusion of the magnetic field from the interior of a bulk superconductor is a unique characteristic of superconductivity.

Now consider a similar experiment where we first apply a field to the sample in the normal state, and then cool it below T_c. The classical result dictates that the field will still be time independent, so that there would be no change in **B**, and the flux will be trapped in the conductor. However, Meissner and Oschenfeld [2] discovered in 1933 that the magnetic field is completely expelled from the interior of the superconductor, in contradiction with the classical expectation. The *Meissner Effect* is a purely quantum mechanical phenomenon.

1.3.1. London Theory

The London brothers [3] provided an explanation for the Meissner effect by proposing that not only must the time derivative of the expression in Eq. (1.8) be zero, but the bracket itself must vanish. Hence we have

$$\nabla^2\mathbf{B} - \frac{4\pi n_s e^2}{m^* c^2}\mathbf{B} = \nabla^2\mathbf{B} - \frac{1}{\lambda_L^2}\mathbf{B} = 0 \quad . \tag{1.10}$$

The specific form of the solution to Eq. (1.10) depends on the particular geometry and boundary conditions, but is typically of the form $\mathbf{B} \propto \exp(-r/\lambda_L)$. Hence the field inside the superconductor decays exponentially with distance, with a characteristic length scale given by the London penetration depth λ_L:[2]

$$\lambda_L = \left\{ \frac{m^* c^2}{4 \pi n_s e^2} \right\}^{\frac{1}{2}} . \tag{1.11}$$

For typical elemental superconductors $\lambda_L \sim 500$ Å, and hence for bulk specimens we have $\mathbf{B} = 0$ in the interior.

In the above expression for λ_L, n_s is the density of "superconducting electrons", and has a natural upper bound which is the total density of conduction electrons. However, since the superconducting transition is a continuous transition, we may anticipate that the density of superconducting electrons will be directly related to the order parameter for the superconductor and hence the London penetration depth will be temperature dependent. In particular, $n_s \to 0$ as $T \to T_c$, and accordingly λ_L will become progressively longer as the transition is approached, and the field will penetrate more and more deeply into the interior of the sample. As $T \to T_c$, $\lambda_L \to \infty$ and the field will penetrate uniformly. Thus there is only flux exclusion for $T < T_c$, and the onset of the diamagnetic susceptibility is gradual (in contrast to the resistance as shown in Fig. 1.1).

We may also obtain a description of the currents induced in the superconductor by the magnetic field. It is straightforward to return to Eq. (1.5) and eliminate $\mathbf{B}$ to obtain an equation for $\mathbf{j}$. The differential equation for the current density analogous to Eq. (1.10) is then

$$\nabla^2 \mathbf{j} - \frac{1}{\lambda_L^2} \mathbf{j} = 0 . \tag{1.12}$$

Eq. (1.12) is identical in form to Eq. (1.10) for the magnetic field, and gives a current density for the induced supercurrents

[2]This analysis assumes that the response of the gas of electrons to an electric field is local, so that the current at any point is determined solely by the value of the electric field at that point. For the spatially nonuniform fields we are considering here, however, the response of the electron gas should be considered nonlocal. Further discussion can be found in the text by Tinkham.

which also decreases exponentially with distance into the bulk of the superconductor. Hence the induced current density is restricted to the surface of the sample.

1.3.2. Magnetic Levitation

One of the most fascinating demonstrations of superconductivity is the levitation of a superconducting particle over a magnet (or vice versa). Typically this is done by dropping a particle of one of the high temperature materials in a dish of liquid nitrogen, with a magnet underneath, and watching the particle jump and hover above the magnet when the temperature drops below T_c. In fact one can arrange the system so that the particle jumps out of the liquid nitrogen, warms above T_c, drops back onto the magnet, and then repeats the cycle. Here we perform a simple analysis to understand the essential features of this phenomenon.

The repulsion of the particle from the magnet is caused by the flux exclusion from the interior of the material. We assume for simplicity that the particle is spherical with radius R and that $R \gg \lambda_L$ so that initially we may neglect surface effects. We will also neglect the (typically very small) susceptibility of the sample in the normal state. Then the difference in free energies (per unit volume) of the normal and superconducting states is

$$F_N - F_S = \frac{B_c^2 - B^2}{8\pi} , \tag{1.13}$$

where B is the average field inside the sample in the normal state and B_c is the critical field. Note that the difference in free energies with no field applied is simply related to the maximum field that the superconductor can expel, and hence B_c is known as the thermodynamic critical field (for both type-I and type-II systems). For a simple type-I system the two free energies become equal as B approaches B_c, and we have a transition from the superconducting to the normal state.

To produce a force on the sphere we of course need a field gradient, so let's assume that B decreases as $1/r$ in the vertical direction as shown in Fig. 1.3.[3] We take h to be the height of the sphere above the magnet, and also assume that the average value of B may be taken at the center of the sphere (i.e. $h \gg R$). Then the magnetic free energy is just

$$F_M = \frac{B^2(a)}{8\pi}\frac{a^2}{h^2}V \tag{1.14}$$

where V is the volume of the sphere, a locates the surface of the magnet, and $B(a)$ is the value of the field at the surface of the magnet. The gravitational potential energy, on the other hand, is just $mgh = \rho Vgh$. Minimizing the total free energy with respect to h then gives the equilibrium position h_E:

$$h_E = \left[\frac{B^2(a)\ a^2}{4\pi\rho g}\right]^{1/3} . \tag{1.15}$$

Note that this result does not depend on the size of the particle, since both energies are directly proportional to V, and in fact the only materials-dependent parameter is the density ρ.

The fact that h_E is independent of R in the above calculation is only valid in the regime that $h \gg R \gg \lambda_L$. If we let R increase, with h_E constant, then the bottom of the sphere will approach the magnet, and eventually it will sit on the magnet. We may also

[3]This is a completely *ad hoc* assumption to simplify the calculation, and we also are not addressing the question of the stability of the sphere parallel to the surface of the magnet. Recall that we must satisfy $\nabla\cdot\mathbf{B} = 0$, and in fact directly over the center of the magnet will be a point of unstable equilibrium.

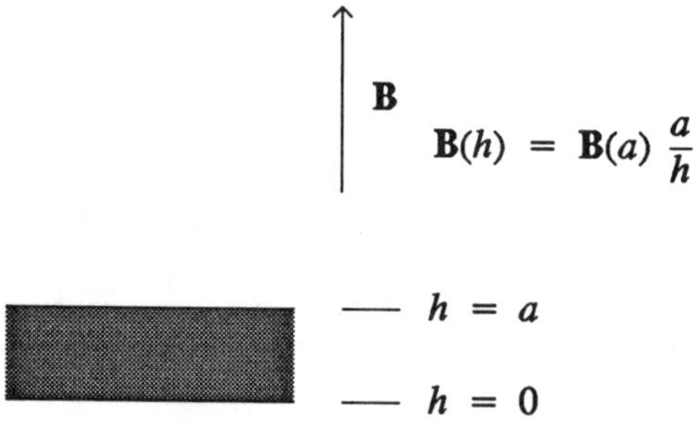

B

$\mathbf{B}(h) = \mathbf{B}(a)\,\frac{a}{h}$

$h = a$

$h = 0$

Fig. 1.3

exceed the critical field near the bottom of the sphere, and hence lose a portion of the energy given in Eq. (1.14) in the region where the sphere is in the normal state.[4] The effective superconducting volume thus will be reduced, and hence h_E will decrease. A similar result will occur in the opposite extreme, when $\lambda_L \gtrsim R$. The flux will penetrate into the sample, and the effective volume will again be reduced, reducing h_E. Hence very small particles will not levitate.

Finally, if we raise the temperature towards T_c, λ_L will increase, and eventually λ_L will become comparable to R. Hence we expect h_E to decrease with increasing temperature, and eventually the sphere will once again rest on the magnet.

1.3.3. Flux Quantization

The current density in any conductor is defined by $\mathbf{j} = nq\mathbf{v}$, where n is the density of carriers, q is the charge, and $\mathbf{v}$ is their average velocity. In the presence of a magnetic field we can write this in terms of the vector potential[5] $\mathbf{A}$ as

$$\mathbf{j} = nq\left[\frac{\mathbf{p}}{m} - \frac{q}{mc}\mathbf{A}\right] \quad . \tag{1.16}$$

If we integrate around a closed path deep inside the superconductor where $\mathbf{j} = 0$ and hence $\oint \mathbf{j}\cdot d\mathbf{l} = 0$, use the relation $\oint \mathbf{A}\cdot d\mathbf{l} = \Phi$, and apply the Bohr-Sommerfeld quantization condition as London did [4], then we find that the magnetic flux is quantized:

$$\Phi_o = \frac{hc}{2e} = 2.07\cdot 10^{-7} \text{ gauss-cm}^2 \tag{1.17}$$

where the charge q for a Cooper pair is (now known to be) $2e$.

Another way to obtain the quantization condition is to employ the theory of Ginzburg and Landau [5], which is a general thermodynamical approach to the theory of phase transitions. To develop a macroscopic theory for the superconducting phase transition they introduced the insightful concept of a complex wave

[4]For a type II superconductor we will also get flux penetration above H_{c1}, although the penetration will not be complete.

[5] $\mathbf{B} = \nabla\times\mathbf{A}$, and in the London gauge $\nabla\cdot\mathbf{A} = 0$.

function to describe the superconducting order parameter, with the density of superconducting electrons being given by $n_s \propto |\Psi|^2$. It is sometimes helpful to think of Ψ as the wave function for a Cooper pair, although the Ginzburg-Landau theory was developed well in advance of the microscopic BCS theory.

If we write the wave function explicitly in terms of a magnitude and a phase as $\Psi = \psi \exp(i\phi)$, then the current density can be written as

$$\mathbf{j} = \left[\frac{2e^2}{mc}\mathbf{A} + \frac{e\hbar}{m}\nabla\phi\right]|\psi|^2 . \tag{1.18}$$

If we now integrate this around a closed loop deep inside a superconductor, where $\mathbf{B} = 0$, we find

$$\oint \mathbf{j}\cdot d\mathbf{l} = 0 = |\psi|^2 \oint \left[\frac{2e^2}{mc}\mathbf{A} + \frac{e\hbar}{m}\nabla\phi\right]\cdot d\mathbf{l} \tag{1.19}$$

$$= \frac{2e^2}{mc}\Phi + \frac{e\hbar}{m} 2\pi\ell$$

where $\oint \nabla\phi \cdot d\mathbf{l} = 2\pi\ell$ and ℓ is an integer to make ψ single valued. Solving for Φ we obtain the flux quantum Φ_o given in Eq. (1.17).

1.4. Energy Gap

One of the central features of a superconductor is that there exists an energy gap in the excitation spectrum for electrons, which was first discovered in specific heat measurements. In a normal metal the specific heat at low temperatures is given by

$$C = \gamma T + BT^3 \tag{1.20}$$

where the linear term is due to electron excitations, and the cubic term originates from phonon excitations. Below the superconducting transition, however, the electronic term was found to be of the form $\exp(-\Delta/k_B T_c)$ which is characteristic of a system with a gap in the excitation spectrum of energy 2Δ. The gap is directly related to the superconducting order parameter, and hence we might expect that $\Delta \to 0$ as $T \to T_c$. The energy gap can be directly observed in tunneling measurements from a normal metal into a superconductor, in the electromagnetic absorption spectrum, and in the lifetime of the phonon excitations.

1.5. Electron-Phonon Interaction and Cooper Pairing

It wasn't until the decade of the 1950's that many of the essential microscopic properties of the superconducting state, such as the energy gap in the electron spectrum discussed in the previous section, were discovered. Another very important discovery [6] was that the superconducting transition temperature depends on the isotopic mass of the material. The simplest form of the *isotope effect* gives

$$T_c \propto (M)^{-\frac{1}{2}} \tag{1.21}$$

where M is the ionic mass. The effect demonstrates that lattice vibrations play an essential role in the formation of the superconducting state. We now discuss how the electron-phonon interaction causes the formation of a bound pair of electrons.

The bare interaction between two electrons is given by Coulomb's law,

$$V_q = \frac{4\pi e^2}{q^2} \tag{1.22}$$

where V_q is the potential strength for scattering a pair of electrons with momentum transfer q. The potential is of course repulsive everywhere, and divergent at $q = 0$. However, in a solid the bare interaction will be screened by the rest of the electrons and ions in the solid. A simple approximation is the Thomas Fermi model, which yields an interaction of the form

$$V_q = \frac{4\pi e^2}{q^2 + \kappa^2} , \tag{1.23}$$

where $\lambda_T = 1/\kappa$ is the characteristic (Thomas-Fermi) screening length. Again this is repulsive for all q. In real space this corresponds to an interaction potential having the familiar Yukawa form $V(r) \propto [\exp(-\kappa r)]/r$.

In order to obtain an interaction which has an attractive part to it, and hence at least a chance of producing a bound state, the dynamics of the system must be taken into account. Considering that the lattice deformation caused by the presence of an electron takes a finite time to relax, and can therefore influence a second electron passing by at a later time, an approximate form for the electron-electron interaction potential can be obtained as [Tinkham, Mahan]

$$V(q,\omega) = \frac{4\pi e^2}{q^2 + \kappa^2}\left[1 + \frac{\omega_q^2}{\omega^2 - \omega_q^2}\right] , \tag{1.24}$$

where ω_q is the phonon frequency at the wave vector q. The first term is the usual screened electron potential, and the second term is the phonon-mediated interaction. The effect of the phonons is to overscreen the Coulomb interaction at low frequencies; for $\omega < \omega_q$ the second term in the brackets is negative and dominates the interaction since the denominator is resonant. Hence the phonon-mediated electron-electron interaction is attractive for $\omega < \omega_q$.

The essential question then is whether Eq. (1.24) has any bound states. If we consider two electrons which are restricted to one spatial dimension, then we know that an attractive interaction will yield at least one bound state no matter how weak. In a three dimensional system, however, Eq. (1.24) is typically too weak to possess any bound states. The crucial point Cooper [7] emphasized is that we are not dealing with a two-electron system, but with an N-electron system. The central result he obtained is that the ground state of an electron gas is unstable to the formation of bound pairs if an attractive electron-electron interaction exists, regardless of the strength of the potential. Hence he showed that the weak electron-phonon interaction could in fact produce bound electron pairs.

Consider then two electrons which are in states just above the Fermi level. Since the interaction is attractive, we expect the bound state to have even spatial symmetry, and thus the spin state must be the antisymmetric singlet $(\uparrow,\downarrow)$, with $S^2 = 0$, to satisfy the Pauli principle. The spatial part of the wave function may be written as

$$\psi(\mathbf{r}_1,\mathbf{r}_2) = \frac{1}{\mathcal{V}} \sum_{k_1,k_2} a_{k_1,k_2}\, e^{i\mathbf{k}_1\cdot\mathbf{r}_1 + \mathbf{k}_2\cdot\mathbf{r}_2} \tag{1.25}$$

$$= \frac{1}{\mathcal{V}} \sum_{k,K} a_{k,K}\, e^{i\mathbf{k}\cdot\mathbf{r} + \mathbf{K}\cdot\mathbf{R}} \tag{1.26}$$

where we have introduced the center of mass coordinates **K** and **R** and $\mathcal{V}$ is the normalization volume. Since the interaction potential does not involve these variables, the only effect is to add a free-particle term in the kinetic energy for the center of mass, and we drop it from the subsequent discussion. The Schrödinger equation for this wave function is then

$$\left[-\frac{\hbar^2\nabla^2}{\mu} + V(\mathbf{r}) - E\right] \frac{1}{\mathcal{V}^{\frac{1}{2}}} \sum_{\mathbf{k}} a_{\mathbf{k}} e^{i\mathbf{k}\cdot\mathbf{r}} = 0 \quad , \tag{1.27}$$

where we have the potential interaction completely general for the moment. Multiplying by $\exp(-i\mathbf{k}'\cdot\mathbf{r})$ and integrating, we obtain

$$(2E_{\mathbf{k}}' - E)a_{\mathbf{k}}' + \frac{1}{\mathcal{V}} \sum_{\mathbf{k}} a_{\mathbf{k}} \int e^{i(\mathbf{k}-\mathbf{k}')\cdot\mathbf{r}}\, V(\mathbf{r})d\mathbf{r} = 0 \quad , \tag{1.28}$$

or

$$(E - 2E_{\mathbf{k}}')a_{\mathbf{k}}' = \frac{1}{\mathcal{V}} \sum_{\mathbf{k}} a_{\mathbf{k}}\, V_{\mathbf{k},\mathbf{k}'} \quad , \tag{1.29}$$

where we have written the energies in terms of the original one-electron energies E_k. To solve this analytically, Cooper assumed a very simple interaction potential of the form[6]

$$V_{\mathbf{k},\mathbf{k}'} = \begin{cases} -F/\mathcal{V} & E_F < \mathrm{E} < \mathrm{E}_F + \hbar\omega_D \\ 0 & \mathrm{E} > \mathrm{E}_F + \hbar\omega_D \end{cases} \quad . \tag{1.30}$$

The interaction potential is then independent of **k** and **k**′, and we

[6]In real space this interaction is a delta function, except that the range of momenta where it is operative is restricted. If all k's were allowed, then this would be a genuine delta function (recall that the Fourier transform of a constant is a delta function).

obtain the simple result

$$a_{\mathbf{k}}' = F \frac{\sum_{\mathbf{k}} a_{\mathbf{k}}}{(2E_{\mathbf{k}}' - E)} . \tag{1.31}$$

Summing over $\mathbf{k}'$, the coefficients a_k cancel, and we obtain the eigenvalue equation

$$1 = \sum_{\mathbf{k}} \frac{F}{(2E_{\mathbf{k}} - E)} . \tag{1.32}$$

$$\simeq FN_F \int_{E_F}^{E_F+\hbar\omega_D} \frac{dE_k}{2E_k - E} , \tag{1.33}$$

and we have assumed the density of states N_F at the Fermi level (for one spin state) is constant over the range of phonon energies $\hbar\omega_D$. Integrating, we find

$$E = \frac{2\hbar\omega_D + (2E_F + E_K)\left(1-e^{1/FN_F}\right)}{1-e^{1/FN_F}} \tag{1.34}$$

$$= 2E_F + E_K + \frac{2\hbar\omega_D}{1 - e^{1/FN_F}}$$

where we have put back in the center of mass kinetic energy E_K for completeness. In the weak coupling limit where $FN_F \ll 1$, which implies a weak potential, we obtain

$$E \simeq 2E_F + E_K - \Delta \tag{1.35}$$

where the energy gap Δ is given by

$$\Delta \simeq 2\hbar\omega_D \, e^{-1/FN_F} \tag{1.36}$$

Thus for two electrons interacting with a weak attraction, in the presence of a Fermi sea, there exists a bound state. Since $E_K = 0$ is the minimum energy, the lowest energy state corresponds to one electron with wave vector **k**, and the other with wave vector -**k**. Thus the full description can be denoted by (**k**↑;-**k**↓). Note also that the Fermi sea plays an essential role, and in fact if we let $E_F \to 0$, then $N_F \to 0$, and hence $\Delta \to 0$.

For the new oxide materials there are some important points to note with regard to Cooper pairing and the BCS theory. First, it is clear that all we need is *any* attractive interaction to form a bound state, not just the electron-phonon interaction. This has lead to the proliferation of theories for the high T_c materials based on other possible excitations in the system: polarons, plasmons, magnons, excitons, ···ons, as reviewed in Chapter 9. Another important point is that in the oxide materials the density of states is low, and the Fermi energy is relatively small, while T_c is relatively high. Hence the assumption that $\hbar\omega_D/E_F \ll 1$ may not be valid.

1.6. BCS Theory

The above calculation shows that the electron gas is unstable to the formation of electron pairs if there is an net attractive interaction. This does not provide a description of the superconducting state, however, or even demonstrate that superconductivity exists at all. In particular, if single electrons can scatter as they move through the lattice, causing resistance, why can't a bound pair of electrons scatter?

Cooper's calculation considers only a single pair of electrons interacting above the Fermi sea: The Fermi sea only enters the calculation in restricting the integrals to $k > k_F$. But if the formation of a bound pair of electrons lowers the energy of the system, then many pairs of electrons can be expected to form, until the system can gain no additional energy by pair formation. Hence the electrons play a dual role, one in belonging to the Fermi sea, and another in selecting a partner for pair formation. To obtain a proper description of the electrons, then, the correlations for all the electrons must be built into the N-electron wave function for the system. This is obviously a very complicated many-body problem.

The solution was provided by Bardeen, Cooper, and Schrieffer [8]. Their theory incorporates the assumption of a weak net attractive interaction via the electron-phonon interaction, as given in Eq. (1.30). The BCS theory provides an explicit N-particle wave function for the ground state, which is then used to calculate properties. Since the electrons are highly correlated, the scattering of a Cooper pair involves disturbing many more electrons than just the two which form the pair, and it turns out that making a transition from one N-electron state to another is extremely unlikely. Hence the Cooper pair can move through the system without hindrance.

The original theory contains a number of approximations, such as for the electron-phonon interaction [Eq. (1.30)], but it works very well for simple metals and contains all the essential elements of the superconducting state. The details are beyond the scope of this short survey, but the interested reader is referred to the discussion by Tinkham or Mahan. Here we will simply review some of the central results of the theory.

There is an energy gap in the one-electron spectrum, which at zero temperature is given by Eq. (1.36). With increasing temperature the energy gap monotonically decreases, and approaches zero as $T \to T_c$. In the vicinity of the transition temperature the theory yields

$$\Delta(T) \approx 1.74\, \Delta(0) \left[1 - \frac{T}{T_c}\right]^{\frac{1}{2}} . \tag{1.37}$$

The continuous decrease of the superconducting order parameter to a value of zero is a characteristic of a second order phase transition. In the present case the behavior is controlled by an exponent value of $\frac{1}{2}$, which is a result of mean field theory. Superconductivity is one of the few phase transitions found in nature where mean field theory works quite well.

Along with the gap collapsing to zero, the critical magnetic field above which the superconductivity is quenched follows

$$H_c(T) = H_c(0)\left[1 - a\left(\frac{T}{T_c}\right)^2\right] \tag{1.38}$$

where a is a constant which is weakly dependent on temperature. Like the energy gap, the critical field approaches zero continuously. The above relation is directly related to the energy difference between the normal and superconducting free energies.

The electronic term in the heat capacity at low temperatures is

$$c_s = 1.34\ \gamma T_c \left[\frac{\Delta(0)}{T}\right]^{3/2} e^{-\Delta(0)/T} \quad , \tag{1.39}$$

which is characteristic of a system which possesses an energy gap in the excitation spectrum. On the other hand at the onset of superconductivity there is a jump in the value of the specific heat which is given by

$$\frac{c_s - c_n}{c_n} = 1.43 \quad , \tag{1.40}$$

where c_n is the specific heat in the normal state, just above T_c, and c_s is the specific heat just below T_c. The discontinuity in the specific heat is also a result of the mean field approximation, which works quite well since the critical regime in a bulk superconductor is exceedingly small.

The electronic term in the specific heat is typically quite small. Fortunately, in conventional superconductors the phonon contribution to the specific heat [Eq. (1.20)] is also small since the ordering temperatures are quite low, and hence accurate measurements for c_s can be made. For the oxide superconductors, on the other hand, T_c is not small, and the phonon term dominates c making measurements of c_s difficult (see Chapter 7). In addition, the simple approximation for the phonon term given in Eq. (1.20) is no longer valid, and the detailed lattice dynamical measurements are needed if unambiguous values for c_s are desired.

The BCS theory also makes a prediction for the ordering temperature in terms of the parameters of the theory. The transition is given by

$$k_B T_c = 1.13\hbar\omega_D e^{-1/FN_F} \tag{1.41}$$

where $\hbar\omega_D$ is the Debye energy for the phonons. Since the phonon frequency is inversely proportional to the square root of the mass for a simple metal, this prediction is in agreement with the isotope effect. However, the electron-phonon coupling will also be dependent on the mass, and hence significant deviations from the Eq. (1.21) can be expected, and accurate results must rely on detailed calculations.

Finally, the theory gives a relationship between the zero temperature energy gap and the ordering temperature:

$$\frac{\Delta(0)}{k_B T_c} = 1.764 \quad . \tag{1.42}$$

This prediction is independent of the parameters of the theory, and agrees quite well with many of the simple elemental superconductors.

1.7. Type-II Superconductors

The behavior of a typical elemental superconductor in an applied magnetic field was shown in Fig. 1.2. The magnetic field is screened from the interior of the sample until a critical field H_c is reached, above which the superconductivity is quenched and the field penetrates uniformly into the material. A qualitatively different behavior is exhibited by type-II superconductors, as shown in Fig. 1.4. At sufficiently small applied fields the system is again in the Meissner state, where the field is screened from the interior bulk of the material. Above a critical field H_{c1}, however, there is partial penetration of the field, while the material remains superconducting. The field increasingly penetrates the system as the applied field is increased, until finally above a second critical field H_{c2} the superconducting state is destroyed. The basic mechanism responsible for the superconductivity is the same for both type-I and type-II materials, and indeed many type-I

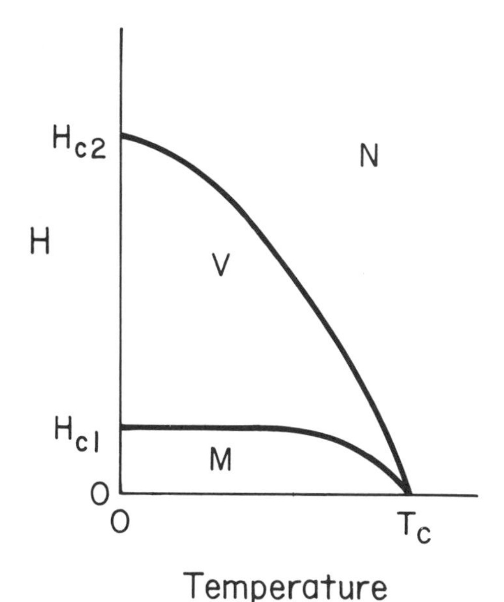

Fig. 1.4. Phase diagram for a type-II superconductor, showing the Meissner (M), Vortex (V), and Normal (N) regions.

materials can be converted to type-II behavior simply by the addition of impurities, but the behavior in a magnetic field is qualitatively different. In the present section we will briefly review the characteristics of type-II systems. Their behavior is discussed in more depth in the next chapter.

The penetration of the field above H_{c1} turns out to be quite nonuniform, and an important question to address is what controls the spatial range over which the properties of the superconducting state, such as the Cooper pair wave function, can vary. A Cooper pair consists of one-electron wave functions with momentum $\pm\mathbf{k}$. If we want to modulate this wave function in the superconducting state, then we must employ states only within $\sim kT_c$ of the Fermi energy in order to avoid excessive kinetic energy costs. For free electrons this gives a spread in energy

$$\Delta E \approx kT_c = \frac{p_f}{m}\Delta p \quad , \tag{1.43}$$

from which we obtain the range of acceptable momentum values

$$\Delta p \sim \frac{kT_c}{v_f} \quad . \tag{1.44}$$

Converting this to a length via the Heisenberg uncertainty principle, we obtain

$$\Delta x \equiv \xi_o = C\,\frac{\hbar v_f}{kT_c} \tag{1.45}$$

where ξ_o is the *coherence length* introduced by Pippard [9]. Basically, ξ_o is the effective 'size' of a Cooper pair. If the superconducting wave function is forced to vary over a region which is smaller than ξ_o then it will be quite costly in energy and hence will adversely affect the superconducting state, while gentle variations over distances much greater than ξ_o can be easily accommodated. The coherence length also depends on the mean free path ℓ of the electrons in the normal state, with ξ given approximately by

$$\frac{1}{\xi} = \frac{1}{\xi_o} + \frac{1}{\ell} \quad , \tag{1.46}$$

where ξ_o is the intrinsic coherence length, and ξ is the effective coherence length. In pure systems where ℓ is large $\xi \approx \xi_o$, while in "dirty" systems (such as alloys) where the mean free path is short ξ can be much smaller than ξ_o.

The coherence length ξ is a fundamental length scale of the superconducting state which is separate from the length scale associated with the London penetration depth λ_L. The ratio of these two length scales plays a central role in the Ginzburg-Landau theory, and defines the dimensionless Ginzburg-Landau parameter κ:

$$\kappa = \frac{\lambda_L}{\xi} \quad . \tag{1.47}$$

It was Abrikosov [10] who showed that the magnetic behavior is controlled by the value of κ. In a pure metal ξ is typically a few thousand angstroms, while λ_L is typically 500 Å, so that we have $\xi \gg \lambda_L$. It is too costly in energy to have the superconducting wave function vary in space on the same length scale as λ_L, and hence we will just have Meissner screening. However, if ξ is small such as in an alloy, with $\xi \ll \lambda_L$, then the superconducting state can be modified and the field can penetrate nonuniformly. Abrikosov showed that the magnetic flux penetrates in discrete vortices, and that the flux in each vortex is quantized in units of the flux quantum Φ_o [Eq. (1.17)]. In the vortex state, then, the field is large in the core of the vortex, and is screened to zero over a length λ_L.

We can obtain an estimate of the critical fields H_{c1} and H_{c2} by considering the energetics of the system. For $H_{applied} = H_{c1}$, the field at the core of the vortex will be H_{c1}, and then it will decrease to zero over a length λ_L. The flux will then be approximately $\pi\lambda_L^2 H_{c1}$, and this must be equal to the flux quantum Φ_o. Hence we have

$$H_{c1} \approx \frac{\Phi_o}{\pi\lambda_L^2} \quad . \tag{1.48}$$

On the other hand, at $H_{applied} = H_{c2}$ the vortices are packed in as tightly as they will go while still maintaining the superconducting state, and hence each will take up an area $\sim\pi\xi^2$. Hence this gives the estimate

$$H_{c2} \approx \frac{\Phi_o}{\pi\xi^2} \quad . \tag{1.49}$$

A quantitative calculation for the critical fields yields

$$H_{c1} \approx \frac{H_c}{\sqrt{2}\,\kappa} \tag{1.50}$$

and

$$H_{c2} = \sqrt{2}\,\kappa\,H_c \tag{1.51}$$

where H_c is the thermodynamic field defined by Eq. (1.13). We see that $H_{c1} = H_{c2}$ when $\kappa = 1/\sqrt{2}$, which defines the crossover from type-I to type-II behavior in terms of the characteristic lengths ξ and λ_L.

Finally, if we take the product of Eq. (1.51) and Eq. (1.50), we obtain

$$H_{c1}\,H_{c2} \cong H_c^2 \quad . \tag{1.52}$$

Hence we see that there is a balance between the lower and upper critical fields: The higher H_{c2} is, the lower H_{c1} is. All the new oxide materials are strong type-II superconductors, with H_{c2} in the megagauss range for some.

1.8. New Topics in Superconductivity

The subject of superconductivity has seen a number of new developments in recent years. One such area of investigation involves superconducting systems which contain magnetic ions. Generally magnetic ions were found to influence the superconducting properties much more strongly than nonmagnetic impurities, and theories have been developed to understand this problem in detail. More recently, systems were discovered where entire magnetic sublattices were contained in superconducting crystal structures. The magnetic sublattices were found to order at low temperatures, and the interaction between the magnetism and the superconducting electrons leads to a rich variety of phenomena including the coexistence of superconductivity and long range magnetic order, long wavelength oscillatory magnetic states, and reentrant superconductivity.

Another pursuit of the superconducting community was to try to find new systems where the conventional BCS-type superconductivity was not applicable. One possibility is in heavy-fermion materials, where the pairing may be based in higher-order angular momentum

($p,d,...$) for the wave function, rather than the s-state (isotropic) Cooper wave function. The other is of course the oxide superconductors, which is the primary subject of this book.

1.8.1. Magnetic Superconductors

A Cooper pair is composed of a time-reversed pair of one-electron states ($\mathbf{k}\uparrow;-\mathbf{k}\downarrow$), and this description provides a good foundation for the theory of superconductivity in pure metals, where the electron mean free path is large. Experimentally it was found that the substitution of one chemical species for another generally does not have a dramatic effect on the superconducting properties, except when the mean free path itself is an essential consideration as we saw in section 1.7. Anderson [11] demonstrated that the pairing of time-reversed states is valid more generally, in particular for so-called "dirty" superconductors where the mean free path is short. Magnetic impurities, on the other hand, represent a perturbation which breaks the time-reversal symmetry, and hence it may have a much more profound effect of the superconducting state.

The interaction of a single magnetic impurity with the conduction electrons can be written as

$$\mathscr{H} = -\mathscr{J}\mathbf{S}\cdot\mathrm{s} \tag{1.53}$$

where $\mathbf{S}$ is the spin of the impurity, $\mathbf{s}$ is the spin of the conduction electron, and $\mathscr{J}$ is the exchange integral. This interaction is not time-reversal invariant, and hence it will split the degeneracy of the Cooper pair in the same way that a magnetic field will. This is the origin of the detrimental effect that magnetic impurities have on the superconducting properties.

For the case of dilute magnetic impurities, Abrikosov and Gorkov [12] worked out a quantitative theory of this spin depairing mechanism. They found that the superconducting transition temperature T_c decreases rapidly with increasing magnetic concentration x. At small x the decrease is linear in x, with a slope which is proportional to the magnitude of the exchange energy $\mathscr{J}$, while at larger values of x the decrease is more rapid and T_c vanishes at a critical concentration x_c. For typical materials $x_c \sim 1\%$, and hence the correlations between magnetic atoms would be quite small in the concentration regime where superconductivity is present. One of the remarkable features that they found was that the energy gap decreased more rapidly with x than T_c, with Δ vanishing at $x_c \cong 0.91\ \mathrm{x}_c$. This is the phenomenon of gapless superconductivity.

In juxtaposition to this typical behavior of magnetic impurities in superconductors, there was another class of materials which behaved quite differently. One of the early systems studied, for example, was (Ce-$\mathscr{R}$)Ru_2, where $\mathscr{R}$ represents one of the heavy rare earth metals such as Gd, Tb, Dy, Er, or Ho. In these systems $\mathscr{J}$ appeared to be zero, and quite large concentrations could be accommodated before there was much effect on the superconducting properties, similar to the alloying behavior observed for the substitution of nonmagnetic impurities. Eventually superconductivity disappeared, but not before magnetic concentrations as large as 30 atomic per cent were present. These concentrations are high enough for substantial magnetic correlations to exist at low temperatures, but the magnetism which developed was of the spin glass variety rather than true long range magnetic order.

It was not until ternary systems such as $HoMo_6S_8$ and $ErRh_4B_4$ were discovered that the coexistence between magnetic order and superconductivity was realized experimentally, and the interaction between these two cooperative phenomena could be explored. In these materials there is a complete sublattice of magnetic ions which is distinct from the sublattice where the superconducting electrons travel. The physical isolation of the magnetic and superconducting electrons yields an exchange energy $\mathscr{J}$ which is very small, and leaves the dipole interaction as the dominant coupling between the magnetism and superconductivity. Since there is no chemical disorder to disrupt the magnetic system, the magnetic entropy compels the system to develop long range magnetic order at sufficiently low temperatures.

If the magnetic interactions are of an antiferromagnetic nature, then antiferromagnetic order will occur in the magnetic ground state. Since there is no macroscopic magnetic (dipolar) field in an antiferromagnet, there is essentially no interaction between the magnetic and superconducting subsystems, and it was found that long range antiferromagnetic order coexisted quite happily with the superconducting state. However, the situation is quite different if the magnetic system prefers ferromagnetic alignment of the moments, since there is a macroscopic magnetization associated with the ferromagnetic state. In essence, the superconductivity cannot tell whether the applied field is being generated externally, or originates from within. It simply knows that it is obliged to screen a magnetic field whose length scale exceeds the London penetration depth λ_L. When the magnetic system initially orders, there is a compromise state in which the magnetization is forced to vary sinusoidally in space, with a characteristic wavelength λ_M given by

$$\lambda_M \sim \left[\gamma\lambda_L\right]^{\frac{1}{2}} \tag{1.54}$$

where γ is a measure of the strength of the ferromagnetism. The larger the value of γ, the longer the magnetic wavelength,[7] while smaller values of λ_L correspond to a stronger superconducting state. This long wavelength sinusoidal magnetism coexists with superconductivity. As the magnetism continues to grow in strength with decreasing temperature, however, the ferromagnetic energy usually dominates, producing a first-order transition where pure ferromagnetism sets in and the superconducting state is destroyed. This is an example of reentrant superconductivity.

Many of the new oxide superconductors will also accommodate the heavy rare earth metals in the lattice, and yield systems which are classic magnetic-superconductor materials. The interaction between magnetism and superconductivity is still an active area of research.

1.8.2. Heavy Fermion Materials

Another class of materials which has been the object of intense research in recent years are f-electron systems such as UBe_{13}, $CeAl_3$, $CeCu_2Si_2$ and URu_2Si_2. At high temperatures (e.g. room temperature and above) these systems consist of f-electron magnetic moments (on the U or Ce) which are disordered, and conduction electrons with electronic properties which are quite similar to normal metals. At low temperatures, however, strong correlations develop between the f-electrons and the conduction electrons. These correlations yield effective masses for the conduction electrons which are one to two orders-of-magnitude larger than the bare electron mass. In addition, strong antiferromagnetic spin correlations exist between the f-electron moments themselves. In some cases these correlations condense into an ordered antiferromagnetic state of the heavy electrons, while in other materials the heavy electrons exhibit superconductivity.[8] There is even evidence that antiferromagnetism and superconductivity coexist under some conditions. Finally, there are materials where no phase

[7]Recall that $\lambda_M \to \infty$ corresponds to true ferromagnetism, while $\lambda_M \sim$ interatomic spacing would correspond to antiferromagnetism.

[8]T_c = 0.9 K for UBe_{13}, 0.65 K for $CeCu_2Si_2$, and 1.5 K for URu_2Si_2. URu_2Si_2 also exhibits an antiferromagnetic phase transition at 17 K. $CeAl_3$, on the other hand, does not undergo any phase transitions.

transitions at all are observed at low temperatures.

It is perhaps not surprising then that many of the low temperature properties of heavy electron systems are unusual. The temperature dependence of the resistivity and Hall effect, for example, are anomalous, and most materials exhibit a *negative* thermal expansion (they shrink on warming) which is very large. Probably the most striking feature, however, is the enormous electronic contribution to the specific heat at low temperatures: Typically specific heat is measured in millijoules/mole-K, while for the heavy fermion systems the appropriate units are joules/mole-K. The magnetic properties are just as remarkable: The susceptibility and magnetostriction, for example, display anomalous variations as a function of temperature and pressure, and are orders of magnitude larger than in ordinary metals.

One of the most revealing characteristics of the heavy fermions systems is their extreme sensitivity to the introduction of impurities into the system. Substituting nonmagnetic Th for U in UBe_{13}, for example, initially depresses the superconducting transition temperature rapidly, much as magnetic impurities were found to depress T_c in "conventional" superconductors as discussed in the previous section. At higher Th concentrations, on the other hand, antiferromagnetism appears, which emphasizes that there is a delicate interplay between the magnetism and superconductivity in these systems. The antiferromagnetism can also be very sensitive to impurities. Substitution of 2% Cu for Zn in UZn_{11}, for example, completely suppresses the antiferromagnetic state.

There is a general consensus now that at least in some of the heavy fermion systems the superconducting pairing is not the conventional isotropic (*s*-wave) state, but rather that the pairs have higher angular momentum components in a manner similar to the superfluid phases of liquid ^{3}He. However, the possible pairing states cannot be classified in terms of their net angular momentum as *s*, *p*, *d*, etc. as they are for liquid ^{3}He, due to crystal symmetry effects and the spin-orbit coupling of the *f*-electrons. In addition, the unusual magnetic and superconducting properties of heavy fermions systems has led to the speculation that the attractive pairing interaction originates from magnetic fluctuations rather than the electron-phonon coupling. The fundamental physics in these materials is still under active investigation, and is every bit as interesting as the oxide superconductors.

1.8.3. Oxides

The sudden escalation in interest in superconductivity is of course due to the dramatic increase in T_c discovered in the oxide superconductors, but it should be noted that the superconducting oxide materials are not really a new phenomenon. Indeed superconductivity was discovered more than twenty years ago in oxygen deficient $SrTiO_{3-x}$ [13]. The pure stoichiometric material ($x = 0$) is a cubic perovskite which is a paraelectric insulator. In fact it is a material which is an incipient ferroelectric, and belongs to a class of materials which exhibit classic soft phonons and structural phase transitions. In the oxygen-deficient state the material is a semiconductor, and becomes superconducting at very low temperatures (below 1 K).

The first "high temperature" oxide superconductor was discovered in the alloy system $BaPb_{1-x}Bi_xO_3$ [14]. Superconductivity exists over the concentration range $0.05 \leq x \leq 0.35$, where the material possesses the tetragonal crystallographic structure. The maximum T_c of $\sim$13 K occurs for $x \approx 0.25$, while for $x > 0.35$ the system is a semiconductor. Very recently the related system $K_{0.4}Ba_{0.6}PbO_3$ also has been found to be superconducting, with a $T_c \sim$ 30 K [15].

The discovery by Bednorz and Müller [16] of superconductivity in the $La_{2-x}Ba_xCuO_4$ was important for two reasons. First, the observed T_c of $\sim$30 K was the highest of any known material at the time, and focussed attention on the oxide superconductors. Second, it unveiled a new class of oxides to be explored. These new ceramic oxides, of course, turned out to be incredibly rich and fertile materials to investigate.

References

[1] H. Kamerlingh Onnes, *Leiden Comm.* **120b, 122b, 124c** (1911).

[2] W. Meissner and R. Ochsenfeld, *Naturwissenschaften* **21**, 787 (1933).

[3] F. London and H. London, *Proc. Roy. Soc (London)* **A149**, 71 (1935).

[4] F. London, *Superfluids*, Vol. I (Wiley, New York, 1950).

[5] V. L. Ginzberg and L. D. Landau, *Sov. Phys. JETP* **20,** 1064 (1950).

[6] E. Maxwell, *Phys. Rev.* **78**, 477 (1950). C. A. Reynolds, B. Serin, W. H. Wright and L. B. Nebsitt, *Phys. Rev.* **78**, 487 (1950).

[7] L. N. Cooper, *Phys. Rev.* **104**, 1189 (1956).

[8] J. Bardeen, L. N. Cooper, and J. R. Schrieffer, *Phys. Rev.* **108**, 1175 (1975).

[9] A. B. Pippard, *Proc. Roy Soc. (London)* **A216**, 547 (1953).

[10] A. A. Abrikosov, *Sov. Phys. JETP* **5**, 1174 (1957).

[11] P. W. Anderson, *J. Phys. Chem. Sol.* **11**, 26 (1959).

[12] A. A. Abrikosov and L. P. Gorkov, *Sov. Phys. JETP* **12**, 1243 (1961).
[13] J. F. Schooley, W. R. Hosler, E. Ambler, J. H. Becker, M. L. Cohen, and C. S. Koonce, *Phys. Rev. Lett.* **14**, 305 (1965).
[14] A. W. Sleight, J. L. Gillson, and P. E. Bierstedt, *Sol. St. Commun.* **17**, 27 (1975).
[15] R. J. Cava, B. Batlogg, J. J. Krajowski, R. Farrow, L. W. Rupp, Jr. A. E. White, K. Short, W. F. Peck and T. Kometani, *Nature* **332**, 814 (1988).
[16] J. G. Bednorz and K. A. Müller, *Z. Phys.* **B64**, 189 (1986).

Introductory References

N. W. Aschroft and N. D. Mermin, *Solid State Physics* (Saunders, 1976), Chap. 34.
C. Kittel, *Introduction to Solid State Physics*, (Wiley, 1976).

Advanced Texts

M. Tinkham, *Introduction to Superconductivity* (McGraw Hill, NY 1975).
P. de Gennes, *Superconductivity of Metals and Alloys*, Benjamin, Menlo Park, CA 1966.
G. D. Mahan, *Many-Particle Physics*, Plenum, New York (1981), Chapter 9.
R. D. Parks, editor, *Superconductivity*, Dekker, New York, 1969.

Magnetic Superconductors

Superconductivity in Ternary Compounds, Vol. II, in *Topics in Current Physics*, edited by M. B. Maple and O. Fischer, Vol. 34 (Springer-Verlag, New York, 1983).

Heavy Fermions

G. R. Stewart, *Rev. Mod. Phys.* **56**, 755 (1984).
Z. Fisk, D. W. Hess, C. J. Pethick, D. Pines, J. L. Smith, J. D. Thompson, and J. O. Willis, *Science* **239**, 33 (1988).
H. R. Ott and Z. Fisk, in *Handbook on the Physics and Chemistry of the Actinides*, ed. by A. J. Freeman and G. H. Lander (North Holland, Amsterdam 1987).
B. R. Coles, *Comp. Phys.* **28**, 143 (1987).
P. Fulde, J. Keller, and G. Zwicknagl, *Solid State Physics* **41**, 1 (1988).

CHAPTER 2

Theory of Type-II Superconductivity

Dietrich Belitz

2.1. Introduction

2.1.1. Motivation

Type-II superconductivity is a prime example of how an old subject, which had been considered closed, at least in principle, many years ago, can not only be revived, but reveal a surprising variety of new physics upon the discovery of a new aspect. In this case, the aspect bearing qualitatively new features was a more thorough approach towards disorder.

Traditionally, the only effect of nonmagnetic disorder had been assumed to be diffusive electron dynamics, with no effect at all on the microscopic interactions. That these assumptions are not always justifiable first became clear in the context of transport in normal metals. In transport theory, if one goes beyond the standard models of non-interacting electrons suffering elastic collisions described in a stochastic approximation, two complications occur. First, elastic collisions are not stochastic. Rather, memory effects and interference phenomena occur, and are more important for higher disorder and lower dimensionality. For non-interacting electrons, this is the realm of localization physics [1,2]. Second, electron-electron interactions have to be taken into account. For clean systems, this is the realm of Fermi liquid theory. If both interactions and disorder are considered simultaneously, qualitatively new phenomena occur due to interplay between disorder and interaction. The first example of this actually dealt with electron-phonon rather than electron-electron interactions, namely Pippard's theory of sound attenuation [3]. For electrons, the first example was given by Schmid [4], who found an anomalous electronic inelastic lifetime. The importance of this was fully appreciated only after Altshuler et al [5] found different manifestations of the same underlying physics.

This new physics, which has not yet been completely worked out even for normal metals, has a particular impact on the subject of type-II superconductivity: Dirty superconductors are of type-II, and the electron-electron interaction, both the direct Coulomb part and the phonon mediated part, is obviously indispensable. Together with recent advances in materials preparation techniques, which have made possible the study of extremely disordered superconductors, and with the fact that many materials of current interest are rather disordered to start with, this has led to the mentioned revival of interest. Furthermore, the new oxide superconductors certainly represent a rather extreme case of type-II materials, if indeed one can apply any of the traditional concepts.[1]

For all these reasons it seems worthwhile to reconsider the corresponding theory. As for the traditional part, it is very well covered in many excellent textbooks and monographs, see e.g. de Gennes [7], Parks [8], Saint-James et al. [9], and Tinkham [10]. For the sake of completeness, this chapter nevertheless reviews the basic concepts and results of the traditional theory. The understanding of the new physics, beyond standard dirty limit theory, is still in its infancy. I will give a brief account of some of the progress which has been made, and discuss what I feel to be some of the most important open questions.

The emphasis of this chapter is on simple physical principles and on the basic structure of the theory, rather than on detailed applications to specific situations, or on materials properties. I will confine myself to three-dimensional systems. Superconducting films are extremely interesting, but they come with complications of their own and form a more or less separate field of research. For lack of space, experimental aspects can only be discussed as far as they are of crucial importance for the theory. For the same reason the presentation is necessarily rather sketchy. Derivations are usually suppressed. They can be found in the literature quoted.

2.1.2. Two Types of Superconductors

F. and H. London showed in 1935 that perfect conductivity together with the Meissner-Ochsenfeld (MO) effect leads to magnetic screening. The magnetic screening length is the London length $\lambda_L = [mc^2/4\pi n_s e^2]^{1/2}$, where n_s is the density of electrons in the superconducting state. At zero temperature, supposedly all

[1]Experimental evidence is emerging that one probably can [6].

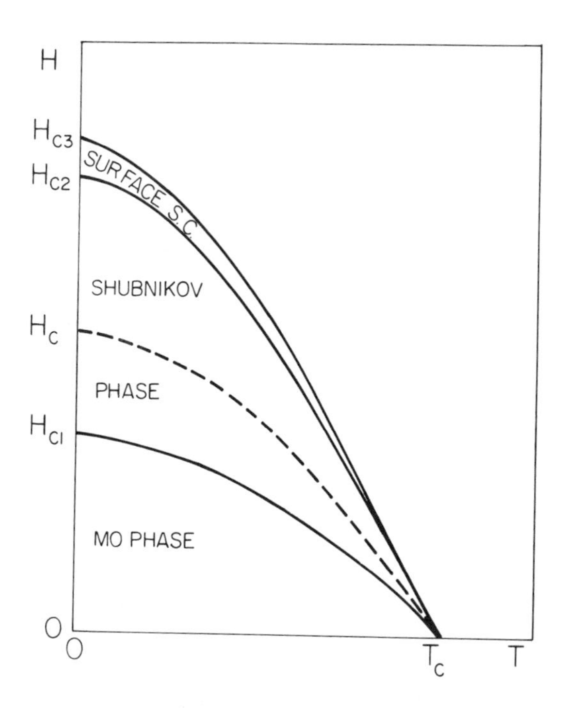

Fig. 2.1. Schematic phase diagram of a type II superconductor.

electrons are superconducting, so $\lambda_L(0) = [mc^2/4\pi ne^2]^{1/2}$. From a simple argument one can see that there is a second length scale in a superconductor. Since the transition temperature sets the energy scale, there is a momentum scale δp of the condensate given by $k_BT_c = \delta(\hbar^2p^2/2m) \simeq \hbar^2(p_F/m)\delta p$. The uncertainty principle then tells us that the condensate density cannot vary on length scales shorter than $\xi_o \sim \hbar v_F/k_BT_c$. ξ_o is called the intrinsic coherence length. Consider now a boundary between a superconductor and a normal metal. Going from the boundary into the superconductor, the system gains condensation energy on a length scale ξ_o. At the same time, it looses magnetic energy on a length scale $\lambda_L(0)$. If $\xi_o \gg \lambda_L(0)$, this leads to an energy maximum at the interface. These superconductors are called "*type-I*". They favor a homogeneous phase with fully developed MO effect. For $\xi_o \ll \lambda_L(0)$, an interface leads to an energy minimum. These are the "*type-II*" superconductors. They have the potential for an interesting structure of alternating normal and superconducting regions. This in turn gives rise to material properties which are the basis of the widespread applications of superconductors.

The experimental result for the phase diagram of a long type-II superconducting cylinder in a longitudinal magnetic field is shown schematically in Fig. 2.1. There are three distinct phases. For fields smaller than the lower critical field H_{c1}, there is complete flux expulsion. This is the MO phase. Between H_{c1} and the upper critical field H_{c2}, magnetic flux is only partially expelled. The material is superconducting, but shows an incomplete MO effect. This is called the Shubnikov phase, mixed state, or vortex state. At H_{c2}, bulk superconductivity disappears, but superconductivity may persist in a surface layer up to an even higher critical field H_{c3}. If one determines the thermodynamical critical field H_c, defined by the magnetic energy being equal to the free energy difference between normal and superconducting state, one finds $H_{c1} < H_c < H_{c2}$ (see chapter 1). All critical fields go to zero linearly as the critical temperature T_c is approached. In the next section we will explain this behavior by means of Ginzburg-Landau theory.

2.2. Ginzburg-Landau Theory

2.2.1. The Ginzburg-Landau Equations

Ginzburg and Landau [11] presented a theory of superconductivity which gives a remarkably complete account of the experimental facts by starting from the following simple postulates:

(1) The superconductor is characterized by
  (i) a complex order parameter $\Psi(\mathbf{r}) = \eta(\mathbf{r}) \exp(i\phi(\mathbf{r}))$
  (ii) a vector potential $\mathbf{A}(\mathbf{r})$, $\mathbf{B} = \nabla \times \mathbf{A}$.[2]

(2) Gauge invariance in the sense of quantum mechanics is postulated. The coupling charge is called q.

(3) The free energy density $f(\mathbf{r})$ is assumed to be a regular function of Ψ, the gauge invariant gradient $(i\nabla/q + \mathbf{A}/\hbar c)\Psi$, and $\mathbf{B}$, all of which are assumed to be small. The equilibrium value of Ψ is the one which minimizes $F = \int d\mathbf{r}\ f(\mathbf{r})$.

(4) Maxwell's theory of electrodynamics is assumed to hold.

Under these assumptions, the free energy density can be expanded as follows:

[2]Here and in the following, **B** denotes the magnetic induction in the material, while **H** denotes the external magnetic field. For normal metals, I neglect the magnetic susceptibility, and put $\mathbf{B} = \mathbf{H}$.

$$f(\mathbf{r})-f_n(\mathbf{r}) = a\,|\,\Psi(\mathbf{r})\,|^2 + \frac{b}{2}\,|\,\Psi(\mathbf{r})\,|^4 + \frac{\hbar^2}{2m}\,|\,[\nabla - \frac{iq}{\hbar c}\,\mathbf{A}(\mathbf{r})]\Psi(\mathbf{r})\,|^2$$

$$+ \frac{1}{8\pi}\,\mathbf{B}(\mathbf{r})^2 \qquad (2.1)$$

Here $a = \alpha(T-T_c)$, f_n is the normal state free energy density, and T_c, α, b, m, and q are the five parameters of the theory. We will soon see that of these five parameters only one is nontrivial. By taking the variation of the free energy as given in Eq. (2.1) with respect to Ψ, Ψ^*, and **A**, one easily obtains the celebrated Ginzburg-Landau (GL) equations:

$$-\frac{\hbar^2}{2m}\,[\nabla - \frac{iq}{\hbar c}\,\mathbf{A}(\mathbf{r})]^2\Psi(\mathbf{r}) + a\Psi(\mathbf{r}) + b\,|\,\Psi(\mathbf{r})\,|^2\Psi(\mathbf{r}) = 0 \quad , \qquad (2.2)$$

$$\nabla\times\mathbf{B}(\mathbf{r}) = \frac{4\pi}{c}\,\mathbf{j}(\mathbf{r}) \quad , \qquad (2.3a)$$

$$\mathbf{j}(\mathbf{r}) = -i\,\frac{\hbar q}{2m}\,[\Psi^*\nabla\Psi - \Psi\nabla\Psi^*] - \frac{q^2}{mc}\,|\,\Psi\,|^2\,\mathbf{A} \qquad (2.3b)$$

$$= \frac{\hbar q}{m}\,|\,\Psi\,|^2\,[\nabla\phi - \frac{q}{\hbar c}\,\mathbf{A}] \quad . \qquad (2.3c)$$

In the derivation surface terms appear. The condition of their being zero imposes boundary conditions on the GL equations. Choosing the correct boundary conditions is nontrivial (see [7]). In this lecture we will mostly deal with infinite systems, where one has the trivial boundary condition $\Psi(x\to\infty)\to 0$.

We see that Eq. (2.2) is the Schrödinger equation for a particle with mass m, and charge q, in a magnetic field and a nonlinear potential $b\,|\,\Psi(\mathbf{r})\,|^2$. The current **j** is given by the usual quantum mechanical expression, Eq. (2.3b). We also see that there is a self-consistency problem: Ψ depends on **B**, **B** is determined by **j**, and **j** in turn depends on Ψ.

2.2.2. Preliminary Analysis, Length Scales

We first consider a homogeneous superconductor without a magnetic field. Then the solution of Eq. (2.2) is $|\,\Psi\,|^2 = -(\alpha/b)(T-T_c)$. From the free energy, $F - F_n = -VH_c^2/8\pi = -V(\alpha^2/2b)(T-T_c)^2$, we obtain

$$H_c = (4\pi\alpha^2/b)^{\frac{1}{2}} \; (T_c - T) \quad . \tag{2.4}$$

The thermodynamical critical field vanishes linearly at T_c. Differentiating the free energy, we also find the discontinuity of the specific heat at T_c:

$$\Delta C = V(\alpha^2/b)T_c \quad , \tag{2.5}$$

where V is the system volume.

Now consider a homogeneous system in a weak magnetic field. Eq. (2.3) yields

$$\left[\Delta - \frac{4\pi q^2}{mc^2} \, |\Psi|^2\right] \mathbf{B}(\mathbf{r}) = 0 \quad . \tag{2.6}$$

If the field is weak, we can use the zero field expression for $|\Psi|$, and obtain the London equation with a temperature dependent penetration depth

$$\lambda(T) = \left[\frac{mc^2 b}{4\pi q^2 \alpha}\right]^{\frac{1}{2}} (T_c - T)^{-\frac{1}{2}} \quad , \tag{2.7}$$

We see that the GL equations describe the MO effect.

To find the temperature dependence of the coherence length, we look at fluctuations of Ψ in zero field. We write $\Psi(\mathbf{r}) = \eta + \delta\eta(\mathbf{r})$, and linearize Eq. (2.2) in the small quantity $\delta\eta$. This way we obtain an Ornstein-Zernicke equation,

$$[\Delta - \xi^{-2}(T)] \; \delta\eta(\mathbf{r}) = 0 \quad , \tag{2.8}$$

where for $T < T_c$,

$$\xi(T) = \frac{\hbar}{(4m\alpha)^{\frac{1}{2}}} \; (T_c - T)^{-\frac{1}{2}} \quad . \tag{2.9}$$

This is called the GL coherence length. Its temperature dependence is the same as that of the penetration depth.

2.2.3. Validity of Ginzburg-Landau Theory

The GL equations have been derived by an expansion in small quantities. Therefore a necessary condition for their validity is that the resulting lengths $\xi(T)$ and $\lambda(T)$ are large compared with ξ_o. From Eqs. (2.7), (2.9) we see that this condition is always fulfilled for temperatures sufficiently close to T_c. There is, however, another restriction. As a mean field theory, GL theory neglects fluctuations, which become important close to T_c. The Ginzburg-Levanjuk criterion states that mean field theory will break down if the fluctuations become of the same order of magnitude as the mean value of the order parameter. The estimate can be carried out in the same way as for the general Landau theory of second order phase transitions, and yields (see Lifshitz and Pitaevskii [12])

$$|T - T_c| >> \frac{b^2(k_B T_c)^2}{2\alpha} (m/\hbar^2)^3 , \tag{2.10}$$

as the condition for fluctuations to be negligible. With the help of Sec. 2.2, we can express this in terms of measurable quantities, and obtain

$$\zeta << |(T - T_c)/T_c| << 1 , \tag{2.11a}$$

$$\zeta = [2\pi(\Delta C/k_B V)\xi_o^3]^{-2} , \tag{2.11b}$$

as the region of validity of GL theory. For a conventional superconductor, typical numbers are $\Delta C/k_B V \cong 3 \cdot 10^{19}$ cm^{-3}, and $\xi_o \cong$ 5000 Å, leading to $\zeta \cong 10^{-15}$. The large coherence length is the reason why mean field theory works extremely well for superconductors. This changes if for some reason the coherence length becomes small. The new oxide superconductors are believed to have coherence lengths as small as 20 Å. If ΔC scales linearly with T, this leads to $\zeta \cong 3 \cdot 10^{-3}$. This means that the critical region will be observable. A recent measurement of the specific heat is consistent with this point of view, and shows that the oxides must have a more complicated order parameter than conventional superconductors [13]. Similarly, the coherence length can be very short in a conventional, but extremely disordered superconductor, see Sec. 2.4. These observations raise the possibility that the transition width seen in these materials are partly of intrinsic nature rather than merely due to material inhomogeneities.

2.2.4 The Ginzburg-Landau Parameter

It is natural to rewrite the GL equations in terms of dimensionless variables. Let us scale Ψ with $\sqrt{(-a/b)}$, $\mathbf{B}$ with $H_c\sqrt{2}$, $\mathbf{r}$ with λ, and $\mathbf{A}$ with $H_c\lambda\sqrt{2}$ (the factor $\sqrt{2}$ is of historic origin). Then we obtain for the free energy density

$$f - f_n = \frac{H_c^2}{4\pi}\left\{-|\Psi|^2 + \frac{1}{2}|\Psi|^4 + \mathbf{B}^2 + |(\nabla/i\kappa - \mathbf{A})\Psi|^2\right\}, \quad (2.12a)$$

where

$$\kappa = \lambda(T)/\xi(T) = \frac{q\sqrt{2}}{\hbar c}\lambda(T)^2 H_c(T). \quad (2.12b)$$

We see that of the five parameters λ, H_c, $-a/b$, T_c, and κ, which are combinations of the original parameters q, m, b, α, and T_c, only κ plays a nontrivial role, while the others set scales only. As a consequence, the thermodynamic state of an infinite system is determined by κ alone. We note that κ is independent of T! From Eq. (2.12a) one can calculate the surface tension δ associated with a normal-superconducting interface. The result is [12] that $\delta > 0$ for $\kappa < 1/\sqrt{2}$, and $\delta < 0$ for $\kappa > 1/\sqrt{2}$. Hence superconductors with $\kappa > 1/\sqrt{2}$ are of type-II, in agreement with the simple argument given in Sec. 2.1.2.

2.2.5. The Upper Critical Fields H_{c2} and H_{c3}

We now can determine the upper critical field H_{c2}, i.e. the field where for $\kappa > 1/\sqrt{2}$ the normal conductor first becomes unstable. If $\Psi \to 0$ for $H \to H_{c2}$,[3] we can neglect the last term in Eq. (2.2), and obtain

$$\frac{\hbar^2}{2m}\left[\nabla - \frac{iq}{\hbar c}\mathbf{A}\right]^2 \Psi = a\Psi \quad . \quad (2.13)$$

This is the Schrödinger equation for a particle with mass m, and charge q, in a magnetic field! We know there is a minimal eigenvalue, so $|a| \geq \hbar\omega_c/2 = \hbar qH/2mc$. Therefore the largest value of H which yields a nonzero solution for Ψ is

$$H_{c2} = \sqrt{2}\,\kappa\, H_c \quad , \quad (2.14)$$

where we have used κ from Eq. (2.12b). We see that for type-II materials ($\kappa > 1/\sqrt{2}$), superconductivity persists up to $H_{c2} > H_c$. H_{c2} vanishes linearly at T_c, cf. Eq. (2.4). For type-I materials, H_{c2} is smaller than the thermodynamic critical field H_c. Its physical meaning in this case is that of a boundary for the

[3]This means one assumes the transition to be of second order. This assumption is not trivial. For instance, the transition of a type-I material in nonzero field is of first order!

metastable normal phase.

In the above consideration, we have tacitly assumed the superconductor to be infinite. That is, we have taken the solution of Eq. (2.13) subject to the boundary condition $\Psi(x \to \infty) \to 0$. If the superconductor has a surface, facing vacuum or an insulator, the appropriate boundary condition is [7] $\partial\Psi/\partial x = 0$ at the surface. This condition simply means that no current is flowing in or out of the superconductor. The problem of the infinite superconductor above was equivalent to that of a particle moving in a harmonic oscillator potential $V(x) = m\omega_c^2(x - x_o)^2$, where x_o is the center coordinate of the Landau orbits. If the field is parallel to the surface, it is easy to see [7] that the appropriate potential is $V(x) = m\omega_c^2(x - x_o \mathrm{sgn}\, x)^2$. This is a double-well potential, and the boundary condition enforces the choice of wave functions with even parity only. The lowest eigenvalue of the double well has even parity, and it is lower than the one for the single well. Therefore, there must be a maximum field $H_{c3} > H_{c2}$. The detailed calculation yields [7]

$$H_{c3} = 2.4\ \kappa\ H_c = 1.7\ H_{c2} \quad . \tag{2.15}$$

In contrast, a field perpendicular to the surface does not change the upper critical field.

2.2.6. Fluxoid Quantization

F. London [14] recognized the quantization of a quantity he called the fluxoid, which later proved to be very important for the structure of type-II superconductors. The argument is very simple. Consider the expression for the current density given in Eq. (2.3c), and integrate it over any closed path $\mathfrak{C}$. Then

$$\oint_{\mathfrak{C}} \mathrm{d}\mathbf{l}\cdot\mathbf{A} + \frac{\hbar c}{q}\frac{m}{\hbar q}\oint_{\mathfrak{C}} \mathrm{d}\mathbf{l}\ \frac{1}{|\Psi|^2}\ \mathbf{j} = \frac{\hbar c}{q}\oint_{\mathfrak{C}} \mathrm{d}\mathbf{l}\ \nabla\phi$$

$$= \frac{\hbar c}{q}\ 2\pi n \qquad (n = 0,1,\ldots). \tag{2.16}$$

Here the far right hand side comes from the condition that the wavefunction must be single valued. Applying Stokes' theorem, and using $\nabla\times\mathbf{A} = \mathbf{B}$, we obtain

$$\Phi' = \Phi + \frac{4\pi}{c} \oint_{\mathfrak{C}} d\mathbf{l}\cdot\mathbf{j}\,\lambda^2 = n\Phi_o \ . \tag{2.17}$$

Here Φ is the flux through the surface $\mathfrak{F}$ surrounded by $\mathfrak{C}$, λ is the penetration depth, and $\Phi_o = hc/q$ is called the flux quantum. Φ' is called the fluxoid. It is important that the flux Φ is in general **not** quantized, even though one often loosely speaks of "flux quantization". However, if the path $\mathfrak{C}$ lies well inside the superconductor, then $\mathbf{j} \cong 0$, and $\Phi \cong \Phi'$.

The fluxoid quantization has been observed in beautiful experiments [15-17], which show that $q=2e$ in accord with BCS theory.

2.2.7. Vortices, and the Structure of the Shubnikov Phase

If $H < H_{c2}$, flux will penetrate the sample. We know that the surface energy is positive, so the system will try to maximize the surfaces of normal conducting regions. The way it does that has been worked out in great detail by Abrikosov [18] (see also [12]). A reasonable guess is that the normal conducting regions will appear in the form of cylindrical filaments in field direction.[4] These filaments are called "vortices". In the extreme type-II limit, $\lambda \gg \xi$, and so the normal conducting core region of a vortex is much smaller than the region of nonzero magnetic induction. This feature facilitates the theoretic treatment substantially.

The energy of a single vortex consists of the magnetic energy, and the kinetic energy of the supercurrent necessary to maintain the vortex,

$$E_V = \int d\mathbf{r} \left\{ \frac{\mathbf{B}^2}{8\pi} + \frac{m}{2}\mathbf{j}^2/n_s q^2 \right\} = \frac{1}{8\pi} \int d\mathbf{r} \left[\mathbf{B}^2 + \frac{4\pi m c^2}{n_s q^2 (4\pi)^2} (\nabla \mathrm{x} \mathbf{B})^2 \right] \tag{2.18}$$

where we have used Eq. (2.3a). With the help of some vector identities and Stokes' theorem, this can be written as

[4]It is important to realize that this is really only an educated guess. Its consequences have been worked out, and they compare very favorably with experiment, in contrast to other possible structures (e.g. a laminar one, see de Gennes [7]).

$$E_V = \frac{1}{8\pi}\int d\mathbf{r}\ [\mathbf{B}^2 - \lambda^2\mathbf{B}\Delta\mathbf{B}] + \frac{\lambda^2}{8\pi}\oint d\mathbf{s}(\mathbf{B} \times curl\mathbf{B}) \quad . \qquad (2.19)$$

Here the second integral is taken over the surface of a hollow cylinder with inner radius ξ, while the outer radius is going to infinity. If $\lambda \gg \xi$, the first integral will contribute very little, since in most of the region the material is superconducting, and hence the integrand is zero. We can therefore neglect this contribution if $\kappa \gg 1$. To complete the calculation we need to calculate the spatial dependence of **B**. We do this by modifying the London equation to allow for the core. Since we have neglected the volume of the core before, we can do that by modeling the core by a delta function. If the vortex contains n flux quanta, we then have

$$\mathbf{B} - \lambda^2\Delta\mathbf{B} = n\Phi_o\delta(\mathbf{r}) \quad . \qquad (2.20a)$$

The solution of this equation which falls off at infinity is

$$B(r) = \frac{\Phi_o n}{2\pi\lambda^2} K_o(r/\lambda) \quad , \qquad (2.20b)$$

where K_o is the modified Bessel function of zeroth order. Using Eq. (2.20b) in Eq. (2.19), one obtains for the vortex energy per unit length

$$\varepsilon_V \equiv E_V/L = [(n\Phi_o)^2/(4\pi\lambda)^2] \log \kappa \quad . \qquad (2.21)$$

The approximation of neglecting the vortex core volume results in Eq. (2.21) being valid only for $\log \kappa \gg 1$. Eq. (2.21) has a number of interesting consequences. First we see that a vortex containing n flux quanta has a higher energy than n vortices containing one flux quantum. Therefore, all vortices are expected to contain one flux quantum only. We can also calculate H_{c1}, the lower critical field where the first vortex appears in the system. To this end we consider the Gibbs free energy in the presence of a vortex,

$$G = F + L\varepsilon_V - \int d\mathbf{r}\ \mathbf{B}\cdot\mathbf{H}/4\pi \quad . \qquad (2.22)$$

H_{c1} is obviously determined by the condition $G = F$, so

$$H_{c1} = (H_c/\kappa\sqrt{2})\ log\ \kappa \quad . \qquad (2.23)$$

We see that apart from the factor log κ, H_c is the geometric mean of H_{c1} and H_{c2}. Again, Eq. (2.23) is valid only for log $\kappa \gg 1$.

For $H > H_{c1}$, there is more than one vortex, so the interaction between vortices has to be taken into account, which turns out to be repulsive. From this it follows that the vortices will form a regular array ("vortex lattice") in order to minimize the repulsion energy [18]. A detailed calculation [19] shows that a triangular lattice has the lowest energy. This prediction has found a spectacular confirmation by experiment [20].

We finally remark that the dynamics of the vortex state is nontrivial, and of great practical importance. The vortices are subject to the Lorentz force, which leads to a motion of the vortices perpendicular to the field ("flux flow"). The resulting $\partial_t \mathbf{B}$ induces an electric field parallel to $\mathbf{j}$, which in experiment shows up as a resistive voltage drop. An ideal vortex state is therefore *not* a superconductor! In practice things are not quite as bad. Material imperfections pin the vortex tubes to local energy minima, which can only be overcome by thermal activation ("flux creep"). To keep the flux creep low is a very important issue in the construction of superconducting magnets.

2.3. Microscopic Theory

2.3.1. Preliminary Remarks

Despite its tremendous success, GL theory has a number of inherent shortcomings. One is that it is valid only close to T_c. Another one, which actually is also an advantage, is that it is valid for clean as well as for disordered systems. In fact, all disorder dependencies are hidden in the parameters of GL theory, and are therefore inaccessible within its framework. To discuss the complete phase diagram, as well as the dependence of T_c and H_{c2} on disorder, we have to resort to the microscopic BCS theory. In doing so, I will restrict myself to the "dirty limit" regime [21], where the electronic mean free path ℓ obeys $\ell << \xi_o$. I will also sketch only those parts of the theory which are absolutely indispensable for the present purpose, and extract the desired results by the simplest method I could find. For a more complete treatment, the reader is referred to the books quoted in the introduction. The theory will be formulated using Anderson's [21] exact eigenstate method (see also [7]). This allows for a straightforward generalization to strong coupling and strong disorder, which is treated in Section 2.4.

2.3.2 The Cooper Instability

Let us consider the BCS model of electrons which interact via an attractive, point-like interaction $g>0$. The interaction part of the Hamiltonian reads

$$H_{int} = -g \int dx\ \Psi^+_\downarrow(x)\Psi^+_\uparrow(x)\Psi_\uparrow(x)\Psi_\downarrow(x)\ . \tag{2.24}$$

Here the Ψ are electron field operators, and the spin structure of the interaction reflects the s-wave pairing envisaged by BCS. The point-like interaction will cause problems at high energies, so Eq. (2.24) has to be supplemented by a cutoff convention: $g \neq 0$ only for electrons within an energy shell of width $\hbar\omega_D$ around the Fermi surface. This cutoff takes care of the fact that the attraction is mediated by phonons whose maximum energy is on the order of the Debye energy. We further allow for a magnetic field, and a random potential to model possible disorder. We now consider the pair propagator

$$\tilde{L}(1,2) = -\langle T_t[\Psi_\downarrow(1)\Psi_\uparrow(1)\Psi^+_\uparrow(2)\Psi^+_\downarrow(2)]\rangle\ , \tag{2.25}$$

where $1 \equiv (x_1,t_1)$, etc. comprises space and imaginary time variables, T_t is the time ordering operator, and the brackets denote the quantum mechanical expectation value. $\tilde{L}$ describes the correlated motion of a pair of electrons with opposite spins moving from point 1 to point 2 in space-time. A simple approximation for $\tilde{L}$ consists of taking only those interaction processes into account where the electrons propagate freely, scatter off each other, propagate freely, scatter again, and so on. This is the ladder approximation shown graphically in Fig. 2.2. The corresponding integral equation is

$$\tilde{L}(1,2) = G(1,2)G(1,2) - ig\int_0^{-i/T} dt_3 \int dx_3 G(1,3)G(1,3)\tilde{L}(3,2)\ , \tag{2.26a}$$

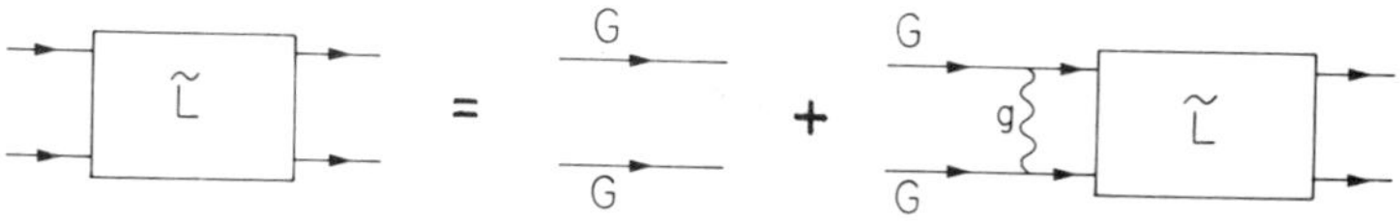

Fig. 2.2. The ladder approximation for the pair propagator.

where

$$G(1,2) = -i \left\langle T_t[\Psi_\downarrow(1)\Psi_\downarrow^+(2)]\right\rangle \qquad (2.26b)$$

is the Green's function for noninteracting electrons. It will turn out that below a certain temperature T_c, $\tilde{L}$ as given by the solution of Eq. (2.26a) is infinite. This signalizes the instability of the system towards the formation of "bound states" of some sort, and it is plausible that this instability might have something to do with superconductivity. It can indeed be shown that the "bound states" are Cooper pairs, and that the system is superconducting below T_c. The ladder approximation used for calculating $\tilde{L}$ is equivalent to the mean field approximation of BCS, so the T_c determined this way is the mean field transition temperature. This route is the fastest for calculating T_c (and H_{c2} as well, as we will see).

It is easy to convince oneself that the first term on the right hand side of Eq. (2.26a) is always finite. Since we are looking for a pole in $\tilde{L}$, we can therefore neglect it. We still have to deal with the disorder in the system. That is, we have to average over the random positions of the impurities. If we denote this impurity average by $<\ldots>_{av}$, we see that Eq. (2.26a) relates $L = <\tilde{L}>_{av}$ to $<\tilde{L}GG>_{av}$. Unfortunately, the only known feasible way to deal with this problem is to factorize $<\tilde{L}GG> \cong L\cdot <GG>_{av}$. Even though one can show [22] that for small disorder the terms neglected this way are less important (in an imprecise sense) than those retained, this approximation is a serious stumbling block if one tries to apply the theory to large disorder. We finally perform a Fourier transform with respect to the time variable. It is easy to see that the first pole in L occurs at zero frequency, so we are only interested in $L(\mathbf{x},\mathbf{y}) \equiv L(\mathbf{x},\mathbf{y};i\Omega=0)$. Then we finally obtain

$$L(\mathbf{x}_1,\mathbf{x}_2) = -ig \int d\mathbf{x}_3\ Q(\mathbf{x}_1,\mathbf{x}_3)\ L(\mathbf{x}_3,\mathbf{x}_2) \quad , \qquad (2.27a)$$

where

$$Q(\mathbf{x}_1,\mathbf{x}_2) = iT \sum_n \left\langle G(\mathbf{x}_1,\mathbf{x}_2;-i\omega_n)\ G(\mathbf{x}_1,\mathbf{x}_2;i\omega_n)\right\rangle_{av} \quad , \qquad (2.27b)$$

and $\omega_n = 2\pi T(n+\frac{1}{2})$ is a Matsubara frequency. It is important that the integral kernel Q is a correlation function for noninteracting electrons. This is a consequence of the mean field approximation. Eq. (2.27) is the central result of the theory as far as the transition point is concerned. It is an implicit equation for T_c or H_{c2}, which we will now analyze. For a detailed derivation of Eq.

(2.27), see e.g. the pedagogical discussion by Ambegaokar [23].

2.3.3 The Integral Kernel

In order to solve Eq. (2.27a), one has to determine the kernel Q. We first observe that in a weak magnetic field, we can employ the quasiclassical approximation, and write [22]

$$Q(\mathbf{x}_1,\mathbf{x}_2) = Q^{B=0}(\mathbf{x}_1,\mathbf{x}_2) \exp\left[i\frac{2e}{c}\int_{\mathbf{x}_1}^{\mathbf{x}_2} d\mathbf{x}\ \mathbf{A}(\mathbf{x})\right] \quad . \tag{2.28}$$

Here $Q^{B=0}$ is the kernel in zero field, and the integral is to be performed along a straight line. The weak field condition necessary for Eq. (2.28) to be valid can be written as $\ell_B \gg \ell$, where $\ell_B = eB/\hbar c$ is the magnetic length, and ℓ is the mean free path. This is the same condition as for the single particle spectrum to remain smooth.

The first task is now to calculate $Q^{B=0}$. Let us assume [21] that the Hamiltonian for the noninteracting electrons in the presence of disorder had been diagonalized. Let us denote the corresponding eigenfunctions by ψ_α, and the eigenenergies by E_α. Then the Green's functions G are diagonal in this basis, and

$$Q^{B=0}(\mathbf{x},\mathbf{y}) = iT\sum_n \int dE d\omega' \frac{1}{E+\omega'+i\omega_n}\frac{1}{E-i\omega_n} F^E_{\mathbf{xyyx}}(\omega') \quad ,(2.29)$$

where

$$F^E_{\mathbf{x}_1\mathbf{x}_2\mathbf{x}_3\mathbf{x}_4}(\omega)$$

$$= \left\langle \sum_{\alpha,\beta} \delta(E+\omega-E_\alpha)\delta(E-E_\beta)\psi^*_\alpha(\mathbf{x}_1)\psi^*_\beta(\mathbf{x}_4)\psi_\alpha(\mathbf{x}_3)\psi_\beta(\mathbf{x}_2)\right\rangle_{av} \quad .(2.30)$$

Of course, neither the ψ_α nor the E_α can be determined explicitly, and so Eq. (2.30) seems to be of little value. However, the special combination needed in Eq. (2.29) can be determined from very general considerations [7,24]. Consider the electron density spectrum ([e.g. [25]),

$$\chi''(\mathbf{q},\omega) = \frac{1}{2}\int dt\; e^{i\omega t}\int d\mathbf{x}d\mathbf{y}\; e^{i\mathbf{q}\cdot(\mathbf{x}-\mathbf{y})}\left\langle [\rho(\mathbf{x},t),\rho(\mathbf{y},0)]\right\rangle_{av}, \qquad (2.31a)$$

where $\rho(\mathbf{x}) = \Psi^{+}(\mathbf{x})\Psi(\mathbf{x})$ is the density operator. In the exact eigenstate basis, the commutator is readily calculated, and one obtains

$$\chi''(\mathbf{q},\omega) = \pi\int dE\; [f(E)-f(E+\omega)]\int d\mathbf{x}d\mathbf{y}\; e^{i\mathbf{q}\cdot(\mathbf{x}-\mathbf{y})}\, F^{E}_{\mathbf{xxyy}}(\omega) \; . \qquad (2.31b)$$

Here f denotes the Fermi function. In the absence of a magnetic field, the wavefunctions in Eq. (2.30) can be chosen to be real, and $F_{xxyy} = F_{xyyx}$. At low temperatures and small frequencies, F will not vary much with E over the integration interval, and we can replace F^{E} by F^{E_F}.[5] We now have expressed $Q^{B=0}$ in terms of the electronic density response χ''. The latter is known to be diffusive at small frequencies and wavenumbers [25],

$$\chi''(q,\omega) = N_F\,\frac{\omega D q^2}{\omega^2 + (Dq^2)^2} \; . \qquad (2.32a)$$

From Eq. (2.29) we then find

$$Q^{B=0}(q) = -2\pi T N_F \sum_n \frac{sgn\,(n)}{4i\pi T(n+\tfrac{1}{2}) + iDq^2\, sgn(n)} \; . \qquad (2.32b)$$

Here N_F is the density of states per spin at the Fermi level, and D is the diffusion constant.

Finally, we cast the full kernel Q, Eq. (2.28), in a manageable form. This can easily be done by realizing that with the small q expression for $Q^{B=0}$, Eq. (2.32), the eigenvalue equation for Q is simply the Schrödinger equation for a particle with charge $2e$ in a magnetic field. Therefore the Landau functions ϕ_η ($\eta = [m,k,k']$) diagonalize Q, and we obtain our final result

[5]This is an approximation whose validity becomes doubtful with increasing disorder. The point seems not to be well understood, though it is of importance for several problems related to transport in disordered systems.

$$Q(\mathbf{x},\mathbf{y}) = \sum_{\eta} \phi^*_\eta(\mathbf{x})\, 4\pi T \sum_{n} \frac{-N_F}{2i\omega_n + iD\nu_\eta}\, \phi_\eta(\mathbf{y}) \ , \qquad (2.33a)$$

with

$$\nu_\eta = 4eB(m+\tfrac{1}{2})/c + k'^2 \ . \qquad (2.33b)$$

We note in passing that the formalism has actually a much greater flexibility than we have made use of here. If for some reason the density response is not diffusive, Eqs. (2.29) - (2.31) are still valid, and one can use the correct density response to calculate $Q^{B=0}$. One can also calculate Q outside the hydrodynamic regime [26].

2.3.4. Anderson's Theorem

We first consider the case B = 0. Then Eq. (2.27a) can be solved by Fourier transform, and yields

$$1 = -ig\, Q^{B=0}(q) \ . \qquad (2.34)$$

The instability first occurs at $q=0$. Remembering the cutoff convention, we restrict the sum in Eq. (2.32) to $n \leq N = \hbar\omega_D/k_B T_c$. We solve for T_c, and find the BCS result

$$T_c = 1.13\, (\hbar\omega_D/k_B)\, e^{-1/gN_F} \ . \qquad (2.35)$$

The remarkable fact about this way of deriving this result is that Eq. (2.35) is valid for arbitrary disorder! The reason is that $Q^{B=0}(q=0)$ as given by Eq. (2.32b) is an exact result which follows from particle number conservation. Hence within the approximations made (the ladder approximation plus the factorization of the impurity average in Sec. 2.3.2), T_c depends on disorder only via N_F. The same is true for some other quantities, e.g. the discontinuity in the specific heat. This fact [21] is sometimes called Anderson's theorem. We will see in Sec. 2.4 that improving on these approximations leads to additional disorder dependencies which can be quite strong.

2.3.5. Dirty Limit Theory

In the presence of a magnetic field, one solves Eq. (2.27a) by going into the Landau basis,

$$1 = gN_F 4\pi T \sum_{n=0}^{N} \frac{1}{4\pi T(n+\frac{1}{2}) + D\nu_\eta} \quad . \tag{2.36}$$

The instability first occurs for $\nu_\eta = \nu_o = 2eH_{c2}/c$, which is given by the implicit equation [7]

$$\ln (T/T_c) = \psi(\tfrac{1}{2}) - \psi\left[\frac{1}{2} + \frac{DH_{c2}}{2\phi_0 T} \right] \quad , \tag{2.37}$$

where ϕ_o is the flux quantum, ψ denotes the digamma function, and T_c is the transition temperature in zero field. We see that in a magnetic field the solution involves a nonzero eigenvalue, and therefore depends directly on the mean free path ℓ via $D = v_F\ell/3$. Consequently, we also have to worry about the region of validity of Eq. (2.37), since its derivation made use of the diffusion pole approximation, which is valid only for small eigenvalues. From Eq. (2.32) we see that $Q^{B=0}(q) = f(q/q^*)$ with $q^* \sim \sqrt{(T/T_c)}/\sqrt{(\xi_o\ell)}$. The diffusion pole approximation is valid for $q\ell \leq 1$. Hence our derivation is valid if $\ell \ll \xi_o$. This region is called the "dirty limit" [21].

Eq. (2.37) can be written in a very simple form [7]. We introduce a reduced temperature $t = T/T_c$, and a reduced magnetic field $h = H_{c2}/(-dH_{c2}/dt)_{t=1}$. Then Eq. (2.37) reads

$$h(t) = \frac{\pi}{8} Un(t) \quad , \tag{2.38a}$$

where the universal function $y = Un(x)$ is defined by

$$ln(x) = \psi(\tfrac{1}{2}) - \psi(\tfrac{1}{2} + y/4\pi x) \quad . \tag{2.38b}$$

This means that a plot of H_{c2} versus temperature in units such that the slope of H_{c2} at T_c is unity results in a universal curve. Experiments are in agreement with this prediction if some complications are taken into account, see Sec. 2.3.6.

We can now obtain the behavior of the coherence length and the penetration depth in the dirty limit by comparison with the results of GL theory. From Eq. (2.14) one obtains $H_{c2} = c/2e\hbar\xi(T)^2$. Using this in Eq. (2.38) yields

$$\xi(T) = 0.85\ (\xi_o \ell)^{\frac{1}{2}}\ (1 - t)^{-\frac{1}{2}} \quad . \tag{2.39}$$

Since $\xi^2 \sim 1/\alpha$, we can then use Eqs. (2.5), (2.7), and Anderson's theorem to obtain

$$\lambda(T) = 0.64\ \lambda_L(0)\ (\xi_O/\ell)^{\frac{1}{2}}\ (1 - t)^{-\frac{1}{2}} \quad . \tag{2.40}$$

Finally, we obtain for the GL parameter

$$\kappa = 0.75\ \lambda_L(0)/\ell \quad . \tag{2.41}$$

Eqs. (2.39) - (2.41) are valid in the dirty limit only. They show that the coherence length shrinks with increasing disorder, while the penetration depth grows. In a disordered system coherence is not quite as good, and hence the magnetic screening is not quite as effective, as in a clean one. As a result, the GL parameter is proportional to $1/\ell$: Sufficiently dirty superconductors are always type-II!

2.3.6. Various Complications

The results of Sec. 2.3.5 have been obtained under various idealizing assumptions. To compare with experiments, some complications have to be taken into account. Here we briefly list some of them.

For smaller disorder than required for the dirty limit, the diffusion pole approximation used in Sec. 2.3.3 is not valid. This can be cured, as mentioned before, by taking the correct density response into account [27]. The results are slight upward deviations from the universal curve.

According to Eq. (2.14), H_{c2} increases with increasing κ without limit. This is an artifact due to the fact that we neglected the effect of the magnetic field on the electronic spin. Obviously, at sufficiently high fields the s-wave superconductor will become unstable against formation of a spin-polarized normal conductor. This is called the paramagnetic limiting effect, or the Clogston-Chandrasekhar limit (see [9]). It results in downward deviations from the universal curve.

Spin-orbit coupling leads to the spin not being a good quantum number. It therefore weakens the paramagnetic effect [28]. Usually neither the paramagnetic effect nor the spin-orbit coupling can be calculated from first principles, so experimental results have to be fitted using two parameters [29]. Recently, however, it has become possible to measure the spin-orbit coupling independently [30]. It might be interesting to use these results for a reanalysis of H_{c2} data.

At high magnetic fields, Cooper pairs with nonzero momentum can be shown to be stable [31]. There is a second order transition from the normal conductor to the Fulde-Ferrell (FF) state, and a first order transition from the BCS state to the FF state. Unfortunately, the FF state is stable only in the clean limit. Therefore its realization needs a clean material with a large κ, which is hard to find.

2.4. Beyond Dirty Limit Theory

2.4.1. Limitations of the Dirty Limit Theory

The dirty limit theory outlined in the last section is subject to various conditions. First of all, it is only valid if the mean free path ℓ obeys $\xi_o \gg \ell \gg 1/k_F$. The upper limit for ℓ is due to the fact that the simple diffusion pole approximation for the density response is invalid at smaller disorder. This can easily be taken into account [27]. The lower limit for ℓ is the usual condition of "weak disorder". It reflects the fact that the standard ensemble averaging technique [32] uses $1/k_F\ell$ as a small parameter. For $k_F\ell \lesssim 1$, perturbation theory breaks down, and localization effects are dominant. This is a major complication, and the nature of superconductivity in this regime is not clear. However, even in the region $k_F\ell \gg 1$ several approximations have been made to derive the dirty limit theory. One was the ladder approximation to obtain the central integral equation (2.26a). Another one was the BCS model for the interaction Hamiltonian, which treats the effective electron-electron interaction as a phenomenological coupling constant g. In a microscopic treatment, $g = \lambda - \mu^*$, where λ depends on the electron-phonon coupling and the phonon spectrum [33], and μ^* is the Coulomb pseudopotential [34]. λ and μ^* are well known for clean materials, but their behavior in the presence of disorder has only recently been investigated.

Within the last few years, several attempts have been made to improve on these points. Some surprising results have been found, some of which agree well with experiment. However, no complete theory for superconductors with an arbitrary degree of disorder has emerged yet. In this last part of this chapter I give a brief account of the progress which has been made, and a brief discussion of some open problems.

2.4.2. Breakdown of Anderson's Theorem

In Sec. 2.3.4 we have seen that within BCS-Gorkov theory, T_c depends on disorder only through the density of states. It turns out that this result depends on the two crucial assumptions of the theory, viz. the ladder approximation, and the phenomenological coupling parameter.

One can improve on the ladder approximation by using the full Green's function in Eq. (2.26a) instead of its counterpart for a noninteracting system. In a clean superconductor, this leads only to Fermi liquid corrections to the density of states. That means, Eq. (2.36) remains unchanged, with N_F denoting the correct density of states. In the presence of disorder, interplay between the electron-electron interaction and the disorder gives rise to novel effects, which lead to a suppression of the single-particle spectrum at the Fermi level (a correlation gap) [5]. This was shown to yield a suppression of T_c with increasing disorder [35,36], even with phenomenological coupling constants. It should be stressed that this is a many-particle effect, as opposed to band structure related density of states effects which are present even in the BCS-Gorkov model. Since its physical roots are the same as those of the Altshuler-Aronov effect, I call this the correlation gap mechanism for T_c degradation. It explains at least qualitatively the sometimes drastic suppression of T_c seen experimentally in some materials. It fails to explain the opposite effect, an increase of T_c with disorder, which is observed in some other materials.

Another source of disorder dependence of T_c is the microscopic coupling constants themselves. To study these effects, one has to work within the framework of strong coupling theory [33,37], which derives the effective electron-electron interaction microscopically from the electron-phonon and the Coulomb interaction. It was realized [38] that the effective electron-phonon interaction λ increases with disorder. This is a mechanism with a tendency to enhance T_c. On the other hand, one would expect the Coulomb pseudopotential to be enhanced as well. This is indeed the case [39], but the effect is small for $k_F\ell \gg 1$. Finally, one has to wonder how the correlation gap mechanism can be incorporated within the framework of strong coupling theory. This turns out to be possible [40] by including the normal self energy part, which here, in contrast to the clean case, is not simply a constant, and therefore cannot be neglected. This way it was possible to include all effects studied before separately within a single unifying theory, and to derive a T_c formula for dirty superconductors [41]:

$$T_c = \frac{\Theta_D}{1.45} \exp\left\{ \frac{-1.04\,[1 + \lambda(\rho) + y(\rho)]}{\lambda(\rho) - \mu^*(\rho)\left(1 + \frac{0.62\,\lambda(\rho)}{1 + y(\rho)}\right)} \right\} . \qquad (2.42)$$

This is a generalization of the well known McMillan formula. Here ρ is the (extrapolated) residual resistivity, which serves as a measure for the disorder, and Θ_D is the Debye temperature. $\lambda(\rho)$ and $\mu^*(\rho)$ are generalizations of the usual parameters λ and μ^*, which now include the effects of disorder on the microscopic coupling constants. $y(\rho)$ results from the normal self energy which vanishes in Eliashberg theory. Accordingly, $y(\rho \to 0) \to 0$. The physics it describes is the correlation gap mechanism. λ, μ^*, and y are all monotonically increasing functions of ρ. We see from Eq. (2.42) that there are many competing disorder effects on T_c. Some tend to enhance T_c, some tend to suppress T_c. They all turn out to be of the same order of magnitude, and T_c depends exponentially on all of them. This makes it clear why no attempt including only one of these effects was able to consistently explain the experimental results.

The theory expresses $\lambda(\rho)$, $\mu^*(\rho)$, and $y(\rho)$ in terms of complicated correlation functions. Up to now they have been calculated only within certain approximations, which restricts the validity of the result to the regime of moderate disorder $\rho \ll \rho_M$, where ρ_M is the Mott resistivity, which for most materials is of the order of 1000 $\mu\Omega$cm. In direct comparison with experiment the theory works well up to $\rho \cong 200$ $\mu\Omega$cm. However, if one restricts oneself to a model calculation using a jellium model for the electrons, and a Debye model for the phonons, the three parameters can be calculated exactly in the limit of small disorder (i.e. to first order in $\rho/\rho_M = 1/k_F\ell$). The result [42] shows that $\delta t/\rho$ with $\delta t = [T_c - T_c(\rho=0)]/T_c(\rho=0)$ for $\rho \to 0$ is large and positive for low T_c, and small and negative for high T_c. Roughly, its behavior is

$$\delta t/(\rho/\rho_M) = \frac{1.04}{\lambda}(a - b\lambda) , \qquad (2.43)$$

where $\lambda = \lambda(\rho=0)$, and a and b are material dependent constants. Eq. (2.43) has been derived by neglecting the Coulomb contribution to δt. The full expression for $\delta t/\rho$ is much more complicated, but its qualitative behavior is the same. The first term is due to the

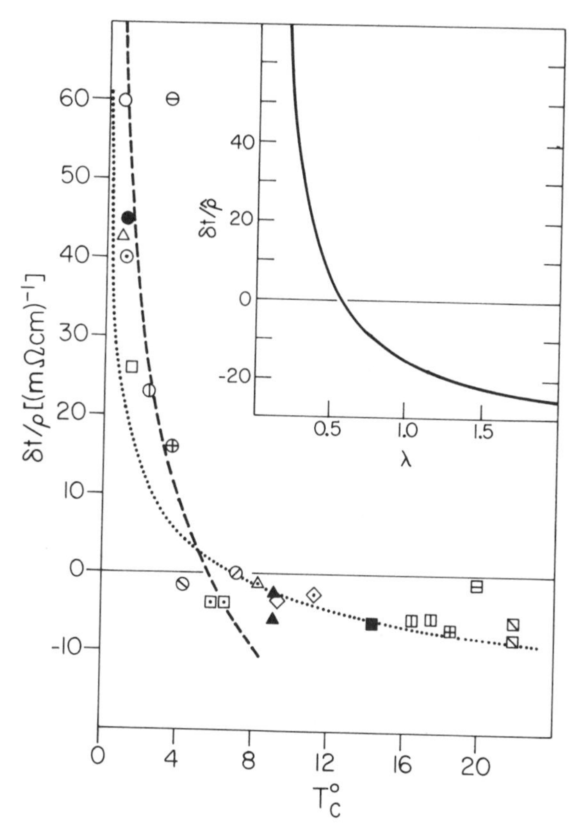

Fig. 2.3. $\delta t/\rho$ plotted versus T_c^0 according to the data shown in Table 2.1. Symbols are the same as those in Table 2.1. Dashed and dotted lines are theoretical results. The dashed (dotted) line corresponds to parameters typical for simple (transition) metals, respectively. The inset shows the simplified result, Eq. (2.43), with the same parameters as those used for the dashed line. For details see [42].

enhancement of the electron-phonon interaction. The second term is due to the correlation gap mechanism. It is relatively more important for larger λ, i.e., for higher T_c. The reason is that the electron-phonon coupling contributes to the correlation gap effect as well as the Coulomb interaction. Equation (2.43) thus states that low T_c superconductors should show an increase of T_c with disorder (at least initially for small disorder), while high T_c materials should show a degradation of T_c. This behavior has indeed been known experimentally for many years. A recent analysis [42] of experimental data has shown that the correlation between $\delta t/\rho$ and T_c

Table 2.1. T_c, and disorder dependence of T_c, for various materials. See Belitz [42] for more parameters of these materials.

Material	$T_c^{o}(K)$	$\delta t/p$ ($10^5 \Omega^{-1} m^{-1}$)	References
Zn	0.85[a]	60±10[b]	[a]McMillan [43]
Ga	1.08[a]	40±5[b]	[b]Buckel and Hilsch [44]
Al	1.16[a]	190+20[b] 45±5[c]	[c]Dynes and Garno [45]
Tl	2.38[a]	23±5[d]	[d]Comberg et al [46]
In	3.40[a]	60±5[b,e]	[e]Bergmann [47]
Sn	3.72[a]	16±4[b,e]	
Hg	4.16[a]	-1.5±0.2[b]	
Pb	7.19[a]	0[b]	
Mo	0.92[a]	43[f,g]	[f]Gurvitch [48]
MoRuP	8.4[h]	-1±0.5[h]	[g]Kimhi and Geballe [49]
Nb	9.22[a]	-5.7±2.3[g] -2.3±1.5[i]	[h]Johnson [50]
			[i]Testardi et al [51]
Mo_3Ge	1.42[j]	26±3[j]	[j]Gurvitch et al [52]
V_3Ge	5.8[k] 6.5[i]	-3.8[i]	[k]Radousky et al [53]
V_3Ga	14.3[k]	-6.2±1[l]	[l]Bending et al [54]
V_3Si	16.5[k,m] 17.5[n]	-5.7±1.2[i,m,n]	[m]Orlando et al [55]
Nb_3Sn	17.9[k]	-1±1[m,o]	[n]Rowell and Dynes [56]
Nb_3Al	18.5[k]	-7±2[p]	[o]Rowell et al [57]
Nb_3Ge	21.8[k]	-6.1±1[n,o] -7.8±1.3[i]	[p]Ghosh and Strongin [70]
$ErRh_4B_4$	9.4[n,o]	-3.6[n,o]	
$LuRh_4B_4$	11.3[n]	-2.5[n]	

is surprisingly well pronounced for a large variety of very different materials, see Fig. 2.3 and Table 2.1. The model calculation explains this correlation quantitatively with very reasonable parameters. These results strongly suggest that the observed overall disorder dependence of T_c is essentially due to very general processes affecting the effective electron-electron interaction, and can be understood neglecting material-specific properties like band structure and phonon spectra. The latter are of course important for a quantitative description of any given material, but not to understand the general trend of many classes of superconductors.

2.4.3. Reentrant H_{C2}

The correlation gap mechanism for T_c degradation described above has an interesting consequence for the phase diagram in a magnetic field [58]. The microscopic processes suppressing T_c include both ordinary diffusion processes (particle-hole ladders), and the Cooper diffusion processes (particle-particle ladders) familiar from localization theory [2,59]. The latter are very sensitive to magnetic fields, and in three dimensions decrease proportional to the square root of the magnetic field. As a result, some of the T_c-suppressing processes are themselves suppressed by the magnetic field. This leads to an *increase* of T_c proportional to $H^{1/2}$, while the pairbreaking is only linear in H, Eqs. (2.14), (2.37). Consequently, at small fields the T_c increase will always win, and there will be a reentrant phase diagram as shown schematically in Fig. 2.4. The characteristic parameters are the magnetic field scale H^*, and the maximum T_c enhancement ΔT. The calculation yields

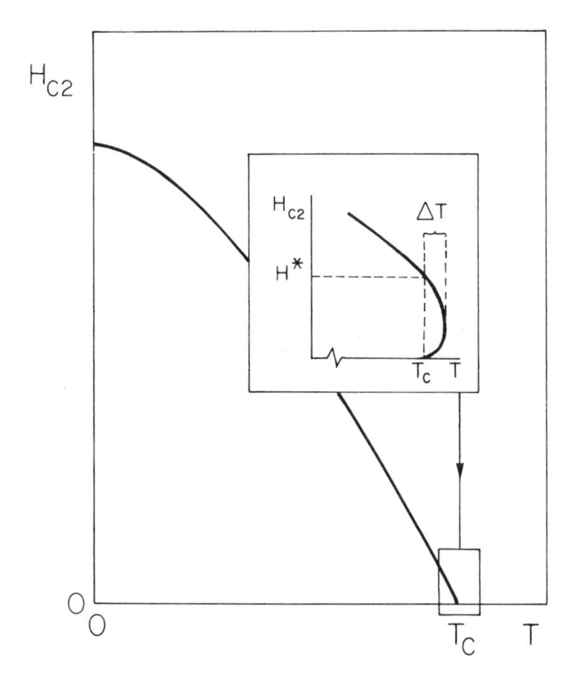

Fig. 2.4. Schematic plot of the reentrant phase diagram predicted by the theory [58].

$H^* \cong 10$ G, $\Delta T \cong 0.1$ mK for common high-field superconductors. The

effect is thus small, and has not been observed so far. A possible way to observe the effect would be to look for anomalies in the fluctuation induced paraconductivity just above T_c, as has been discussed in reference [58]. An experimental check of these ideas would be valuable since it would provide a check for the theoretical claim that the correlation gap mechanism underlies the observed T_c degradation.

2.4.4. Breakdown of Mean Field Theory

We have seen in Sec. 2.3.5 that the coherence length decreases with decreasing mean free path ℓ. Since there is no relevant length scale between ξ and $1/k_F$, one expects this decrease to continue until $k_F\ell \cong 1$. In this regime the mean free path looses its meaning, and one has to use the diffusion constant instead. In terms of the diffusion constant, Eq. (2.39) becomes

$$\xi^2 \sim (\hbar D/k_B T_c)(1-t)^{-1} \quad . \tag{2.44}$$

This is a general expression: The diffusivity D is meaningful even for $k_F\ell < 1$, where the electrons undergo a localization transition, but it becomes scale dependent. From the theory of Anderson localization, we known what happens to D[6] (see, e.g., [60]):

$$D = \frac{1}{m} \begin{cases} k_F\ell \quad , \quad k_F\xi_{loc} \sim 1 & \text{weak disorder} \quad (2.45a) \\ \dfrac{1}{k_F\xi_{loc}} \quad , \quad 1 \ll k_F\xi_{loc} < k_F\ell_T & \text{critical disorder} \quad (2.45b) \\ \dfrac{1}{k_F\ell_T} \quad , \quad 1 \ll k_F\ell_T < k_F\xi_{loc} & \text{saturation at } T \neq 0. \quad (2.45c) \end{cases}$$

Here $\xi_{loc} \sim (1/k_F)|1 - n_i/n_c|^{-\nu}$ is the localization length, which serves as a measure of disorder. $n_i \sim \ell^{-3}$ is the impurity density (or any other measure of disorder), and n_c denotes the critical value of n where the Anderson transition takes place. ν is an exponent whose value is not exactly known in $d = 3$. Extrapolation of an ε-expansion around $d = 2$ yields $\nu = 1$. ℓ_T is the inelastic

[6]This knowledge so far is restricted to the case of non-interacting electrons, which is not quite appropriate for superconductors.

length. One has $\ell_T \sim (N_F T)^{-1/3}$, where the exponent can be shown to be exact [61]. Equation (2.45a) is the usual Boltzmann result. Equation (2.45b) describes the onset of localization: as $n_i \to n_c$, ξ_{loc} diverges, and $D \to 0$. Equation (2.45c) expresses the fact that at nonzero temperature there never is strict localization, but rather D eventually crosses over to a nonzero value. If we use Eq. (2.45) in Eq. (2.44), we obtain the result of Kotliar and Kapitulnik [62],

$$\xi^2(T) \sim \frac{1}{1-t} \begin{cases} \xi_0 \ell \quad , \quad k_F \xi_{loc} \sim 1 \quad , & (2.46a) \\ \ell_{T_c}^3 / \xi_{loc} \quad , \quad 1 \ll k_F \xi_{loc} < k_F \ell_{T_c} \quad , & (2.46b) \\ \ell_{T_c}^2 \quad , \quad 1 \ll k_F \ell_{T_c} < k_F \xi_{loc} \quad . & (2.46c) \end{cases}$$

With typical values for the density of states one obtains $\ell_{Tc} \sim 10$ Å for $T_c \sim 10$ K. This yields for the Ginzburg-Levanjuk parameter of Sec. 2.2.3 $\zeta \sim 10^{-3}$ at $k_F \ell \sim 1$. This means that the mean field theory will break down at the same time when localization effects become important. This opens the possibility to observe the critical region of dirty superconductors the same way as it has recently been done for a superconducting oxide [13]. The critical behavior of a superconductor is actually not known. Possibilities which have been discussed include an xy-model like transition [63] and a first order transition [64]. One should also keep in mind that the mean field theory will break down whenever ξ becomes as short as ~ 10 Å, regardless of the reason for ξ being short. This may be important in connection with other materials with an intrinsically short coherence length, e.g. like Chevrel phases.

2.4.5. Concluding Remarks

The theory of type-II superconductors has recently experienced a renewed surge of interest, which will probably continue in the face of the substantial experimental progress that has been made. Theory has not quite kept up with this development, there are many beautiful experiments on disordered superconductors which are not understood (this is true even if one excludes the superconducting oxides). It is not clear, for instance, how superconductivity eventually is destroyed at extreme disorder, nor if this destruction takes place above or below or at the localization transition. The main problem for a theoretical exploration of this regime is that the interplay between interaction and disorder discussed in Sec. 2.4 gets extremely complicated near a mobility edge. The problem of interaction and localization is not completely understood even in the absence of superconductivity, even though some progress has been made [65,66]. It has become clear, however, that the Coulomb interaction is substantially modified by localization effects, and in turn severely changes the localization behavior. The same is true for the electron-phonon coupling. The latter has only been explored in the context of sound attenuation [67,68], but by analogy the Eliashberg function will show critical behavior. This leads to the problem of anomalously enhanced coupling strengths with different signs. This is a continuation of the phenomenon we have encountered already at moderate disorder in Sec. 2.4.2. On the other hand, superconductivity near the Anderson transition has so far only been studied within models where the Coulomb interaction is neglected, and the electron-phonon coupling is kept fixed [62,69]. This leads to superconductivity even in the localized regime, an extreme manifestation of Anderson's theorem. The construction of a theory combining these approaches poses a major challenge for the future.

There are many more topics which have hardly or not at all been discussed in this chapter. One is the new high-T_c oxide superconductors. The more these materials are explored, the more it seems that they behave more or less like ordinary superconductors in surprisingly many respects. It is remarkable, for instance, that the structure of the vortex phase is the same [6]. The most spectacular feature distinguishing the oxides from ordinary materials is still the high T_c (and possibly the isotope effect). T_c, however, as any phase transition point, is notoriously hard to calculate, since in contrast to the critical behavior it depends on all the details of the system. It seems possible that on the phenomenological level, where T_c is a parameter, existing theory needs only few modifications. For instance, in view of the short coherence length of these materials a more thorough consideration of fluctuations might be appropriate, cf. Sec. 2.4.4. Also, the order parameter may be more complicated than in GL theory [13]. Of course, a microscopic theory for the new materials is a different issue altogether. Still any such theory will have to predict properties other than T_c which are different from those of ordinary superconductors.

All in all the theory of type-II superconductors appears to me as a building under construction. The top is far from being finished, but the foundations have survived thirty years of challenge, and are certainly strong enough to further build on them. Doubtlessly a lot of exciting physics is still to be found in the process.

Acknowledgements

I would like to thank M. R. Beasley, G. Bergmann, S. Das Sarma, R. C. Dynes, R. E. Glover, W. Götze, M. Gurvitch, A. F. Hebard, H. Kinder, V. Korenman, R. E. Prange, M. Weger, and K. I. Wysokinski, who all have contributed to my understanding of this subject. This work was supported by the Materials Science Institute at the University of Oregon.

References

[1] P. W. Anderson, *Phys. Rev.* **109** 1492 (1958).
[2] E. Abrahams, P. W. Anderson, D. C. Licciardello and T. V. Ramakrishnan, *Phys. Rev. Lett.* **42** 673 (1979).
[3] A. B. Pippard, *Phil. Mag.* **46** 1104 (1955).
[4] A. Schmid, *Z. Phys.* **271** 251 (1974).
[5] B. L. Altshuler and A. G. Aronov, *Sol. State. Commun.* **30** 115 (1979); B. L. Altshuler, A. G. Aronov and P. A. Lee, *Phys. Rev. Lett.* **44** 1288 (1980).
[6] L. P. Gammel, D. J. Bishop, G. J. Dolan, J. R. Kwo, C. A. Murray, C. F. Schneemeyer and J. V. Waszczak, *Phys. Rev. Lett* **59** 2592 (1987).
[7] P. G. de Gennes, *Phys. Kondens. Materie* **3** 79 (1964); *Superconductivity in Metals and Alloys* (Benjamin New York, 1966).
[8] R. D. Parks, (ed) *Superconductivity* (Dekker, New York, 1969).
[9] D. Saint-James, G. Sarma and E. J. Thomas, *Type II Superconductivity* (Pergamon, Oxford, 1969)
[10] M. Tinkham, *Introduction to Superconductivity* (McGraw-Hill, New York, 1975).
[11] V. L. Ginzburg and L. D. Landau, *Zh. Eksp. Teor. Fiz.* **20** 1064 (1950). An English translation of this paper can be found in D. ter Haar, *Men of Physics: L D Landau* (Pergamon Oxford, 1965).
[12] E. M. Lifshitz and L. P. Pitaevskii, *Statistical Physics* Part 2, Vol 9 of *Course of Theoretical Physics*, by L. D. Landau and E. M. Lifshitz (Pergamon New York, 1980)
[13] S. E. Inderhees, M. B. Salamon, N. Goldenfeld, J. P. Rice, B. G. Paizol, D. M. Ginsberg, J. Z. Liu, and G. W. Crabtree, *Phys. Rev. Lett.* **60**, 1178 (1988). *ibid.* **60**, 2445 (1988).
[14] F. London, *Superfluids* (Wiley New York 1950).
[15] R. Doll and M. Näbauer, *Phys. Rev. Lett.* **7** 51 (1961).
[16] B. S. Deaver and W. M. Fairbanks, *Phys. Rev. Lett.* **7** 43 (1961).
[17] W. A. Little and R. D. Parks, *Phys. Rev. Lett* **9** 9 (1962).
[18] A. A. Abrikosov, *Zh. Eksp. Teor. Fiz.* **32** 1442 [*Sov. Phys. JETP*

5 1174] (1957).
[19] W. H. Kleiner, L. M. Roth and S. H. Autler, *Phys. Rev.* **133** A1226 (1964).
[20] U. Essmann and H. Träuble, *Phys. Lett.* **24A** 526 (1967).
[21] P. W. Anderson, *J. Phys. Chem. Solids* **11** 26 (1959).
[22] L. P. Gorkov, *Zh. Eksp. Teor. Fiz.* **37** 1407 [*Sov. Phys. JETP* **37(10)** 998] (1959).
[23] V. Ambegaokar, in *Superconductivity*, ed. by R. D. Parks (Marcel Dekker New York, 1969)
[24] E. Abrahams, P. W. Anderson, P. A. Lee and T. V. Ramakrishnan, *Phys. Rev.* **B24** 6783 (1981).
[25] D. Forster, *Hydrodynamic Fluctuations, Broken Symmetry, and Correlation Functions* (Benjamin, New York 1975).
[26] P. A. Lee and M. G. Payne, *Phys. Rev.* **B5** 923 (1972).
[27] E. Helfand and N. R. Werthamer, *Phys. Rev.* **147** 288 (1966).
[28] K. Maki *Physics* **1** 127 (1964).
[29] N. R. Werthamer, E. Helfand and P. C. Hohenberg, *Phys. Rev.* **147** 295 (1966).
[30] G. Bergmann, *Phys. Rep.* **107** 1 (1984).
[31] P. Fulde and R. A. Ferrell, *Phys. Rev.* **135A** 550 (1964).
[32] A. A. Abrikosov, L. P. Gorkov and I. E. Dzyaloshinski, *Methods of Quantum Field Theory in Statistical Physics* (Prentice Hall Englewood Cliffs, 1963).
[33] G. M. Eliashberg *Zh. Eksp. Teor. Fiz.* **38** 966 [*Sov. Phys. JETP* **11** 696] (1960).
[34] P. Morel and P. W. Anderson, *Phys. Rev.* **125** 1263 (1962).
[35] H. Fukuyama, H. Ebisawa and S. Maekawa, *J. Phys. Soc. Japan* **53** 1919 (1984).
[36] D. Belitz, *J. Phys.* F **15** 2315 (1985).
[37] D. J. Scalapino, J. R. Schrieffer and J. W. Wilkins, *Phys. Rev.* **148**263 (1969).
[38] B. Keck and A. Schmid *J. Low Temp. Phys.* **24** 611 (1976).
[39] P. W. Anderson, K. A. Muttalib and T. V. Ramakrishnan, *Phys. Rev.* **B28** 117 (1983).
[40] D. Belitz, *Phys. Rev.* **B35** 1636 (1987).
[41] D. Belitz, *Phys. Rev.* **B35** 1651 (1987).
[42] D. Belitz, *Phys. Rev.* **B36** 47 (1987).
[43] W. L. McMillan, *Phys. Rev.* **167** 166 (1968).
[44] W. Buckel and R. Hilsch, *Z. Phys.* **138** 109 (1954).
[45] R. C. Dynes and J. P. Garno, *Phys. Rev. Lett.* **46** 137 (1981).
[46] A. Comberg, S. Ewert and G. Bergmann, *Z. Phys.* **271** 317 (1974).
[47] G. Bergmann, *Z. Phys.* **228** 25 (1969).
[48] M. Gurvitch, *Phys. Rev.* **B28** 544 (1983).
[49] D. B. Kimhi and T. H. Geballe, *Phys. Rev. Lett* **45** 1039 (1980).
[50] W. L. Johnson, *J.Appl. Phys.* **50**, 1557 (1979).
[51] L. R. Testardi, J. M. Poate and H. J. Levinstein, *Phys. Rev.*

B15 2570 (1977).

[52] M. Gurvitch, A. K. Ghosh, B. L. Gyorffy, H. Lutz, O. F. Kammerer, J. S. Rosner and M. Strongin, *Phys. Rev. Lett* **41** 1616 (1978).

[53] H. B. Radousky, T. Jarlborg, G. S. Knapp and A. J. Freeman, *Phys. Rev.* B **26** 1208 (1982).

[54] S. J. Bending, M. R. Beasley and C. C. Tsuei, *Phys. Rev.* **B30** 6342 (1984).

[55] T. P. Orlando, E. J. McNiff, S. Foner and M. R. Beasley, *Phys. Rev.* **B19** 4545 (1979).

[56] J. M. Rowell and R. C. Dynes, *Bad Metals, Good Superconductors* (unpublished, 1980).

[57] J. M. Rowell, R. C. Dynes and P. H. Schmidt, in *Superconductivity in d- and f-Band Metals*, ed. by H. Suhl and M. B. Maple (Academic New York, 1980).

[58] D. Belitz, *Phys. Rev. Lett.* **56** 1175 (1986).

[59] L. P. Gorkov, A. I. Larkin and D. E. Khmelnitskii,*Pisma Zh Eksp Teor Fiz* **30** 248 [*JETP Lett.* **30** 228] (1979).

[60] P. A. Lee and T. V. Ramakrishnan, *Rev. Mod. Phys.* **57** 287 (1985).

[61] D. Belitz and K. I. Wysokinski, *Phys. Rev.* **B36** 9333 (1987).

[62] G. Kotliar and A. Kapitulnik, *Phys. Rev.* **B33** 3146 (1986).

[63] C. Dasgupta and B. I. Halperin, *Phys. Rev. Lett.* **47** 1556 (1981).

[64] B. I. Halperin, T. C. Lubensky and S. K. Ma, *Phys. Rev. Lett* **32** 292 (1974).

[65] A. M. Finkelshtein, *Zh. Eksp. Teor. Fiz.* **84** 168 [*Sov. Phys. JETP* **57** 1] (1983).

[66] C. Castellani, C. Di Castro, P. A. Lee and M. Ma, *Phys. Rev.* **B30** 527 (1984).

[67] T. R. Kirpatrick and D. Belitz, *Phys. Rev.* B **34** 9008 (1986).

[68] C. Castellani and K. G. Kotliar, *Phys. Rev.* **B34** 9012 (1986).

[69] M. Ma and P. A. Lee, *Phys. Rev.* **B32** 5658 (1985).

[70] A. K. Ghosh and M. Strongin, in *Superconductivity in d- and f-Band Metals*, ref. 57.

CHAPTER 3

The Josephson Effect

Richard A. Ferrell

3.1. Phenomenology

3.1.1. Thermodynamics

Consider a pair of superconductors separated by an insulating barrier some 10 or 20 atomic layers thick. In each superconductor, the electron-phonon interaction leads to the formation of Cooper pairs. In 1962, Josephson realized that these electron pairs can tunnel through the barrier, resulting in an electric current in the absence of any potential difference across the barrier [1]. Because no energy is required to transfer an electron pair from one side of the barrier to the other, the quantum ground states of the junction in which different numbers of pairs have been transferred are degenerate. Let Φ_ν be the wave function representing the state in which ν pairs have been transferred. Quantum mechanics requires, in general, that one consider a superposition of these degenerate wave functions Φ_ν, of the form

$$\Psi = \sum_\nu A_\nu \Phi_\nu \quad , \tag{3.1}$$

where the numerical coefficients A_ν depend upon the details of the problem at hand.

With an approach somewhat different from that of Josephson and of Anderson [2], Ferrell and Prange [3] deduced the desired superposition by drawing an analogy to a familiar solid-state problem, the tight-binding model of an electron bound in a one-dimensional crystal lattice. In the tight-binding model, ions placed at equally spaced sites present to the electron an attractive periodic potential. The strongly bound lowest states in each well are all degenerate so that the degeneracy is equal to the number of ions in the lattice. Each bound state has small tails which extend through the classically forbidden regions into the neighboring wells, allowing the electron to tunnel from one well to another. The translational symmetry of the lattice requires, by the Floquet-Bloch theorem, that the correct zeroth-order ground state wavefunction be a superposition, in which the lowest bound state at each site is multiplied by a phase factor e^{ikna}, where n is an integral site coordinate (a is the lattice spacing) and k is the crystal momentum (we are using units in which Planck's constant, h, equals 2π). In the Josephson problem, the number of pairs transferred, ν, is the analog of the site coordinate, n, in the lattice problem. With a set equal to one, the desired superposition is

$$\Psi_\alpha = \sum_\nu e^{i\nu\alpha} \Phi_\nu \ , \tag{3.2}$$

where α, the analog of k, is a constant of the motion.

To determine the energy, we write the Hamiltonian as

$$H = H_o + H' \ , \tag{3.3}$$

where, in addition to the zero-order part H_o, we include the perturbing term H', which splits the degeneracy. In the tight-binding model, H' represents the tunneling of the electron from one site to another. In the Josephson problem, H' transfers a pair across the barrier. The energy splitting is given by the expectation value of H':

$$E(\alpha) = (\Psi_\alpha, H'\Psi_\alpha)/(\Psi_\alpha, \Psi_\alpha)$$

$$= \frac{\sum_{\nu,\nu'} e^{i\alpha(\nu-\nu')} (\Phi_{\nu'}, H'\Phi_\nu)}{\sum_{\nu,\nu'} e^{i\alpha(\nu-\nu')} (\Phi_{\nu'}, \Phi_\nu)} \ . \tag{3.4}$$

We will examine the precise nature of the transfer Hamiltonian, H', when we discuss the microscopic theory. For now, we write the tunneling matrix element as

$$(\Phi_{\nu'}, H'\Phi_\nu) = -\frac{E_J}{2}(\delta_{\nu',\nu+1} + \delta_{\nu',\nu-1}) \ , \tag{3.5}$$

corresponding to the two possibilities of transferring an electron pair in either direction. In the denominator of Eq. (3.4), orthogonality requires

$$(\Phi_{\nu'},\Phi_{\nu}) = \delta_{\nu',\nu} \quad . \tag{3.6}$$

Thus

$$E(\alpha) = -\frac{E_J}{2}\frac{\sum_{\nu}(e^{i\alpha}+e^{-i\alpha})}{\sum_{\nu} 1} = -E_J\cos\alpha \quad . \tag{3.7}$$

The simple form of the dependence of the energy on α in Eq. (3.7) is familiar from the tight-binding model. By analogy to this problem, α, the analog of the crystal wave number, is canonically conjugate to ν, the analog of the lattice coordinate. Postponing the consideration of intrinsic quantum mechanical effects, which become important for very small junctions, we treat ν and α as canonically conjugate *classical* variables. Hamilton's equation for the time rate of change of the coordinate ν is then

$$\dot{\nu} = \frac{\partial E}{\hbar\partial\alpha} = \frac{1}{\hbar} E_J \sin\alpha, \tag{3.8}$$

where $2\pi\hbar$ is Planck's constant. (In a fully quantum-mechanical treatment, this equation is valid for the operators in the Heisenberg picture, or for the expectation values in the Schrödinger picture.) The electric current across the barrier, J, is obtained by multiplying the pair transfer rate, $-\dot{\nu}$, by the charge, $-2e$, of each pair, so that

$$J = 2e\dot{\nu} = I_J \sin\phi, \tag{3.9}$$

where $I_J = 2eE_J/\hbar$ is the amplitude of the Josephson current. (We have now changed to the more conventional notation and are writing ϕ for the Cooper pair phase α. Here, $e = |e|$ is the absolute value of the electronic charge.) This is the equation describing the DC Josephson effect: A constant current, determined by the phase, flows across the junction in the absence of any applied voltage. The current vanishes for $\phi = 0$ and reaches its extremal values of $\pm I_J$ for $|\phi| = \pi/2$. A further increase of $|\phi|$ reduces the magnitude of the current, a phenomenon that we will examine when discussing the AC Josephson effect.

Because ϕ is the Cooper pair phase, it is possible to "current bias" the junction and to control the DC Josephson current by connecting the junction in series with a superconducting inductor of inductance L. The energy of the inductor is given by the usual formula $LI^2/2$, where the current, I, is proportional to ϕ_L, the phase drop across the inductor. The total phase accumulated by going around the series circuit of inductor and junction is $\phi_L + \phi = \phi_T = 2\pi\ell$, where ℓ, the "winding number", is an integer, i.e. a quantum number of the system. ϕ_T is therefore a constant of the motion. The energy stored in the inductor is proportional to

$$\phi_L^2 = (\phi_T - \phi)^2 = \phi_T^2 - 2\phi_T\phi + \phi^2 \quad , \tag{3.10}$$

the first term being a constant, which can be dropped. If the phase change across the inductor is large compared to ϕ, the term quadratic in ϕ may also be dropped, so that the energy of the coupled system is given by

$$U = - E_J \cos\phi - C_1\phi \quad , \tag{3.11}$$

where C_1 is a constant proportional to $L\phi_T$. In effect, we have a classical problem in which the phase, ϕ, adjusts itself to minimize this total energy. (Later, we will consider capacitive charging effects which bring in the canonically conjugate variable ν and which can lead to appreciable quantum fluctuations of the phase.) Minimizing U, we obtain

$$E_J \sin\phi = C_1 \quad , \tag{3.12}$$

so that the junction current is

$$J = I_J \sin\phi = \frac{2eC_1}{\hbar} \quad . \tag{3.13}$$

The tilted sinusoidal variation of the potential energy, as a function of ϕ, has given rise to the name "washboard potential". The minimization fixes the phase at a value off the usual Josephson coupling energy minima, viz., for ϕ equal to an integral multiple of 2π, so that a non-vanishing current flows through the junction. The more the washboard is tilted, the greater is the current, until a point of maximal biasing, at $\sin\phi = 1$, is reached. Further tilting brings about an unstable, runaway situation. Because of quantum mechanical tunneling, under certain conditions runaway can set in even before the classical condition of $\sin\phi = 1$ is reached.

A mechanical analog for the Josephson junction is the pendulum because, here, too, the potential energy has a $\cos\phi$ dependence, with ϕ now being the angle that the pendulum has swung away from the vertical. The pendulum can be imagined as a massless wheel, of radius ℓ and pivoted at its center. A mass m is attached to the rim of the wheel. In addition, in order to simulate the effect of the current-biasing inductor, we imagine that the rim serves also as a spool around which a string is wound. A mass m' is attached to the loose end of the string and hangs vertically down, thereby exerting a torque $m' g\ell$ on the wheel, g being the acceleration of gravity. The resulting total energy,

$$U = mg\ell(1 - \cos\phi) - m' g\ell\phi \quad , \tag{3.14}$$

is of the desired form. This mechanical system will stabilize, with m tilted away from the vertical direction at the angle

$$\phi = \sin^{-1} \frac{m'}{m} \quad , \tag{3.15}$$

provided $m' < m$. For $m' > m$, m will pass beyond the horizontal orientation of the rod and then tumble over and accelerate. Such mechanical analogies may be exploited to develop an intuitive understanding of the behavior of Josephson junctions.

A Josephson junction may also be biased with a source of voltage V in series with a resistor of resistance R so as to call for the current $I = V/R$ from the junction. For some purposes, this is less desirable than using an inductor because the resistor introduces Johnson noise.

3.1.2. Time Dependence

If a voltage V is applied across a Josephson junction, taking a pair across will change the energy by $-2eV$. The time-dependent Schrödinger equation then requires the pair phase to change at the rate

$$\dot{\phi} = \frac{2eV}{\hbar} \quad . \tag{3.16}$$

If V is a constant, say V_o, it follows that

$$\phi = \frac{2eV_o t}{\hbar} + \text{const.} \quad , \tag{3.17}$$

so that the current flowing through the junction is

$$J = I_J \sin\phi = I_J \sin\left[\frac{2eV_o t}{\hbar} + \text{const.}\right] \quad . \tag{3.18}$$

This is the AC Josephson effect: An applied DC voltage leads to an alternating current of frequency proportional to the voltage.

Another kind of oscillation is the so-called Josephson plasma oscillation where, by virtue of the junction capacitance, C, the changing phase results from the voltage that is produced by the Josephson current itself. From Eq. (3.16), the rate of change of the charge transferred through the junction on open circuit is

$$J = -\dot{Q} = -C\dot{V} = -\frac{C}{2e\hbar}\ddot{\phi} \quad . \tag{3.19}$$

As before,

$$J = I_J \sin\phi \quad , \tag{3.20}$$

so that

$$\ddot{\phi} = -\omega_J^2 \sin\phi \quad , \tag{3.21}$$

where the "Josephson plasma frequency" is

$$\omega_J = \left(\frac{2eI_J}{\hbar C}\right)^{\frac{1}{2}} \quad . \tag{3.22}$$

For small oscillations of ϕ about zero, Eq. (3.21) can be linearized and ω_J is the actual frequency of vibration. This "Josephson plasmon" is a natural mode of oscillation of the system, just as the ordinary plasmon is a mode of oscillation of the mobile electrons in a metal.

On the other hand, if ϕ undergoes small oscillations, $\Delta\phi$, about a nonzero bias value ϕ_o, Eq. (3.21) can be linearized by

$$\sin\phi = \sin(\phi_o + \Delta\phi) \simeq \cos\phi_o \cdot \Delta\phi \quad , \tag{3.23}$$

which lowers the "plasmon" frequency by the factor $(\cos\phi_o)^{\frac{1}{2}}$. In problems of this kind it is convenient to characterize the Josephson junction as a phase-dependent inductor. In general, instead of integrating Eq. (3.16), we can differentiate Eq. (3.23) to obtain

$$\dot{J} = I_J\left[\cos\phi\right]\dot{\phi} = \frac{2eI_J}{\hbar}\cos\phi \; V \quad . \tag{3.24}$$

The coefficient of V is, by definition L^{-1}, the reciprocal of the inductance L:

$$L = \frac{\hbar}{2eI_J} \cdot \frac{1}{\cos\phi} \quad . \tag{3.25}$$

The natural resonant frequency of the junction, regarded as an "LC" circuit, is therefore

$$\omega = \frac{1}{\sqrt{LC}} = \omega_J(\cos\phi)^{\frac{1}{2}} \quad , \tag{3.26}$$

with the dependence on bias phase that was discussed above. We note in passing the intrinsic instability of a Josephson junction for phases such that $\cos\phi < 0$, as manifested by the unphysical negative inductance that is then predicted by Eq. (3.25).

Consider now the situation in which both a static voltage, V_o, and an oscillating voltage, $V_\omega \sin \omega t$, are applied to the junction. To understand the effect of the modulation of V_o by the oscillating voltage, one can visualize the motion of the unit vector in the complex plane, $\exp(i\phi) = \cos\phi + i\sin\phi$. The tip of the unit vector moves around the unit circle at the rate $\dot{\phi}(t) = 2eV_o/\hbar + (2eV_\omega/\hbar)$ $\sin\omega t$. When $\omega t = \pi/2, 5\pi/2, 9\pi/2$, etc., the passage occurs more quickly, and when $\omega t = 3\pi/2, 7\pi/2, 11\pi/2$, etc., the passage is slower. If the synchronization condition $\omega = 2eV_o/\hbar$, is satisfied, the effect is cumulative and a DC current can result. This occurs when the time average of the projection of the unit vector along the imaginary axis is nonzero, or, in other words, when the unit vector repeatedly traverses the upper half of the unit circle faster than the lower half. Quantitatively, integrating the voltage yields

$$\phi(t) = \frac{2eV_o t}{\hbar} - \frac{2e}{\hbar\omega} V_\omega \cos\omega t + \phi_o \quad , \tag{3.27}$$

where ϕ_o is a constant of integration. Substituting into Eq. (3.9) gives

$$J(t) = I_J \sin\phi(t) = I_J \sin\left[\frac{2eV_o t}{\hbar} + \phi_o - \frac{2eV_\omega}{\hbar\omega}\cos\omega t\right] \quad . \tag{3.28}$$

For $eV_\omega \ll \hbar\omega$ we can expand Eq. (3.28) in a Taylor's series. To first order, it becomes

$$J(t) \simeq I_J \sin\left[\frac{2eV_o t}{\hbar} + \phi_o\right] - \frac{2eV_\omega}{\hbar\omega} I_J \cos\left[\frac{2eV_o t}{\hbar} + \phi_o\right] \cos\omega t \quad . \tag{3.29}$$

Because of the nonlinear response of the junction, the oscillating voltage produces sidebands in the current at the frequencies $2eV_o/\hbar \pm \omega$. When $\omega = 2eV_o/\hbar$ one of the sidebands comes down to zero frequency, giving a DC current. The strength of this current is evidently, from Eq. (3.29), equal to

$$\overline{J(t)} = \frac{2eV_\omega}{\hbar\omega} I_J \cos\phi_o \overline{\cos^2\omega t} = \frac{e}{\hbar\omega} V_\omega I_J \cos\phi_o \quad . \tag{3.30}$$

If the term quadratic in eV_ω/ω in Eq. (3.28) is retained, a second pair of sidebands appears,

$$J_2(t) = I_J \left[\frac{eV_\omega}{\hbar\omega}\right]^2 \sin(2eV_o t + \phi_o) \cos 2\omega t \quad . \tag{3.31}$$

Thus, a DC current can be obtained for $2eV_o/\hbar = 2\omega$. Indeed, by expressing this "frequency modulation" of the current in terms of Bessel's functions, one sees that there are sidebands at $2eV_o/\hbar \pm n\omega$, where n is any integer. Therefore, the general condition for a DC current and voltage steps is

$$V_o = \frac{\hbar\omega}{2e} \text{ x integer} \quad , \tag{3.32}$$

with a spacing between the steps of $\hbar\omega/2e$. It follows that a measurement of the frequency determines the voltage in terms of the ratio of the fundamental constants, $2e/\hbar$. This feature has been exploited for constructing a voltage standard.

3.1.3. Large Junctions

While the insulating barrier separating the two superconductors is only tens of Ångstroms thick, a junction may be a good fraction of a millimeter or more in its lateral dimensions. The spatial variation of the phase in the tangential directions x and y must therefore be taken into account. A space-dependent phase implies, according to Eq. (3.9), the spatially varying current density

$$J(x,y) = I_J \sin\phi(x,y) \quad , \tag{3.33}$$

where, now, I_J and J are defined per unit area. In practice, such static phase variations are obtained by the application of a steady magnetic field,

$$\mathbf{B} = \mathrm{curl}\mathbf{A} \quad , \tag{3.34}$$

where **A** is the vector potential. From the usual expression relating velocity and momentum for electron pairs of charge $q = -2e$ and mass $m' = 2m$,

$$\mathbf{v} = \frac{1}{m'}(\mathbf{p} - \frac{q}{c}\mathbf{A}) \quad , \tag{3.35}$$

where c is the velocity of light, we obtain

$$\mathbf{p} = -\frac{2e}{c}\mathbf{A} \quad . \tag{3.36}$$

This holds at points more deeply inside the superconductors than λ, the penetration depth, where $\mathbf{v} = 0$. The change in pair momentum across the junction is

$$\Delta\mathbf{p} = -\frac{2e}{c}\Delta\mathbf{A} = -\frac{2e}{c}\nabla\Phi(x,y) \quad , \tag{3.37}$$

where, by Stoke's theorem, $\Phi(x,y)$ is the amount of magnetic flux within the junction that passes by one side of the point (x,y). Substituting from Eq. (3.37), we find

$$\nabla\phi(x,y) = \frac{1}{\hbar}\Delta\mathbf{p} = -\frac{2\pi}{\Phi_o}\nabla\Phi(x,y) \quad , \tag{3.38}$$

or

$$\phi(x,y) = -2\pi\frac{\Phi(x,y)}{\Phi_o} + \text{const.} \quad , \tag{3.39}$$

where $\Phi_o = hc/2e$ is the flux quantum. Substitution into Eq. (3.9) yields the spatially dependent Josephson current density

$$J(x,y) = -I_J\sin\left(2\pi\frac{\Phi(x,y)}{\Phi_o} + \text{const.}\right) \quad . \tag{3.40}$$

For a magnetic field B in the y direction,

$$\Phi(x,y) = h_J Bx \quad , \tag{3.41}$$

where $h_J = 2\lambda + \tau$ is the effective magnetic thickness of the barrier, λ, the penetration depth, and τ, the actual geometrical thickness. In this case, Eq. (3.40) assumes the simple form

$$J(x) = I_J\sin(kx + \text{const.}) \quad , \tag{3.42}$$

where

$$k = \frac{2\pi B h_J}{\Phi_o} \quad . \tag{3.43}$$

For a junction of rectangular shape, of length L in the x-direction, the total current (per unit length in the y-direction) is

$$J_T = \int_{L/2}^{L/2} J(x)\mathrm{d}x \quad . \tag{3.44}$$

Choosing the constant of integration to be $\pm\pi/2$, so as to maximize J_T, gives, by substitution from Eq. (3.43),

$$J_T^{max} = I_J \left| \int_{-\frac{L}{2}}^{\frac{L}{2}} \cos(kx)\mathrm{d}x \right| = I_J \frac{|\sin kL/2|}{k/2}$$

$$= I_J L \frac{|\sin(\pi\Phi_T/\Phi_o)|}{\pi\Phi_T/\Phi_o} \quad , \tag{3.45}$$

where $\Phi_T = Bh_J L$ is the total flux passing through the junction. If the magnetic field is very large, so that $\Phi_T \gg \Phi_o$, $J_T^{max} \rightarrow 0$. This expression for J_T^{max} is mathematically equivalent to that describing the diffraction pattern of light passing through a slit of width L. For $B = 0$, the entire junction has the same phase and is contributing coherently. For $B \neq 0$ the phase uniformity is destroyed and destructive interference sets in. The first minimum of the "diffraction pattern" occurs for $\Phi_T = \Phi_o$, or for $k = 2\pi/L$. The current distribution for this case is

$$J(\mathrm{x}) = I_J\cos(2\pi x/L) \quad . \tag{3.46}$$

Thus the current density at the edges, at $x = \pm L/2$, is equal in magnitude and opposite in sign to that at the center. The integrated contributions from the center and sides of the junction exactly cancel, in this case, to give $J_T = 0$.

3.1.4. Self-Field Effects

The magnetic flux that produces the dephasing of the Josephson current that was discussed in the previous section is not necessarily set up exclusively by electrical currents external to the junction. Visualizing the barrier as lying horizontally, parallel to the x-y plane, the Josephson current, $J(x,y)$, that flows upward in the z-direction through the barrier gives rise to a surface current $\mathbf{S}(x,y)$ in the upper superconductor (and to $-\mathbf{S}$ in the lower superconductor). These currents are related by the continuity equation,

$$\nabla \cdot \mathbf{S} = -J = -I_J \sin\phi(x,y) \quad . \tag{3.47}$$

By Ampere's law, the magnetic field that is set up by the surface current is given by

$$\hat{\mathbf{z}} \times \mathbf{B} = -\frac{4\pi}{c}\mathbf{S} \quad , \tag{3.48}$$

where $\hat{\mathbf{z}}$ is the unit vector normal to the barrier. The differential form of Eq. (3.41),

$$\nabla \Phi = -h_J \, \hat{\mathbf{z}} \times \mathbf{B} \quad , \tag{3.49}$$

substituted into Eq. (3.38) and (3.48) yields

$$\nabla \phi = \frac{2\pi h_J}{\Phi_o} \hat{\mathbf{z}} \times \mathbf{B} = -\frac{8\pi e h_J}{\hbar c^2} \mathbf{S} \quad . \tag{3.50}$$

Eliminating $\mathbf{S}$ from Eqs. (3.47) and (3.50) gives

$$\nabla^2 \phi(x,y) = \frac{1}{\lambda_J^2} \sin\phi(x,y) \quad , \tag{3.51}$$

where λ_J, the Josephson penetration depth, is determined by

$$\lambda_J^{-2} = \frac{8\pi e h_J}{\hbar c^2} I_J \quad . \tag{3.52}$$

If $\phi(x,y)$ is everywhere small, it decays exponentially away from the edges of the junction, with λ_J being the characteristic decay length.

Because of the nonlinear nature of Eqs. (3.9) and (3.51), the current cannot be increased indefinitely by increasing ϕ. For a junction of lateral dimensions much larger than λ_J, only the edges, of effective width of $O(\lambda_J)$, will contribute to the total current passing through the junction, with the interior remaining inactive. Thus, for such a very large junction, of perimeter P, the maximum Josephson current is of order

$$J_T^{max} \sim O(P\lambda_J I_J) \propto P I_J^{\frac{1}{2}} \ . \tag{3.53}$$

The effect of increasing the Josephson strength, I_J, is partially frustrated by the shrinkage in the Josephson penetration depth that is required by Eq. (3.52).

In the Josephson plasma oscillations of Section 3.1.2 above, we have already encountered one type of self-field effect. Combining the considerations that lead to Eqs. (3.21) and (3.51), one arrives at the Josephson wave equation

$$\nabla^2\phi - \frac{1}{c_J^2}\ddot{\phi} = \frac{1}{\lambda_J^2}\sin\phi \ , \tag{3.54}$$

where

$$c_J^2 = \omega_J^2\lambda_J^2 = \frac{c^2}{4\pi C h_J} \ . \tag{3.55}$$

The capacitance per unit area of a parallel-plate capacitor containing a dielectric slab of thickness τ and dielectric constant ε is

$$C = \frac{\varepsilon}{4\pi\tau} \ . \tag{3.56}$$

From Eqs. (3.55) and (3.56) the ratio of c_J to $c_\varepsilon = c/\sqrt{\varepsilon}$, the velocity of light in an unbounded medium of dielectric constant ε, is

$$\frac{c_J}{c_\varepsilon} = \left(1 + 2\frac{\lambda}{\tau}\right)^{-\frac{1}{2}} \ . \tag{3.57}$$

Because λ is generally much larger than τ, $c_J \ll c_\varepsilon$.

3.2. Microscopic Theory

3.2.1. The BCS Ground State

We now apply the BCS theory of superconductivity to justify our phenomenological treatment of the Josephson effect. This will give us a connection between the amplitude of the Josephson current and the normal-state conductivity of the junction. We begin by recalling the form of the ground state wave function of a single superconductor. An attractive interaction leads to the pairing of electrons of opposite spins and antiparallel momenta. If $a^{\dagger}_{p\sigma}$ denotes the creation operator for an electron of momentum **p** and spin σ, the BCS ground state is

$$\Psi_{BCS} = \prod_{\mathbf{p}} \left(u_{\mathbf{p}} + v_{\mathbf{p}} a^{\dagger}_{\mathbf{p}\uparrow} a^{\dagger}_{-\mathbf{p}\downarrow} \right) \Psi_{VAC} \quad , \tag{3.58}$$

where Ψ_{VAC} is the vacuum state. Thus, the ground state is a product of pair wavefunctions in which a pair state is occupied, with amplitude v_p, or empty, with amplitude u_p. The sum of the probabilities for these two cases is

$$|u_p|^2 + |v_p|^2 = 1 \quad . \tag{3.59}$$

The coefficients can be taken to be real and positive without loss of generality (any relative phase can be absorbed in the creation operators). A variational calculation yields

$$v_p^2 = \frac{1}{2}\left(1 - \frac{\varepsilon_p}{E_p}\right) \quad , \tag{3.60}$$

where

$$\varepsilon_p = \frac{\hbar^2 p^2}{2m} - \mu \tag{3.61}$$

and

$$E_p = \left(\varepsilon_p^2 + \Delta^2 \right)^{\frac{1}{2}} \tag{3.62}$$

is the quasiparticle energy. μ is the Fermi energy. The energy gap, Δ, sets the scale for the smoothing of the normal-state step function behavior of u_p and v_p, i.e., $v_p^2 = \theta(-\varepsilon_p)$. From Eqs. (3.59, 60, and 62) follows the important identity

$$u_p v_p = \frac{\Delta}{2E_p} \quad . \tag{3.63}$$

In order to understand the Josephson effect from the point of view of the microscopic theory, rather than simply on the basis of the phenomenological approach presented in Section 3.1.1 above, it is necessary to study single-electron tunneling in the BCS theory. In this way, one can arrive at a detailed description of pair tunneling. The basic ingredient in the microscopic formalism concerns the addition to, or the removal of, an electron from the BCS ground state. Because of the Pauli exclusion principle, it is evident that the addition of an electron will block one of the pair states in the product in Eq. (3.58). It follows that the application of the creation operator $a^\dagger_{\mathbf{p}\uparrow}$ yields

$$a^\dagger_{\mathbf{p}\uparrow}\Psi_{BCS} = u_{\mathbf{p}}\Psi_{\mathbf{p}\uparrow} \;, \tag{3.64}$$

where the normalized one-quasiparticle state is

$$\Psi_{\mathbf{p}\uparrow} \equiv \prod_{\mathbf{p}' \neq \mathbf{p}} (u_{\mathbf{p}'} + v_{\mathbf{p}'}a^\dagger_{\mathbf{p}'\uparrow}a^\dagger_{-\mathbf{p}'\downarrow})a^\dagger_{\mathbf{p}\uparrow}\Psi_{VAC} \;. \tag{3.65}$$

For removing an electron from the time-reversed state of momentum -**p** and spin down we need to employ in succession the anticommutation relations

$$a_{-\mathbf{p}\downarrow}a^\dagger_{\mathbf{p}\uparrow} = -\,a^\dagger_{\mathbf{p}\uparrow}a_{-\mathbf{p}\downarrow} \tag{3.66a}$$

and

$$a_{-\mathbf{p}\downarrow}a^\dagger_{-\mathbf{p}\downarrow} = 1 - a^\dagger_{-\mathbf{p}\downarrow}a_{-\mathbf{p}\downarrow} \;. \tag{3.66b}$$

Applied to Ψ_{VAC}, the last term vanishes. Thus we find

$$a_{-\mathbf{p}\downarrow}\Psi_{BCS} = -\,v_{\mathbf{p}}\Psi_{\mathbf{p}\uparrow} \;, \tag{3.67}$$

the same quasiparticle state as defined by Eq. (3.65). Denoting the single-electron state $\mathbf{p}_\downarrow$ by s, and its time inversion by $\bar{s}$, we obtain Eqs. (3.64 and 67) in the more compact form

$$a^\dagger_s\Psi_{BCS} = u_s\Psi_s \tag{3.68}$$

and

$$a_{\bar{s}}\Psi_{BCS} = -\,v_s\Psi_s \;. \tag{3.69}$$

Similarly, we find

$$a^\dagger_{\bar{s}}\Psi_{BCS} = u_s\Psi_{\bar{s}} \tag{3.70}$$

and

$$a_s\Psi_{BCS} = v_s\Psi_{\bar{s}} \;. \tag{3.71}$$

3.2.2. The Tunneling Hamiltonian

To motivate the application of BCS theory to the problem of two superconductors separated by an oxide barrier, we first recall the familiar problem of a single particle moving in a double potential well consisting of two deep minima. The energy eigenfunctions of this latter problem are generally, to a good approximation,

$$\psi_e = \frac{1}{\sqrt{2}} (\psi_r + \psi_\ell) \tag{3.72}$$

and

$$\psi_o = \frac{1}{\sqrt{2}} (\psi_r - \psi_\ell) \quad , \tag{3.73}$$

where ψ_r and ψ_ℓ are the wave functions corresponding to states localized in the right-hand or left-hand well, respectively. If the small energy splitting between the even and odd states is $2T$, we can write the splitting in operator form as

$$H' = T(a_o^\dagger a_o - a_e^\dagger a_e) \quad , \tag{3.74}$$

where the operators $a_o^\dagger$ and a_o create and annihilate a particle in the odd state, respectively. Similarly for $a_e^\dagger$ and a_e. We can alternatively express H' in terms of the operators that create or destroy a particle in the left or right well. By substituting into Eq. (3.74) the operator transformations corresponding to Eqs. (3.72, 73), namely,

$$a_e^\dagger = \frac{1}{\sqrt{2}} (a_r^\dagger + a_\ell^\dagger) \tag{3.75}$$

and

$$a_o^\dagger = \frac{1}{\sqrt{2}} (a_r^\dagger - a_\ell^\dagger) \quad , \tag{3.76}$$

(and similarly for the annihilation operators), we obtain

$$H = - T(a_r^\dagger a_\ell + a_\ell^\dagger a_r) = - Ta_r^\dagger a_\ell + \text{h.c.} \quad . \tag{3.77}$$

Evidently, H' is an operator that effects particle tunneling from one well to the other with the transition matrix element $-T$. For the many-electron problem of two metals separated by a barrier, H' is generalized to a sum over all possible electron states on the left and on the right of the barrier:

$$H' = \sum_{r\ell} T_{r\ell} a_r^{\dagger} a_{\ell} + \text{h.c.} \quad . \tag{3.78}$$

Eq. (3.78) can, alternatively, be written in terms of the time-inverted states as

$$H' = \sum_{r\ell} (T_{r\ell}\, a_{\bar{\ell}}^{\dagger} a_{\bar{r}} + \text{h.c.}) \quad . \tag{3.79}$$

An intermediate step in the tunneling of an electron pair is described by the wave function $\Psi^{(\nu)}_{r\bar{\ell}}$ corresponding to the excited state of a quasiparticle in both the right-hand and left-hand superconductors. Starting with Φ_{ν}, as defined in Section 3.1.1, and moving one electron from the left to the right, we find from Eqs. (3.68-71),

$$a_r^{\dagger} a_{\ell} \Phi_{\nu} = u_r v_{\ell} \Psi^{(\nu)}_{r\bar{\ell}} \quad , \tag{3.80}$$

which defines the phase of $\Psi^{(\nu)}_{r\bar{\ell}}$. The essence of the Josephson effect is now captured in the observation that the very same two-quasiparticle state can be reached by starting from $\Phi_{\nu+1}$ and transferring an electron in the opposite direction, i.e. from right to left, according to

$$a_{\bar{\ell}}^{\dagger} a_{\bar{r}} \Phi_{\nu+1} = v_r u_{\ell} \Psi^{(\nu)}_{r\bar{\ell}} \quad . \tag{3.81}$$

(For the correct relative sign in Eq. (3.81) it is necessary to respect the anticommutator $\{a_r^{\dagger}, a_{\bar{\ell}}^{\dagger}\} = 0$.) Alternatively, which brings us even closer to the actual tunneling of a pair, we consider the two-step process of repeated application of H' so as to pass from Φ_{ν} to $\Phi_{\nu+1}$. By means of Eqs. (3.78) and (3.81, 82), we find for the matrix elements describing these two steps,

$$(\Psi^{(\nu)}_{r\bar{\ell}} \, , \, H' \Phi_{\nu}) = T_{r\ell} u_r v_{\ell} \tag{3.82}$$

$$(\Phi_{\nu+1}, H' \Psi^{(\nu)}_{r\ell}) = T^{*}_{r\ell} v_r u_\ell \quad . \tag{3.83}$$

In the next section, by applying perturbation theory, we will obtain from these equations an explicit expression for the strength of Josephson tunneling.

3.2.3. Josephson Coupling Energy

The effective pair tunneling matrix element, $-E_J/2$, that was introduced phenomenologically in Section 3.1.1, can now be written down explicitly in terms of the two single-electron tunneling matrix elements, Eqs. (3.82, 83), and the excitation energy of the two-quasiparticle intermediate state, $E_r + E_l$. Second-order perturbation theory yields

$$-\frac{E_J}{2} = -\sum_{r\ell} \left(E_r + E_\ell\right)^{-1} \left(\Phi_{\nu+1}, H' \Psi^{(\nu)}_{r\ell}\right) \left(\Psi^{(\nu)}_{r\ell}, H' \Phi_\nu\right) \quad . \tag{3.84}$$

Substitution from Eq. (3.82, 83) and (3.63) gives

$$E_J = 2 \sum_{r\ell} \left|T_{r\ell}\right|^2 \frac{u_r v_r u_\ell v_\ell}{E_r + E_\ell} = \frac{T^2}{2} \sum_{r\ell} \frac{\Delta_r \Delta_\ell}{E_r E_\ell (E_r + E_\ell)} \quad , \tag{3.85}$$

where we have abbreviated the average value of $|T_{r\ell}|^2$ by T^2. Restricting ourselves to the equal gap case, $\Delta_r = \Delta_\ell = \Delta$, including a factor of two for spin degeneracy, and introducing $N(0)$, the density of states at the Fermi surface (for given spin), we can reduce Eq. (3.85) to an integration:

$$E_J = T^2 \Delta^2 N(0)^2 \int_{-\infty}^{\infty} \int_{-\infty}^{\infty} \frac{d\varepsilon_1 \, d\varepsilon_2}{E_1 E_2 (E_1 + E_2)} \quad . \tag{3.86}$$

The integral is readily evaluated as π^2/Δ by substituting

$$\frac{\varepsilon_{1,2}}{\Delta} = \sinh u_{1,2} \tag{3.87}$$

and

$$\frac{E_{1,2}}{\Delta} = \cosh u_{1,2} \quad , \tag{3.88}$$

and by changing to the variables $u_1 \pm u_2$ following the application of standard trigonometric identities. Thus the Josephson coupling energy parameter is

$$E_J = \pi^2 T^2 N(0)^2 \Delta \quad , \tag{3.89}$$

and the Josephson current amplitude is

$$I_J = \frac{2e}{\hbar}\pi^2 T^2 N(0)^2 \Delta \quad . \tag{3.90}$$

As I_J is the maximum zero-voltage current that can flow across the junction, it is natural to ask what will happen when the junction is inserted into a circuit that calls for a current greater than I_J. The answer (at zero-temperature) is that a voltage $V > 2\Delta/e$ will be set up and that actual energy-conserving quasiparticle excitations will take place. By Fermi's "Golden Rule" of time-dependent perturbation theory, the magnitude of the quasiparticle current is

$$\begin{aligned} I_{Q.P.}(V) &= \frac{2\pi e}{\hbar}\sum_{r\ell} \left|(\Psi^{(v)}_{r\ell}, H'\Phi_v)\right|^2 \delta(E_r + E_\ell - eV) \\ &= \frac{2\pi e}{\hbar} T^2 \sum_{r\ell} u_r^2 v_\ell^2 \delta(E_r + E_\ell - eV) \\ &= \frac{4\pi e}{\hbar} T^2 N(0)^2 \int_0^\infty\int_0^\infty d\varepsilon_1 d\varepsilon_2 \delta(E_1 + E_2 - eV) \quad , \end{aligned} \tag{3.91}$$

by substitution of Eqs. (3.82), (3.60), and (3.59). The coherence factors can be replaced by unity upon summing over both branches of the quasiparticle spectrum. Let us carry out the integration near threshold, for $0 < eV - 2\Delta \ll \Delta$. For this we need

$$\frac{d\varepsilon}{dE} = \frac{E}{\varepsilon} = \frac{E}{(E^2 - \Delta^2)^{\frac{1}{2}}} \simeq \left(\frac{\Delta}{2}\right)^{\frac{1}{2}} (E - \Delta)^{-\frac{1}{2}} \quad , \tag{3.92}$$

approximated for $0 < E - \Delta \ll \Delta$. The integral then becomes

$$\int_\Delta^\infty\int_\Delta^\infty dE_1 dE_2\left(\frac{d\varepsilon_1}{dE_1}\right)\left(\frac{d\varepsilon_2}{dE_2}\right)\delta(E_1+E_2-eV)$$

$$\simeq\frac{\Delta}{2}\int_\Delta^\infty\int_\Delta^\infty\frac{dE_1 dE_2}{(E_1-\Delta)^{\frac{1}{2}}(E_2-\Delta)^{\frac{1}{2}}}\delta(E_1+E_2-eV)$$

$$=\frac{\Delta}{2}\beta(\tfrac{1}{2},\tfrac{1}{2})=\frac{\pi}{2}\Delta\ . \tag{3.93}$$

Substituting Eq. (3.93) into Eq. (3.91) and comparing with Eq. (3.90) yields the following identity for the quasiparticle current at voltage threshold:

$$I^{thres.}_{Q.P.}=I_J\ . \tag{3.94}$$

Thus, in a sense, the quasiparticle current picks up where the Josephson current leaves off.

It remains to relate I_J to σ_N, the conductivity of the junction in its normal state. In this case the coherence factors become step functions and Fermi's Golden Rule gives for the strength of the current,

$$I_N(V)=\sigma_N V=\frac{2\pi e}{\hbar}T^2\sum_{r\ell}\theta(\varepsilon_r)\theta(-\varepsilon_\ell)\delta(\varepsilon_r-\varepsilon_\ell-eV)\ . \tag{3.95}$$

With the sums converted to integrals and a factor of two included for spin degeneracy, Eq. (3.95) yields

$$\sigma_N=\frac{4\pi e}{\hbar V}T^2N(0)^2\int_0^{eV}d\varepsilon_r=\frac{4\pi e^2}{\hbar}T^2N(0)^2\ . \tag{3.96}$$

This equation makes it possible to eliminate the tunneling parameter T^2 in favor of σ_N, a more directly measurable quantity. By comparison of Eq. (3.96) with Eq. (3.90), the desired connection is

$$I_J=\frac{\pi\Delta}{2e^2}\sigma_N\ , \tag{3.97}$$

which can be expressed in the following way: I_J (and $I^{thres.}_{Q.P.}$) are equal to $\pi/4$ times the current that would flow in the normal state upon application of the "gap voltage", $2\Delta/e$.

In closing this section we remind the reader that all of our work has been at zero temperature. As calculated by Ambegaokar and Baratoff [4], the strength, expressed by Eq. (3.97) in the zero-temperature limit, is, in fact, a monotonically decreasing function of temperature and vanishes at the transition temperature.

3.3. Quantum Effects

While the unique electrodynamic properties of the Josephson junction are ultimately attributable to the essentially quantum mechanical phenomena of superconductivity and tunneling, we have seen that the junction may be treated by regarding ϕ, the quantum mechanical phase, as itself a classical parameter. The behavior of the junction can be modeled as a classical particle moving in the potential $V = -E_J\cos\phi$. We now discuss the limitations imposed on such a description by the Heisenberg uncertainty principle and by the quantum fluctuations arising from the non-commuting nature of the canonically conjugate variables, ϕ and ν (the pair number).

For a junction of finite capacitance C, the charging energy due to the transfer of ν pairs is $(-2e\nu)^2/2C$, so that the junction Hamiltonian is

$$H = \frac{2e^2\nu^2}{C} - E_J\cos\phi = \frac{4e^2}{C}\left(\frac{1}{2}\nu^2 - g\cos\phi\right) , \qquad (3.98)$$

with

$$g = \frac{CE_J}{4e^2} . \qquad (3.99)$$

From the tight-binding analogy, we have the commutation relation

$$[\nu,\phi] = i , \qquad (3.100)$$

so that the junction has a ϕ-dependent wave function which obeys a Schrödinger equation which is the same as for a pendulum in a gravitational field of strength g. If the junction is biased so that the mean value of the phase is ϕ_o, the Josephson current is given by

$$I = I_J \langle\sin(\phi_o + \Delta\phi)\rangle$$

$$= I_J[\langle\cos\Delta\phi\rangle\sin\phi_o + \langle\sin\Delta\phi\rangle\cos\phi_o] = I_J\langle\cos\Delta\phi\rangle\sin\phi_o \quad ,(3.101)$$

where we have denoted the expectation value by angular brackets and have assumed that positive and negative values of $\Delta\phi$, the fluctuation in phase, are equally likely. Thus, because of the quantum fluctuations, the strength of the Josephson current is reduced below its "classical" value by the factor

$$<\cos\Delta\phi> \; < 1 \; . \tag{3.102}$$

In order to calculate the above reduction factor, it is necessary to solve Schrödinger's equation for the pendulum, i.e.

$$\left(-\frac{1}{2}\frac{d^2}{d\phi^2} - g\cos\phi\right)\psi(\phi) = \varepsilon\psi(\phi) \quad , \tag{3.103}$$

which is Mathieu's differential equation. In the "weak-field" limit, $g \ll 1$, ε can be calculated in perturbation theory. A variational calculation can be used, in lieu of numerical tables, to interpolate between this limit and the $g \gg 1$ harmonic-oscillator limit, for which the approximation $\cos\phi \simeq 1 - \phi^2/2$ can be substituted. From the Feynman-Hellman theorem, $<\cos\phi>$ is determined, if $\varepsilon(g)$ is known, by

$$<\cos\phi> = -\frac{\partial\varepsilon}{\partial g} \quad . \tag{3.104}$$

While such fluctuation effects are interesting from a physicist's point of view, they are undesirable for practical applications. They correspond to a weakening and to an uncontrollable noise, which reduces a junction's reliability.

As the Josephson technology proceeds to the fabrication of junctions of ever smaller dimensions, it may become necessary to take the quantum noise into consideration. In this connection, a simplification that sets in is worth mentioning. From Eq. (3.99), it would appear that g varies from junction to junction depending on the values of C and E_J. Surprisingly, for the case of a very small junction, g depends only upon the junction normal state resistance, σ_N^{-1}, and is given by the universal formula

$$g = \frac{3}{2^{11}} r^{-2} \quad , \tag{3.105}$$

where $r = e^2\sigma_N^{-1}/h$ is the resistance measured in units of $h/e^2 = 26$ KΩ. This gap-independent formula for g arises from the fact, noted by Larkin and Ovchinnikov [5] and Ekern et al [6], that the junction has an additional effective capacitance, due to the virtual tunneling of quasiparticles across it, equal to

$$\Delta C = \frac{3}{64}\frac{h}{\Delta} \sigma_N \quad . \tag{3.106}$$

This formula can also be derived from the Kramers-Kronig relations [7]. As ΔC is inversely proportional to the gap, the gap cancels from the combination $\Delta C E_J$. Substituting Eqs. (3.97), (3.106), and the expression for I_J following Eq. (3.9) into Eq. (3.99) yields Eq. (3.105).

One application of the theory of quantum fluctuations, which tends to confirm their physical reality, can be mentioned here. By means of Eq. (3.105) and mean field theory, Ferrell and Mirhashem [8] derived the universal threshold criterion

$$g_c = 2^{-5} \tag{3.107}$$

for phase coherence in a regular square array of Josephson junctions. According to Eq. (3.105), this signifies that the array will not be superconducting unless $r < \sqrt{3}/8$, i.e. not unless the normal state resistance is less than 5.7 $K\Omega$. In view of the approximations involved in the theory, this would seem to be in satisfactory agreement with the universal threshold value of 6.5$K\Omega$ observed by Jaeger et al [9] for the occurrence of superconductivity in thin granular metal films in the zero-temperature limit. Kobayashi and Komori [10] have recently reported a somewhat smaller experimental threshold resistance.

3.4. Summary

The phenomenology that has been presented in Sections 3.1.1-3.1.4 provides an adequate basis for understanding most of the applications of the Josephson effect. These include voltage standards, based on the AC effect, and the extremely sensitive magnetometers or "SQUIDS" (superconducting quantum interference devices), which, in principle, only require the DC effect. In all such applications, which are described in detail elsewhere in this book, the quantum mechanical phase of the Cooper electron pairs plays a key and vital role.

The microscopic theory that has been reviewed in Sections 3.2.1-3.2.3 has been limited, for the sake of simplicity, to zero temperature. It provides further insight into the underlying mechanism for the tunneling of a Cooper pair by the successive tunneling of single electrons. The connection that was established there between the critical (or maximum) Josephson current and the quasiparticle current at the gap voltage is useful in understanding the operation of a Josephson junction as a logic gate, or switching device.

Compared to semiconductor logic elements, the Josephson junction is superior, by orders of magnitude, in both speed and power consumption. Nevertheless, the cryogenic requirement and the concomitant thermal incompatibility with the semiconductors has restricted the incorporation of Josephson junctions into computers. With the recent discovery of high temperature superconductors, the situation can be expected to change. As miniaturization proceeds, it may eventually be necessary to take quantum fluctuations of the pair phase into account, as discussed in Section 3.3 above.

In this brief chapter it has been possible to discuss only a few of the salient features selected from the large body of knowledge that has grown up around the Josephson effect during the past one-quarter of a century. The interested reader will find that the list of references collected below provides an entrée into this vast domain. In closing it is a pleasure to acknowledge the assistance of Mr. Behzad Mirhashem in preparing these lecture notes.

References

[1] B. D. Josephson, *Phys. Lett.* **1**, 251 (1962).

[2] P. W. Anderson, "Special Effects in Superconductivity" in *Lectures on the Many-Body Problem*, Ravello 1963, Vol. II (E. R. Caianiello, ed.), Academic Press, New York, 1964, p. 113.

[3] R. A. Ferrell and R. E. Prange, *Phys. Rev. Lett.* **10**, 479 (1963).

[4] V. Ambegaokarand A. Baratoff, *Phys. Rev. Lett.* **10**, 486 (1963); erratum **11**, 104 (1963).

[5] A. I. Larkin and Yu. N. Ovchinnikov, *Phys. Rev.* B **28**, 6281 (1983).

[6] U. Ekern, G. Schon, and V. Ambegaokar, *Phys. Rev.* B **30**, 6419 (1984).

[7] R. A. Ferrell, *Physica* C **152**, 10 (1988); 231 (1988).

[8] R. A. Ferrell and B. Mirhashem, *Phys. Rev.* B **37**, 648 (1988); B. Mirhashem and R. A. Ferrell, *Physica* C **152**, 361 (1988).

[9] H. M. Jaeger, D. B. Haviland, A. M. Goldman, and B. G. Orr, *Phys. Rev.* B **34**, 4920 (1986).

[10] S. Kobayashi and F. Komori, *J. Phys. Soc. Japan* **57** No. 6, in press (1988).

Additional References on the Josephson Effect

A. Barone and G. Paterno, *Physics and Applications of the Josephson Effect*, John Wiley & Sons, New York, 1982.

R. A. Ferrell, Phys. Rev. B **25**, 496 (1982).

B. D. Josephson, Rev. Mod. Phys. **36**, 216 (1964).

B. D. Josephson, Adv. Phys. **14**, 419 (1965).

B. D. Josephson, "Weakly Coupled Superconductors" in *Superconductivity* (R. D. Parks, ed.), Marcel Dekker, New York, 1969, p. 423.

J. E. Mercereau, "Macroscopic Quantum Phenomena" in *Superconductivity* (R. D. Parks, ed.), Marcel Dekker, New York, 1969, p. 393.

J. E. Mercereau, "DC Josephson Effects" in *Tunneling Phenomena in Solids* (E. Burstein and S. Lundqvist, eds.), Plenum Press, New York, 1969, p. 461.

D. J. Scalapino, "The Theory of Josephson Tunneling" in *Tunneling Phenomena in Solids* (E. Burstein and S. Lunqvist, eds.), Plenum Press, New York, 1969, p. 477.

T. Van Duzer and C. W. Turner, *Principles of Superconductive Devices and Circuits*, Elsevier, New York, 1981.

CHAPTER 4

Crystallography

Anthony Santoro

4.1. Introduction

The high temperature superconductors discovered so far belong to five chemical systems having the following general formulas:

(i) $BaPb_{1-x}Bi_xO_3$

(ii) $La_{2-x}M_xCuO_{4-y}$ (M = Ba, Sr)

(iii) $Ba_2MCu_3O_\delta$ (M = Y, Gd, Eu, etc.)

(iv) $Ba_{2-x}La_{1+x}Cu_3O_\delta$, and

(v) $Bi_2CaSr_2Cu_2O_\delta$.

The structures of the superconducting compounds, and those of the related phases, have been studied by both x-ray and neutron diffraction techniques, and in this chapter we will review the atomic arrangement and the crystal chemistry of the most important materials analyzed up to the present time.

Almost all the structural refinements of superconductors have been carried out using neutron powder diffraction data for the reasons briefly outlined in what follows. The atomic numbers of the elements involved in the systems range from 8 (oxygen) to 83 (bismuth), and in all compounds light atoms like oxygen are combined with atoms of heavy metals. Under these conditions it is difficult to precisely determine the location of the oxygen atoms in a structure with an x-ray experiment because the atomic scattering factors for x-rays are proportional to the atomic number Z and the diffracted intensities are dominated by the heavy elements. This disadvantage does not exist in the case of neutrons since the neutron scattering amplitudes of light and heavy atoms are comparable.

A second and equally important reason to prefer neutrons over x-rays is absorption. The mass absorption coefficients μ/ρ (in cm^2/g) of the elements present in the superconducting materials are shown in Table 1 for x-rays and neutrons of wavelengths commonly used in diffraction experiments. From these values we may easily calculate the mass absorption coefficients of the compounds of interest with the formula:

$$\frac{\mu}{\rho} = \sum_i g_i \left(\frac{\mu}{\rho}\right)_i \tag{4.1}$$

where g_i is the mass fraction contributed by element i to the molecular mass, μ is the linear absorption coefficient (cm^{-1}) of the compound, and ρ is its density. If ρ is known, we may evaluate μ, and from this the transmission factor

$$\frac{I}{I_o} = \exp(-\mu t) \tag{4.2}$$

Table 1. Absorption coefficients for the elements in the most important superconducting compounds

	μ/ρ†			μ‡	
Element	Neutrons (1.08 Å)	X-rays (1.54 Å)	Compound	Neutrons (1.08 Å)	X-rays (1.54 Å)
O	$1\cdot10^{-5}$	11.5	$La_{1.85}Ba_{0.15}CuO_4$	0.13	1717
Cu	$2.1\cdot10^{-2}$	52.9	$La_{1.85}Sr_{0.15}CuO_4$	0.13	1642
Sr	$5\cdot10^{-3}$	125	$Ba_2YCu_3O_7$	0.05	1091
Y	$6\cdot10^{-3}$	134	$Ba_2YCu_3O_6$	0.05	1074
Ba	$2.6\cdot10^{-3}$	330	$Ba_{1.5}La_{1.5}Cu_3O_{7.25}$	0.09	1394
La	$2.3\cdot10^{-2}$	341	$BaBiO_3$	0.07	1941
Pb	$3\cdot10^{-4}$	232			
Bi	$6\cdot10^{-5}$	240			

†in cm^2/g. Data taken from [3], p. 72. The density ρ used to calculate the linear absorption coefficient μ has been evaluated with the formula $\rho = (nM)/(NV)$ where M is the molecular weight, N the Avogadro number ($N = 6.02257\text{x}10^{23}$ molecules per mole), and n the number of molecules in the unit cell of volume V (cm^3) [$V = abc(1-\cos^2\alpha-\cos^2\beta-\cos^2\gamma+2\cos\alpha\cos\beta\cos\gamma)^{1/2}$, where a, b, c, α, β, and γ are the lattice parameters].
‡in cm^{-1}

of a homogeneous material of thickness t (cm). In Eq. (4.2), I_o and I are the incident and transmitted intensities. If we consider as an example the case of $La_{1.85}Ba_{0.15}CuO_4$, we have from Table 1 μ = 1717 cm^{-1} for x-rays (λ = 1.54 Å) and μ = 0.13 cm^{-1} for neutrons (λ = 1.08 Å), From these values we may easily calculate that in the case of x-rays, 4/100 mm of the material is sufficient to attenuate the incident beam by a factor of 10^{-3}, while in the case of neutrons a thickness of 1 cm gives a transmission factor I/I_o = 0.88. Two problems are connected with the study of highly absorbing materials. In the first place, for absorptions as severe as those reported in Table 1 only the surface of the sample is involved in the diffraction process, and this may well be in a physical state different from that of the interior. In addition, the correction of the observed intensities becomes very difficult, and even the determination of lattice parameters is imprecise, unless precautions and extrapolation techniques are used in the experiment and in the analysis of the data [1].

It is not surprising, therefore, that neutron diffraction has played a major role in the study of superconductors. The powder technique, combined with the Rietveld method of profile analysis [2], has been used in all cases not only because of the difficulty involved in the preparation of single crystals sufficiently large for neutron diffraction, but also because the presence of complex twinning in many of the superconducting compounds significantly complicates the measurement of the diffracted intensities.

4.2. The Rietveld Method

Neutron powder intensities can be collected with two techniques. The more traditional of the two consists of fixing the wavelength of the incident beam and measuring the intensities over a range of scattering angles. The other method consists of fixing the scattering angle and measuring the intensities over a range of wavelengths. In the first case a monochromatic beam of neutrons is used, in the second a polychromatic pulsed source is needed. Figure 1 shows schematically the layout of a diffractometer operating with constant wavelength, and in what follows we will limit our discussion to this technique, although our conclusions can easily be extended to the other case as well.

The intensity y_i observed in a neutron powder pattern at the angular position $2\theta_i$ is given by [4]:

$$y_i = f_i + \sum_k I_k G_{ik} + e_i \quad , \tag{4.3}$$

where f_i is the contribution from the background, I_k is the

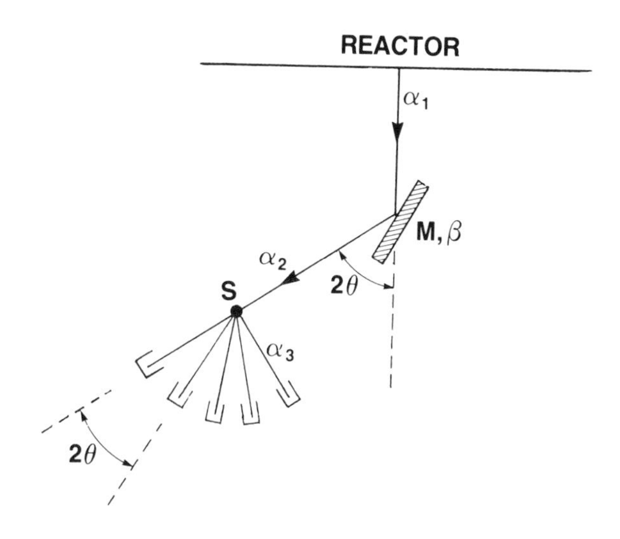

Fig. 1. Schematic view of a multicounter neutron powder diffractometer. M is a monochromating single crystal of mosaic spread β, and α_1, α_2, and α_3 are the horizontal angular divergencies of the three collimators of the diffractometer. The monochromatic beam, emerging from M, is diffracted by the sample S and its intensity is measured by a bank of counters that can be rotated over any angular interval of interest.

integrated intensity of reflection k, G_{ik} is a function describing the peak shape,[1] and e_i is a random variable drawn from a population with zero mean. The sum is extended over all the reflections whose contribution to y_i is significant. The value of I_k depends on the crystal structure of the material under study and it is given by the equation

$$I_k = K j_k L_k (F_{nk}^2 + F_{mk}^2) \quad , \tag{4.4}$$

where K is a scale factor, j_k is the multiplicity of reflection k and depends on the symmetry of the compound, L_k is the Lorentz factor [5], and F_{nk} and F_{mk} are the nuclear and magnetic structure factors, respectively. We will be concerned only with nuclear structures, and for this case we have for the structure factor

[1]The function G_{ik} is normalized so that $\int G(x)\, dx = 1$

$$F(hkl) = \sum_{i=1}^{N} \left\{ (b_i \exp\left[-B_i \left(\frac{\sin\theta}{\lambda}\right)^2\right] \exp\left[2\pi i(hx_i + ky_i + lz_i)\right] \right\} , \quad (4.5)$$

where the sum extends over all the N atoms contained in the unit cell, b_i is the scattering amplitude of atom i located at the position x_i, y_i, z_i, hkl are the indices of the reflection occurring at the angular position θ, and B_i is the isotropic thermal factor of the atom. Equations (4.3-4.5) show that the intensities diffracted by a powder can be used to refine the structural parameters of a material. Two methods may be used to do it. The first consists in minimizing the function

$$M_I = \sum_{k}^{N} w_k \left[\Sigma_i (y_i - f_i) - I_k \right]^2 . \quad (4.6)$$

In this expression N is the total number of Bragg reflections used in the refinement, w_k is a weight associated with reflection k, and the sum $\sum(y_i - f_i)$ is the observed integrated intensity of the reflection which is compared with the theoretical prediction I_k calculated on the basis of a structural model by means of Eqs. (4.4) and (4.5). The second procedure, developed by Rietveld [2], consists of minimizing the function

$$M_p = \sum_{i=1}^{\mathcal{N}} w_i \left[\left[y_i - f_i \right] - \sum_{k=1}^{k} I_k G_{ik} \right]^2 , \quad (4.7)$$

where $(y_i - f_i)$ is the net intensity observed at the angular position θ_i, w_i is a weight associated with this intensity, and $\mathcal{N}$ is the number of points used in the refinement. Here again the net intensity $(y_i - f_i)$ is compared with the theoretical prediction $\Sigma_k I_k G_{ik}$ calculated from a structural model and assuming for the peak shape a known function G_{ik}.

A comparison of Eqs. (4.6) and (4.7) shows the basic difference between the two methods. The use of M_I does require the measurement of the integrated intensities $\Sigma_i(y_i - f_i)$ of single reflections, and this cannot be done for moderately complex structures giving powder patterns with severe overlapping. The function M_p, on the other hand, does not have this limitation, since it only requires the measurement of net intensities $(y_i - f_i)$ at every point of the profile, irrespective of overlap. This function, therefore, makes much more efficient use of the diffracted intensities than M_I. The disadvantages of M_p, on the other hand, are that the background must be known everywhere and that a function G_{ik} describing the peak shape must be selected.

The first problem can be circumvented by noting that the background is a slowly varying function of the diffraction angle θ. This function can be included in the model and can be therefore estimated also in regions not free from diffracted intensity [6]. In this case the function to be minimized is

$$M_p' = \sum_{i=1}^{\mathcal{N}} w_i \left[y_i - \left(f_i + \sum_{k_1}^{k_2} {}_k \, I_k \, G_{ik} \right) \right]^2 \quad . \tag{4.8}$$

The first function proposed to describe the peak shape in neutron diffraction was a Gaussian with full width at half maximum calculated from the instrumental parameters and without taking into account the possibility of broadening due to the physical conditions of the sample [7]. The assumption of Gaussian peak shape is generally valid to a good approximation. There are cases, however, where the peaks are better described by other functions, such as the

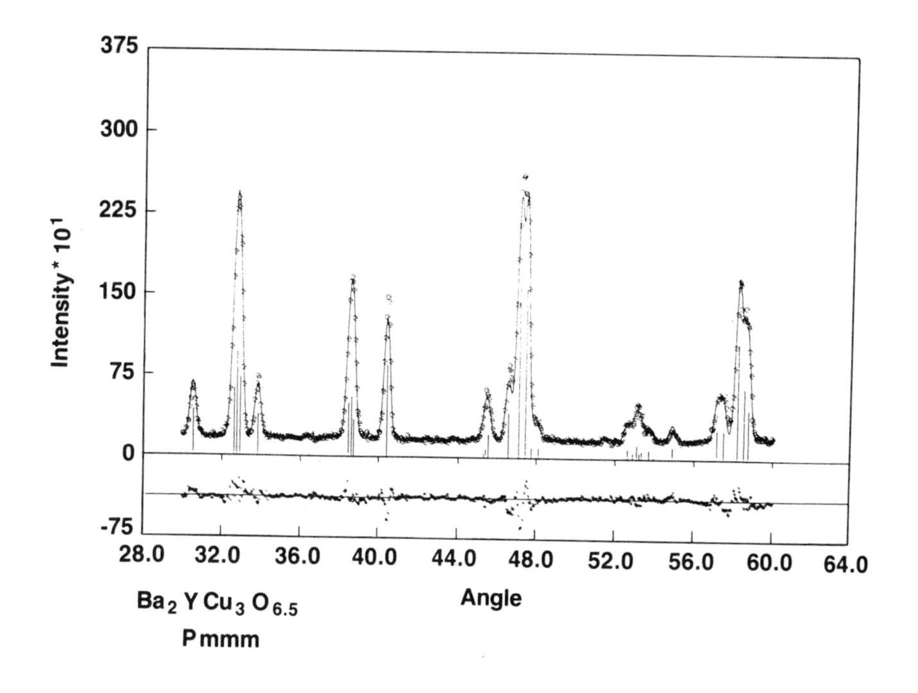

Fig. 2. Plot of the observed and calculated intensities over the angular 2θ interval 30° - 60°, for the compound $Ba_2YCu_3O_{6.5}$. The circles represent the observed intensities, and the continuous line those calculated from a model of the structure. The lower part of figure is the plot of the difference I(calc) - I(obs).

Pearson type-VII function [8]:

$$G_{ik} = 2\left[\frac{2^{1/m}-1}{\pi}\right]^{\frac{1}{2}} \frac{\Gamma(m)}{\Gamma(m-1/2)}\frac{I}{H_k}\left[1+4\left[\frac{2\theta_i-2\theta_k}{H_k}\right]^2\left(2^{1/m}-1\right)\right]^{-m} . \quad (4.9)$$

In this expression, H_k is the full width of half maximum of a peak occurring at the angular position θ_k and m is a parameter which fixes the peak shape. It can be shown that varying the value of m from 1 to ∞ changes the shape of the function continuously from a Lorentzian to a Gaussian.

In the Rietveld method the agreement between observed and calculated intensities is evaluated at every point of the powder pattern, and in general is expressed by means of appropriate plots and by so-called agreement factors. Figure 2 shows the calculated and observed profiles after the final refinement of the structure of $Ba_2YCu_3O_{6.5}$. The agreement factors most commonly employed in profile analysis are defined by the following equations:

$$R_N = 100 \times \left[\frac{\sum |I_{obs} - I_{cal}|}{\sum I_{obs}}\right] , \quad (4.10)$$

$$R_p = 100 \times \left[\frac{\sum |y_{obs} - y_{cal}|}{\sum y_{obs}}\right] , \quad (4.11)$$

$$R_w = 100 \times \left[\frac{\sum w\, [y_{obs} - y_{cal}]^2}{\sum w\, y_{obs}^2}\right]^{1/2} \quad (4.12)$$

$$R_E = 100 \times \left[\frac{N - P + C}{\sum w\, y_{obs}^2}\right]^{1/2} . \quad (4.13)$$

In these expressions N is the number of independent observations, P is the number of refined parameters, C is the number of constraints, y is the intensity at the angular position θ, I is the integrated Bragg intensity, and w is the statistical weight. The factor R_E expresses the agreement that one should theoretically expect on the basis of counting statistics alone, and the ratio R_W/R_E is the well known quantity "chi". Well refined structures typically have a value of χ between $\sim$1.3 and 2.0. The agreement factors previously defined, as well as χ, will be used extensively in the following sections.

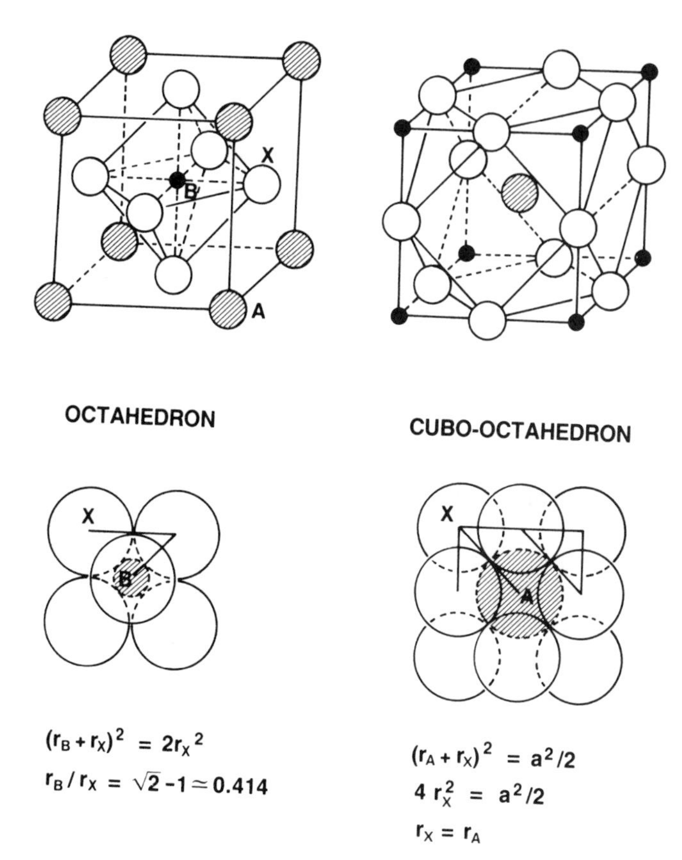

Fig. 3. Idealized structure of perovskite ABX_3. (a) Unit cell with origin at the center of atom A, showing the octahedral coordination of atom B; (b) unit cell with origin at the center of atom B, showing the cube-octahedron of atoms X surrounding A; (c) and (d). Projections of the structures of Figs. (a) and (b), respectively, on the plane (001), showing the conditions that the ionic radii of the atoms must satisfy in order to give a perovskite-type structure.

4.3. The Structure of Perovskite

All the superconductors discussed in this chapter have a close relationship with the structure of perovskite ($CaTiO_3$). It seems logical, then, to begin our study with an analysis of this important structural type common to a great variety of compounds of general formula ABX_3. The structure of perovskite is illustrated in Fig. 3 in its idealized cubic form (space group Pm3m). From Figs. 3a and 3b it is clear that the A and B atoms have cubo-octahedral and octahedral coordination, respectively. Such a highly symmetrical arrangement of atoms imposes rather severe limitations on the size of the ions present in the structure. If r_A, r_B, and r_X are the ionic radii [9] of A, B, and X, respectively, then from Fig. 3(a) we have $r_A + r_X = (r_B + r_X)\sqrt{2}$. In practice it has been found that the structure of perovskite exists whenever the condition $0.75 \leq p \leq 1.0$ is satisfied, where

$$p = \frac{(r_A + r_X)}{\sqrt{2}(r_B + r_X)} \quad . \tag{4.14}$$

However, distortions in the structure may be present, with consequent lowering of the symmetry (tetragonal, trigonal, and orthorhombic perovskite structures are known to occur). The A cations, with cubo-octahedral coordination, are larger than the B cations, and from Figs. 3(c) and 3(d) it is easy to derive that $r_A \simeq r_X$ and that the ratio r_B/r_X must be approximately equal to 0.4. If these conditions are not satisfied, at least approximately, other structural types are formed. Since the ions A and X are nearly equal in size, the structure of perovskite contains close packed

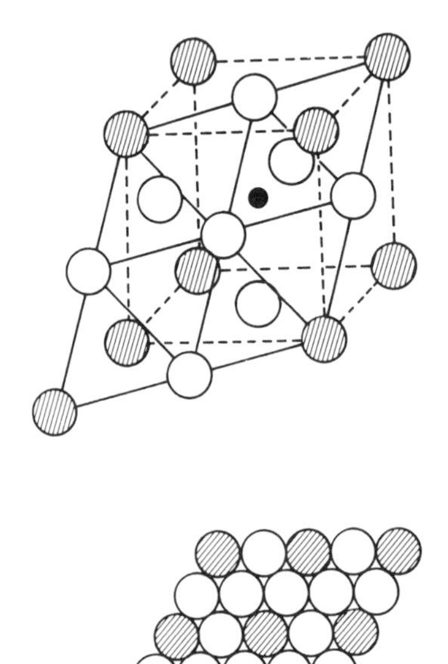

Fig. 4. (a) The perovskite structure showing the planes (111) filled with atoms A and X; (b) View of the planes (111) with the closed packed configuration of atoms A and X.

layers of composition AX_3. These are parallel to the (111) plane and are shown in Fig. 4. The atoms X are linked to four ions A and two ions B (Fig. 3a) and are close to eight other ions X.

As we have mentioned previously, many compounds have the perovskite type structure. This atomic arrangement is not restricted to bivalent and tetravalent atoms, as in $CaTiO_3$. In fact, compounds such as $KNbO_3$ and $LaAlO_3$ have the same structure. It appears, then, that the size of the ions is a more important factor than the valence of the atoms in determining the atomic arrangement of these materials. The A and/or B sites may even be occupied by cations of different kinds as in $Sr(Ga_{0.5}Nb_{0.5})O_3$ and in $(Ba_{0.5}K_{0.5})(Ti_{0.5}Nb_{0.5})O_3$, or some of the A sites may be vacant, as occurs in the sodium tungsten bronze, Na_xWO_3, in which $x \leq 1.0$. In this case, of course, to preserve neutrality some of the W^{5+} ions must convert to W^{6+}.

4.4. The System $BaPb_{1-x}Bi_xO_3$

High-temperature superconductivity in ceramic materials was first detected in the system $BaPb_{1-x}Bi_xO_3$ [10], and soon after this discovery structural studies were carried out over the entire range of composition, including the end members $BaPbO_3$ and $BaBiO_3$ [11-16]. The solid solution in this system extends from $x = 0$ to $x = 1$, and although the general structural features do not change dramatically with composition, the degree and type of distortions present in the basic structure do vary with x. At room temperature, the symmetry of the material changes according to the following sequence [11, 17]:

Orthorhombic	$0 \leq x \lesssim 0.05$
Tetragonal	$0.05 \lesssim x \lesssim 0.35$
Orthorhombic	$0.35 \lesssim x \lesssim 0.90$
Monoclinic	$0.90 \lesssim x \leq 1.0$

Superconductivity exists only for values of x between 0.05 and 0.35. The value of the critical temperature T_c increases with x, reaches a maximum $T \simeq 13$ K for $x \simeq 0.25$, and then decreases in the range $0.25 < x < 0.35$. For $x > 0.35$, the material becomes a semiconductor.

The crystal data and the structural parameters of the compound $BaPb_{0.75}Bi_{0.25}O_3$ are shown in Table 2 and the structure and its relationship to perovskite are illustrated in Fig 5.[2] From the

[2]The relationship between two unit cells is usually expressed by means of the axes transformation required to go from one reference system to the other. In Fig. 5 we can see that

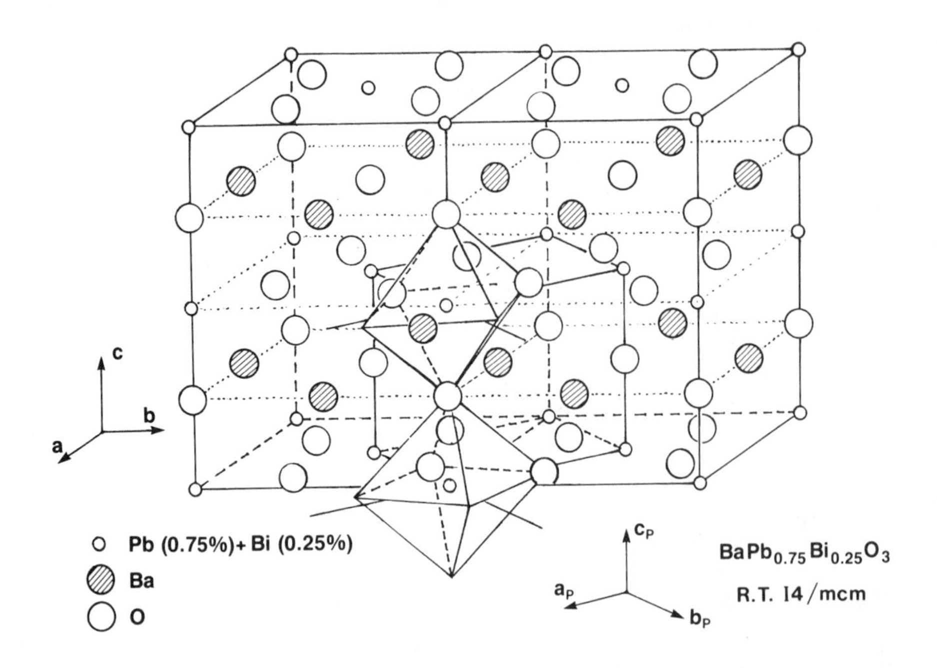

Fig. 5. Crystal structure of the compound $BaPb_{0.75}Bi_{0.25}O_3$. Two unit cells are joined together to show the close relationship with the unit cell of perovskite, defined by the translations $\mathbf{a}_p$, $\mathbf{b}_p$, and $\mathbf{c}_p$. Two octahedra of oxygen atoms have been drawn to show the type of distortion present in the structure (the two octahedra are rotated in opposite senses about the c-axis).

figure it is apparent that the oxygen atoms located on the layers at $z = 0$ and $z = \frac{1}{2}$ are displaced from the positions that they occupy in perovskite. The distortion responsible for this shift, and for the consequent lowering of symmetry from cubic to tetragonal and doubling of the volume of the unit cell, can be described as a rotation of the oxygen octahedra of about 8° around the [001] axis of the pseudo-cubic unit cell.[3] A consequence of the displacement

$$(\mathbf{abc}) = (110/\text{-}110/002)(\mathbf{a}_p\mathbf{b}_p\mathbf{c}_p)$$

where $(\mathbf{a}_p,\mathbf{b}_p,\mathbf{c}_p)$ is the column vector of the three vectors defining the unit cell of perovskite, and $(\mathbf{a},\mathbf{b},\mathbf{c})$ is the column vector of the triplet defining the unit cell of the compound.

[3]All the phases previously listed in this system are generated in a similar way. Thus, the orthorhombic and monoclinic modifications

Table 2. Refined structural parameters of $BaPb_{0.75}Bi_{0.25}O_3$
(C. Chaillout, private communication)

Space group I4/mcm, n = 4*, ρ_c = 8.27 g./cm^3,
a = 6.0496(1) Å,# c = 8.6210(2) Å, V = 315.5 Å^3

Atom	Position†		x≢	y	z	B(Å^2)	Occupancy‡
Ba	4b	$\bar{4}$2m	½	0	¼	0.81(3)	1.0
Pb	4c	4/m	0	0	0	0.30(2)	0.75
Bi	4c	4/m	0	0	0	0.30(2)	0.25
0(1)	4a	42	0	0	¼	1.24(4)	1.0
0(2)	8h	mm	0.2182(1)	0.7182(1)	0	1.17(2)	1.0

R_N = 4.94, R_p = 6.14, R_W = 7.88, R_E = 5.04, χ = 1.56

Symmetry_operations¶
$xy\bar{z}$ → xyz, $xy\bar{z}$, $\bar{y}xz$, $\bar{y}x\bar{z}$, $x\bar{y}$ ½+z, $x\bar{y}$ ½-z, yx ½+z, yx ½-z, plus center and centering (½½½).

*n represents the number of formula units per unit cell.
†The three symbols of an atomic position represent, respectively, the multiplicity of the position (i.e. the number of symmetry related locations or sites belonging to the position), a letter which is a coding scheme to identify the position, and a symmetry symbol which describes the symmetry at the position. The symbols and their terminology are taken from [21, 22].
‡The value of the occupancy indicates the fraction of the sites that are occupied. The number of atoms in the unit cell, then, is given by the occupancy times the multiplicity of the position. Thus, in our case there are four Ba, three Pb, one Bi, four 0(1), and eight 0(2) atoms per unit cell.
¶The symmetry operations give all the symmetry related sites of a position. These will all be distinct from one another only for the general position, i.e. for the position with site symmetry 1. When the site symmetry is higher than 1, duplications will result. In any case the number of distinct locations is equal to the multiplicity of the position.
#Numbers in parentheses are standard deviations on the last decimal place.
≢Atomic coordinates are given as fractions of the corresponding lattice parameters.

---

that appear for certain compositions are generated by 8-10° rotations of the octahedra about the [110] axes of the pseudo cubic cell.

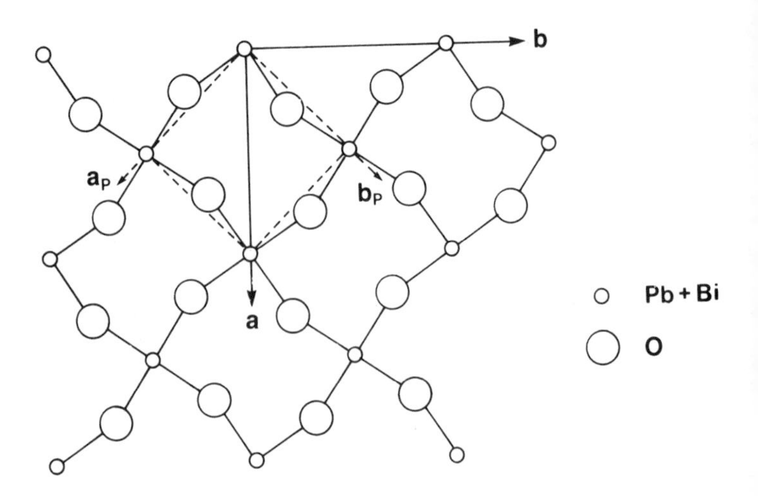

Fig. 6. Atomic configuration of oxygen and *R* atoms on the (001) plane for z = 0. The effect of the octahedra rotations indicated in Fig. 5 is to produce a typical buckling of the two-dimensional *R*-O chains. The axes **a**, **b** of the original perovskite unit cell and the axes **a** and **b** of the actual structure are also indicated.

of the oxygen atoms is that the interconnected *R*-O (*R* = 0.75 Pb + 0.25 Bi) chains in the layers at z = 0 and z = ½ are buckled, as illustrated in Fig. 6, and we will see later that this appears to be a feature common to all the superconductors discovered so far.

The relationship between the structures of $BaPb_{0.75}Bi_{0.25}O_3$ and perovskite is so simple and direct that there is no change in the coordination of the Ba and *R* cations, the first being twelvefold coordinated with a distorted cubo-octahedron as coordination polyhedron, and the second octahedrally coordinated. Also, the oxygen anions remain bonded to four Ba and two *R* cations in the same way as the X anions of the perovskite-type structure.

Two formulations are possible for $BaPb_{1-x}Bi_xO_3$, depending on the oxidation state assigned to the bismuth cations, i.e.,

$$Ba^{2+}\,Pb^{4+}_{1-x}\,Bi^{4+}_{x}\,O_3 \quad \text{and} \quad Ba^{2+}\,Pb^{4+}_{1-x}\,(Bi^{3+}_{0.5}\,Bi^{5+}_{0.5})_x\,O_3 \quad ,$$

and a number of structural studies have been carried out to determine the valence of bismuth in this important system. Most of this research has been done on the compound $BaBiO_3$ which is monoclinic [14, 16]. The Bi cations in the structure are located in two crystallographically different positions of space group I2/m.

Powder neutron diffraction results have shown that the distances Bi(1)-0 and Bi(2)-0 for the two inequivalent Bi cations are significantly different (2.283 Å and 2.126 Å, respectively) and from these values it was readily concluded that the Bi(1) position is occupied by Bi^{3+} and the position Bi(2) by Bi^{5+}. Consequently, the correct formulation for this compound should be $Ba_2(Bi^{3+}Bi^{5+})O_6$. X-ray photoemission spectroscopy on single crystals of $BaBi0_3$ [18] confirmed that there are two crystallographically independent positions for bismuth. However, the difference in electron density was found to be too small to justify a complete disproportionation into 3+ and 5+ valence states for Bi. This conclusion is in agreement with electronic structure calculations [19]. Studies on samples of $BaBi0_3$ obtained from single crystals synthesized in oxygen atmosphere [20] showed that these apparent contradictions may be explained by assuming a different degree of ordering of the Bi^{3+} and Bi^{5+} cations in the two inequivalent positions, in the sense that one position is occupied by more Bi^{3+} cations than Bi^{5+}, and the other is occupied in the opposite way. It was also shown that different methods of preparation may alter the ratio Ba^{3+}/Bi^{5+} in each position.

4.5. The System $La_{2-x}M_xCuO_{4-y}$ (M = Ba, Sr)

The discovery of superconductivity in the system La-Ba-Cu-O was made on a sample consisting of a mixture of phases [23], and only in subsequent experiments was the superconducting material identified as having the stoichiometry $La_{2-x}Ba_xCuO_{4-y}$ and the tetragonal K_2NiF_4 type structure at room temperature [24, 25]. This structural type can be described as containing alternate layers of perovskite (ABX_3) and rock salt (AX) units, as shown in Fig. 7. In this description, the composition A_2BX_4 may be considered as a member of structurally related phases having the general formula $AX(ABX_3)_N$, with N = 1,2,... [26]. Compounds of composition A_2BX_4 generally exhibit the K_2NiF_4 atomic arrangement if the ionic radii of the component atoms obey the condition $1.0 < r_A < 1.9$ Å and $0.5 < r_B < 1.2$ Å.

The value of the critical temperature T_c in the system $La_{2-x}Ba_xCu0_{4-y}$ is a function of x and reaches its maximum value of about 35 K for $x \simeq 0.15$. The detailed structure for this composition was determined by powder neutron diffraction [27], and the results of this study are shown in Table 3. The structures at 295 K and 10 K are practically identical and no phase transition was detected in this study. The agreement factor χ, however, has a reasonable value for the structure at room temperature (1.72) while it becomes unacceptably high (2.78) for the structure at 10 K, indicating the possibility of a subtle structural distortion.

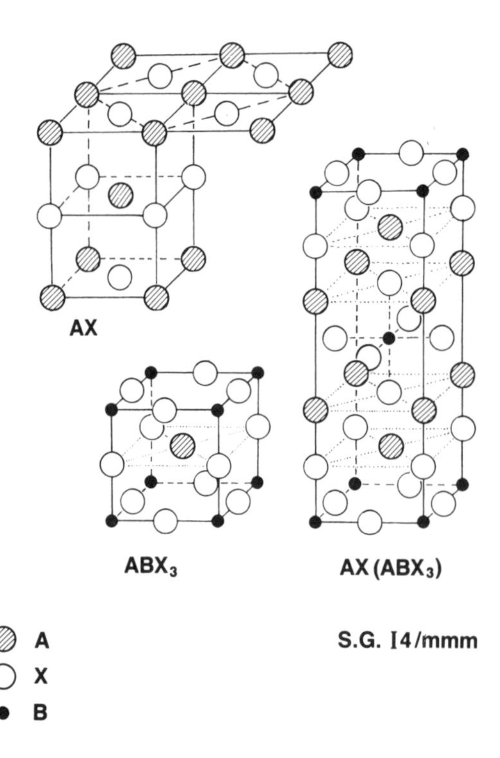

Fig. 7. The K_2NiF_4-type structure is a member of a series of phases of general formula $AX(ABX_3)_n$. The phase with n = 1 is shown together with perovskite and rock salt. Note that the unit cell of the rock salt structure adopted in the figure is unconventional (the conventional cubic axes are those indicated by the two heavy broken lines in the horizontal plane on the top and by the vertical axis).

High-resolution neutron diffraction experiments, done on a powder of the same compound with the same composition, did in fact show that a tetragonal to orthorhombic transition occurs in the vicinity of 180 K [28]. In this study some structural anomalies were observed at low temperature, and, more specifically, it was observed that the splitting of the diffraction lines due to the lowering of the symmetry increases with decreasing temperature, reaches a maximum around 75 K, and then decreases sharply and saturates at about 35 K, in coincidence with the onset of superconductivity. The space group of the orthorhombic phase was found to be Abma (the same as the conventional Cmca expressed in a different reference system), but no structural parameters were reported. Very high resolution x-ray diffraction data, taken with synchrotron radiation and a sample of

Table 3. Refined structural parameters of $La_{1.85}Ba_{0.15}CuO_{4-y}$ at 295 K and 10 K [27].

Space group I4/mmm, n = 2

a = 3.7873(1) Å, c = 13.2883(3) Å, V = 190.598(4) Å
3.7817(1) Å, 13.2487(3) Å, 189.478(5) Å*

Atom	Position	x	y	z	B(Å^2)†	Occupancy
La + Ba	4e 4mm	0	0	0.36063(9)		1
	4e 4mm	0	0	0.36075(9)		1
Cu	2a 4/mmm	0	0	0		1
	2a 4/mmm	0	0	0		1
0(1)	4c mmm	0	½	0		0.99(1)
	4c mmm	0	½	0		0.97(1)
0(2)	4e 4mm	0	0	0.1828(2)		1.02(1)
	4e 4mm	0	0	0.1824(2)		1.00(1)

R_w =7.54 R_E =4.38 χ =1.72
8.43 3.03 2.78

Symmetry operations
xyz, $\bar{x}\bar{y}z$, $\bar{x}yz$, $x\bar{y}z$, $\bar{y}xz$, $y\bar{x}z$, yxz, $\bar{y}\bar{x}z$ plus center and centering (½½½).

*For each parameter, the value given in the first line refers to the structure at 295 K and that in the second line to the structure at 10 K.

†The values of the temperature factors are not given in the original paper.

composition $La_{1.8}Ba_{0.2}CuO_{4-y}$ as a function of temperature, showed at about 10 K a broadening of the diffraction lines which is not consistent with an orthorhombic structure, but rather with monoclinic or triclinic symmetry [29]. It must be noted, however, that the starting material in this study was a mixture of two phases, both tetragonal at room temperature and with slightly different lattice parameters. The presence of shoulders of the second phase in the vicinity of the main peaks represents a serious obstacle to the unambiguous interpretation of the experimental results and the existence of a monoclinic or triclinic phase in this system is still doubtful. Recent neutron diffraction work [30] on a sample of $La_{1.85}Ba_{0.15}Cu0_4$ has shown that in the powder pattern of the compound at 10 K there are several weak peaks (Fig. 8) which can be indexed in terms of an orthorhombic unit cell related to the tetragonal cell of the structure at room temperature by the relations $a_o \simeq c_o \simeq a_t\sqrt{2}$ and $b_o \simeq c_t$, where a_o, b_o, c_o are the lattice

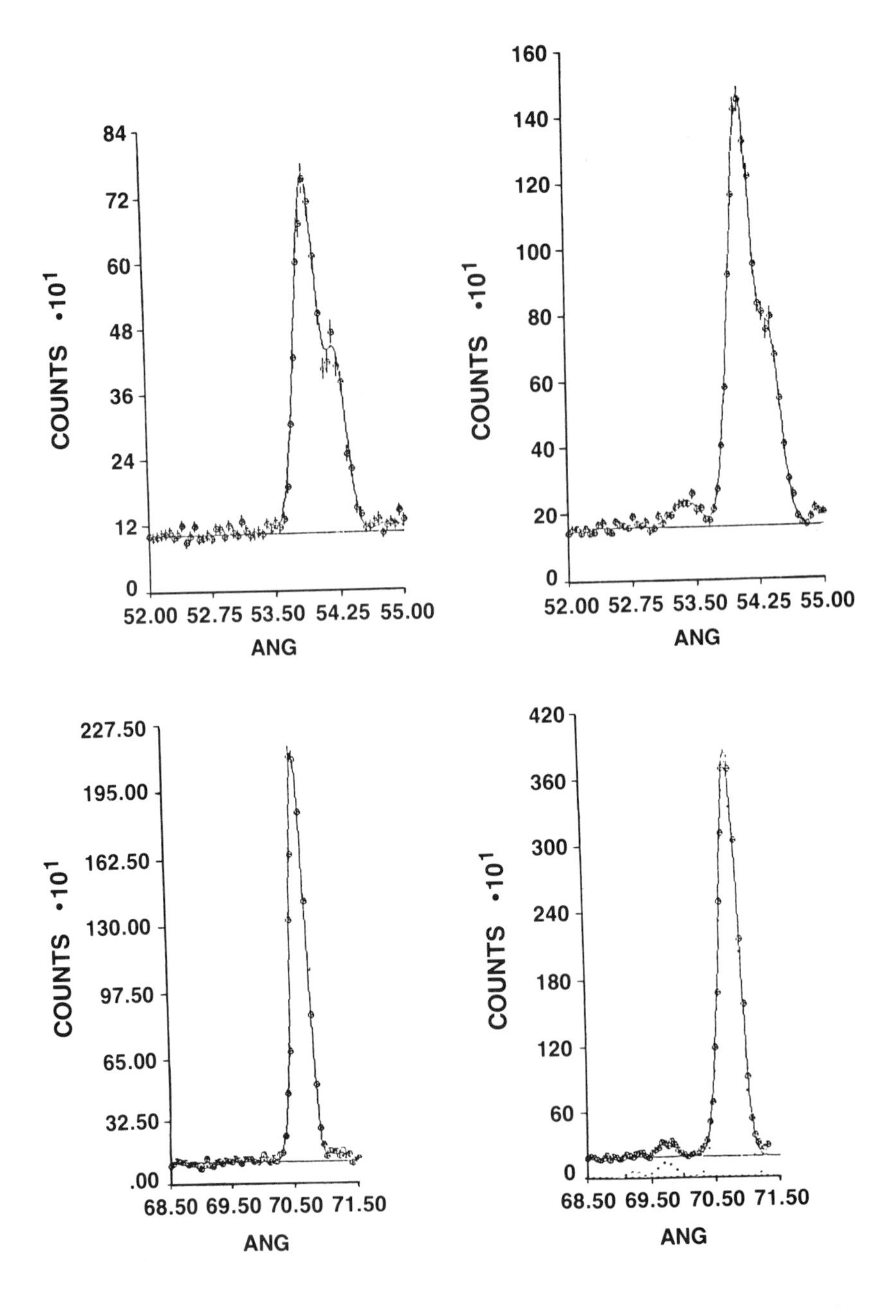

Fig. 8. Diffraction lines showing the phase transition in the compound $La_{1.85}Ba_{0.15}CuO_4$. Peaks on the left are those observed at room temperature, and those on the right at 10 K. The weak reflections on the left side of the main peaks are due to the lowering of symmetry from tetragonal to orthorhombic.

parameters of the orthorhombic phase, and a_t and c_t those of the tetragonal phase. Systematic extinctions seem to indicate the symmetry of space group Cmca, and in fact, a refinement of the structure in this space group gave excellent results, as indicated in Table 4.

In view of the differences between the results obtained in the previously described studies, we must conclude that the structural phase diagram of the system $La_{2-x}Ba_xCuO_{4-y}$ is not yet fully understood, and certainly more work is needed to clarify some of the problems still unsolved. We may note, however, that in none of the refinements done so far has any oxygen deficiency been found.

Structural analyses of the system $La_{2-x}Sr_xCuO_{4-y}$ have shown that the compounds of Sr and Ba are isomorphic at room temperature [31, 32]. The composition corresponding to $x \simeq 0.15$ has a T_c of about 40 K and undergoes a phase transition from tetragonal to orthorhombic symmetry at approximately 200 K [33]. The transformation is quite evident in this system, as many of the reflections are clearly split in the powder neutron diffraction patterns taken at low temperature. The refined parameters of the compound $La_{1.85}Sr_{0.15}CuO_4$ at room temperature, 60 K, and 10 K are given in Table 5 and comparison with the data reported in Table 4 shows that the Sr and Ba compounds are also isomorphic at low temperature [the transformation of coordinates in the two settings is $(xyz)_{Bmab}$ = $(100/00\text{-}1/010)(xyz)_{Cmca}$]. Thus, all the considerations made on the Sr compound do apply also to the Ba analog.

At all temperatures, the lanthanum and strontium cations were

Table 4. Refined structural parameters of $La_{1.85}Ba_{0.15}CuO_4$ at 10 K (Santoro, Beech, Miraglia, unpublished results)
Space group Cmca, n = 4, ρ_c = 7.06 g./cm^3
a = 5.3430(3) Å, b = 13.2504(2) Å, c = 5.3479(3) Å, V = 378.61 Å^3

Atom	Position	x	y	z	B(Å^2)	Occupancy
Ba	8f m	0	0.3607(1)	0.0052(4)	0.15(2)	0.075
La	8f m	0	0.3607(1)	0.0052(4)	0.15(2)	0.925
Cu	4a 2/m	0	0	0	0.23(3)	1.0
O(1)	8e 2	¼	0.0052(2)	¼	0.42(3)	1.0
O(2)	8f m	0	0.1821(1)	-0.0206(5)	0.90(3)	1.0

R_N = 3.68, R_p = 5.87, R_W = 7.95, R_E = 4.71, χ = 1.69
Symmetry operations
xyz, $x\bar{y}\bar{z}$, x ½-y ½+z, x ½+y ½-z, plus center and centering (½½½).

Table 5. Refined structural parameters of $La_{0.85}Sr_{0.15}CuO_{4-y}$ at room temperature, 60 K, and 10 K [31].

Space group* I4/mmm, n =2, ρ_c =6.99 g/cm^3
Bmab† 4 7.03

a =3.7793(1) Å, b =3.7793(1) Å, c =13.2260(3) Å, V =188.91 Å^3
5.3252(1) 5.3546(1) 13.1844(1) 375.94
5.3240(1) 5.3547(1) 13.1832(1) 375.83

Atom	Position	x	y	z	B(Å^2)	Occupancy
La + Sr	4e 4mm	0	0	0.3605(1)	0.48(2)	1.0
	8f m	0	-0.00496(1)	0.36072(3)	0.278(4)	1.0
	8f m	0	-0.004964(9)	0.36077(2)	0.281(5)	1.0
Cu	2a 4/mmm	0	0	0	0.42(3)	1.0
	4a 2/m	0	0	0	0.223(6)	1.0
	4a 2/m	0	0	0	0.214(6)	1.0
0(1)	4e 4mm	0	0	0.1824(1)	1.33(3)	1.0
	8f m	0	0.0256(1)	0.18257(7)	0.662(7)	1.0
	8f m	0	0.0255(1)	0.18260(6)	0.656(7)	1.0
0(2)	4c mmm	0	½	0	0.78(3)	1.0
	8e 2	¼	¼	-0.00560(6)	0.435(9)	1.0
	8e 2	¼	¼	-0.00573(6)	0.399(9)	1.0

R_N	R_p	R_W	R_E	χ
3.71	7.15	9.75	7.40	1.32
3.42	5.20	7.60	4.27	1.78
3.41	5.22	7.59	4.23	1.79

Symmetry operations (for Bmab)
xyz, $x\bar{y}\bar{z}$, x ½+y ½-z, x ½-y ½+z, plus center and centering (½0½).

*For each parameter, the values given in the first, second, and third line refer to room temperature, 60 K, and 10 K, respectively.

†Space group Bmab is not the conventional space group, which is Cmca, and is used here to permit direct comparison of the parameters with the room temperature phase. The axes transformation necessary to go from Bmab to Cmca is

$$\{abc\}_{Cmca} = (100/001/0\bar{1}0)\{abc\}_{Bmab}$$

found to be distributed at random over the equivalent sites of the same position, although the possibility of short range ordering cannot be ruled out on the basis of powder data. A schematic picture of the tetragonal structure is shown in Fig. 9. On the scale of the figure, the tetragonal and orthorhombic phases are not distinguishable, except that the orthorhombic unit cell is twice as large as the tetragonal cell. In Fig. 9, the perovskite-like planes are emphasized to illustrate the two-dimensional nature of the

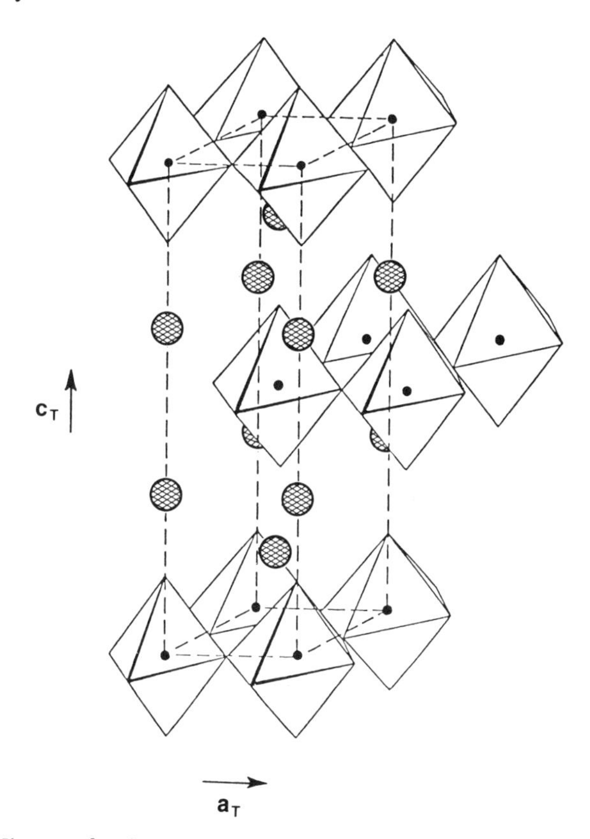

Fig. 9. View of the room temperature tetragonal structure of $La_{1.85}Sr_{0.15}CuO_4$. The large shaded circles represent the La/Sr atoms, the small filled circles the Cu atoms. The oxygen atoms are located at the vertices of the polyhedra. The Cu-O perovskite layers are perpendicular to the c-axis.

copper-oxygen layers perpendicular to the c-axis. These layers are separated by the La/Sr-oxygen planes with a rock salt type of atomic arrangement. The copper atoms in one plane do not share oxygen atoms with copper atoms in other planes. Each oxygen atom of the perovskite layers (0(2)) is bonded to two copper atoms in the same plane and to four R atoms (R = 0.925 La + 0.075 Sr) in adjacent planes, while each oxygen atom of the rock salt layers (0(1)) is linked to five R atoms and one copper atom in a distorted octahedral configuration. The Cu-O distances within the perovskite type planes are short (1.889 Å) and those in a perpendicular direction are rather long (2.411 Å). The coordination polyhedron around the copper atoms is a bipyramid, schematically shown in Fig. 10, with the relevant bond distances given for the structure at 10 K. The shape of this bipyradmid changes in a subtle way when the symmetry

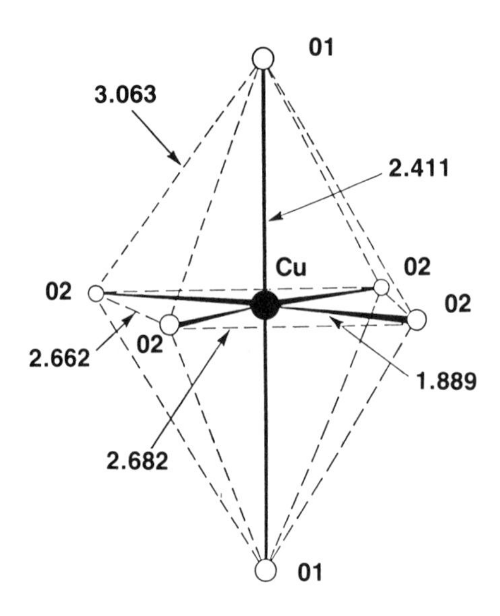

Fig. 10. The copper-oxygen coordination in $La_{1.85}Sr_{0.15}CuO_4$ at 10 K. The solid lines are the copper-oxygen bonds, the dashed lines outline the oxygen polyhedron. The atom separations are in Angstrom units.

is lowered from tetragonal to orthorhombic. The base becomes a rectangle and the copper-oxygen planes become buckled as a consequence of a small but significant rigid rotation of the bipyramids about axes parallel to the a-axis of the unit cell and passing through the shared oxygen atoms of the bipyramids. This important distortion is illustrated in Fig. 11, and a comparison with the configuration in Fig. 6 shows the similarity of buckled chains in the structures of $BaPb_{0.75}Bi_{0.25}O_3$ and $La_{1.85}Sr_{0.15}CuO_4$. The *R* atoms in this structure are nine-coordinated, so that their coordination is the average of what it would be in perovskite (12) and in rock salt (6). The shape of the coordination polyhedron is that of a square antiprism with a capped face and with the *R* atom shifted towards this face. This coordination is shown in Fig. 12 together with the relevant bond distances for the structure at 10 K. The *R* atoms are strongly bonded to both the oxygen atoms located on the same plane and those of the perovskite-like layers. Thus, the nature of the atoms forming *R* may strongly influence the Cu-O bonding. From a crystallographic point of view no change can be detected between 60 K and 10 K, that is, due to the superconducting transition.

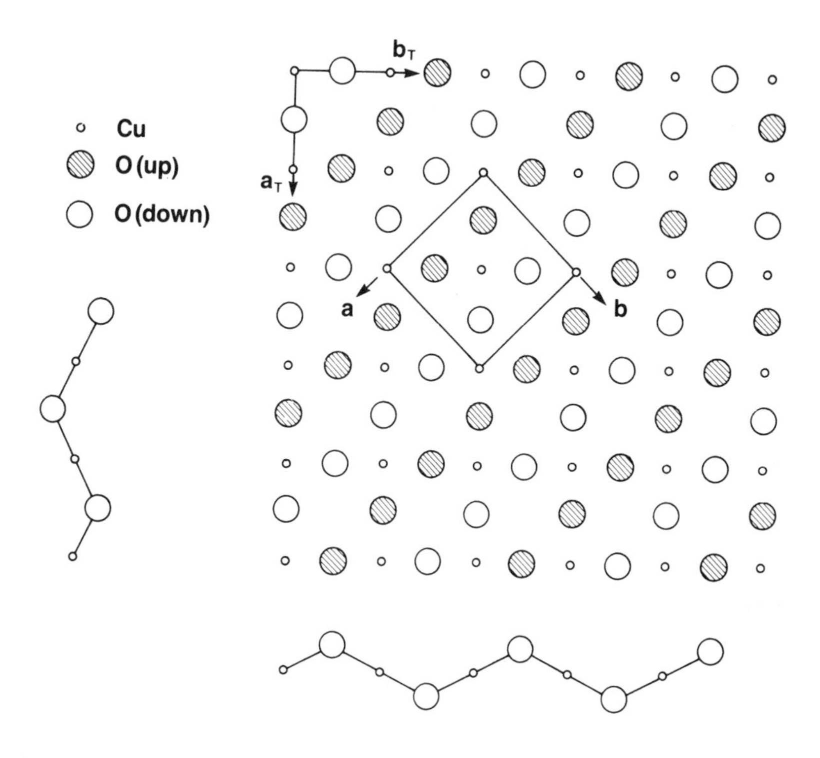

Fig. 11. The (001) plane at z = 0 in the structure of $La_{1.85}Sr_{0.15}CuO_4$ at 10 K. The rotation of the oxygen polyhedra at the phase transition causes the buckling of the Cu-O chains as indicated in the figure. The tetragonal and orthorhombic axes are also indicated.

As noted before, no oxygen deficiency has been observed in the Ba and Sr compounds. This means that charge compensation for the substitution of Sr or Ba for La must be accomplished entirely by oxidation of copper from Cu^{2+} to Cu^{3+}. By assuming a simple ionic model, the oxidation state of copper in $La_{1.85}M_{0.15}CuO_4$ would be $Cu^{2.15+}$, i.e. about 17 Cu^{2+} and 3 Cu^{3+}.

4.6. The System $Ba_2MCu_3O_\delta$ (M = Y, Gd, Eu, etc.)

Superconductivity with $T_c \simeq 94$ K was discovered in the system Ba-Y-Cu-O in a sample which was a mixture of phases [34] and, again, the superconducting compound was identified later in several

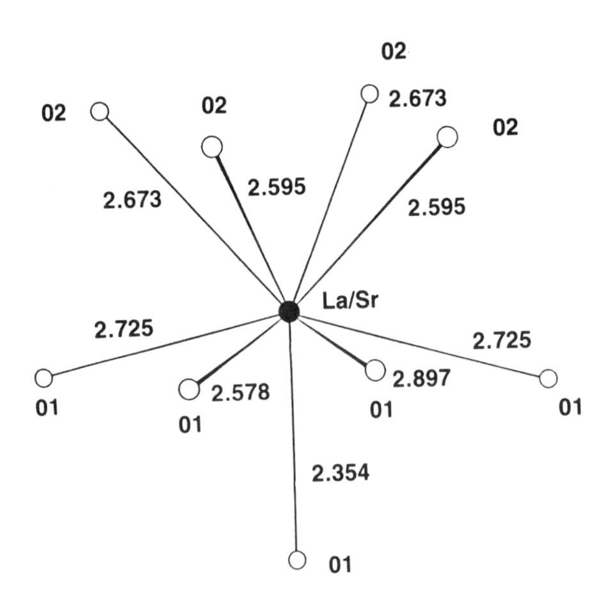

Fig. 12. The ninefold coordination of the La/Sr atoms in $La_{1.85}Sr_{0.15}CuO_4$. Atom separations are in Angstrom units. The oxygen atoms are labeled consistently with Table 5.

laboratories as $Ba_2YCu_3O_\delta$, with $\delta \simeq 7.0$ [35-39].

Immediately following this discovery, a number of structural studies were carried out by means of single crystal and powder x-ray diffraction techniques [35-43]. These investigations indicated that the structure of $Ba_2YCu_3O_7$ can be obtained from that of perovskite by tripling the c-axis, by eliminating all the oxygen atoms at $00\frac{1}{2}$ and half of those at $0\frac{1}{2}0$ and $\frac{1}{2}00$, by ordering the metal atoms, with the yttrium at $\frac{1}{2}\frac{1}{2}\frac{1}{2}$ and the barium at $\frac{1}{2}\frac{1}{2}{\sim}1/6$ and $\frac{1}{2}\frac{1}{2}{\sim}5/6$, and by shifting some of the atoms from the ideal positions they occupy in perovskite. Using single crystal x-ray data, this model was refined in space groups Pmmm [40], P4/mmm [43], and P$\bar{4}$m2 [42]. As mentioned before, in all cases it was assumed that both the sites at $0\frac{1}{2}0$ and $\frac{1}{2}00$ were occupied by oxygen atoms, with an occupancy factor of 50%.

Profile analysis of neutron powder data [44-49] immediately revealed that the correct space group of $Ba_2YCu_3O_\delta$ (with $\delta \simeq 7.0$) is Pmmm and that the model obtained in the x-ray experiments is not entirely correct because the sites at $0\frac{1}{2}0$ are completely filled by oxygen atoms and the sites at $\frac{1}{2}00$ are completely vacant. The refined structural parameters obtained in one of these neutron diffraction studies are given in Table 6 and the structure of $Ba_2YCu_3O_7$ is illustrated in Fig. 13 where it is compared with the parent structure of perovskite. No phase transitions have been

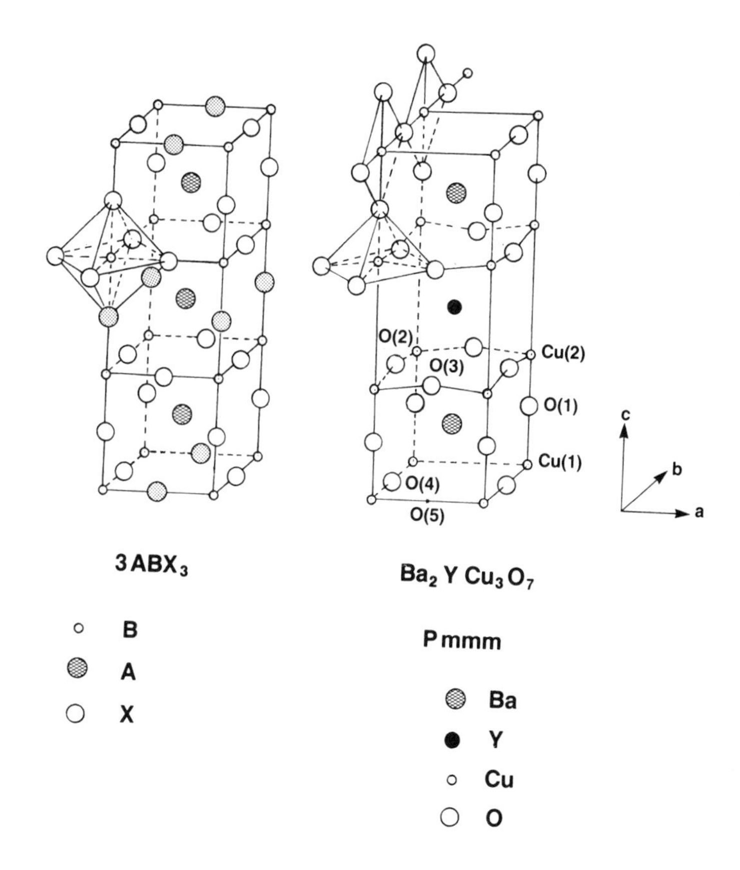

Fig. 13. Structure of $Ba_2YCu_3O_7$ (right side) compared with that of perovskite (left side). The structure of the superconductor can be derived from that of perovskite by eliminating the dotted oxygen atoms, ordering the cations, and shifting slightly the arrangement of the oxygen atoms. The pyramidal and square-planar coordinations of copper are indicated. The chains along the b-axis and the buckling of the Cu-O(2) and Cu-O(3) chains are also indicated in the figure.

observed in going from room temperature down to 10 K.

The ordered elimination of some of the oxygen atoms from the atomic arrangement of perovskite has significant effects on the features of the resulting compound. The copper atoms are located on two inequivalent positions of space group Pmmm. The first (Cu(2) in Fig. 13), at $00\sim 1/3$, has a pyramidal, almost square-planar coordination, rather than an octahedral one; the second (Cu(1)), located at the origin, has square planar coordination in which the

Table 6. Refined structural parameters of $Ba_2YCu_3O_7$ at room temperature [48].
Space group Pmmm, n = 1 , ρ_c = 6.38 g/cm^3,
a = 3.8198(1) Å, b = 3.8849(1) Å, c = 11.6762(3) Å, V = 173.27 Å^3

Atom	Position	x	y	z	B(Å^2)	Occupancy
Ba	2t mm	½	½	0.1839(2)	0.65(5)	1.0
Y	1h mmm	½	½	½	0.56(4)	1.0
Cu(1)	1a mmm	0	0	0	0.55(4)	1.0
Cu(2)	2q mm	0	0	0.3547(1)	0.49(4)	1.0
O(1)	2q mm	0	0	0.1581(2)	0.78(5)	1.0
O(2)	2s mm	½	0	0.3779(2)	0.57(5)	1.0
O(3)	2r mm	0	½	0.3776(2)	0.55(5)	1.0
0(4)	le mmm	0	½	0	1.73(9)	1.0

R_N = 4.84 R_P = 6.10 R_W = 8.40 R_E = 5.42 χ = 1.55

Symmetry operations
xyz, $\bar{x}\bar{y}z$, $x\bar{y}\bar{z}$, $\bar{x}y\bar{z}$, plus center of symmetry.

near square Cu-O_2 units share one corner and form chains along the b-axis of the unit cell. The atoms Cu(2) are strongly bonded to the four oxygen atoms O(2) and O(3), forming the basis of the pyramid, and are weakly bonded to the oxygen atom O(1) at the apex. Because of this feature, there exists in the structure two-dimensional layers of copper and oxygen atoms perpendicular to the c-axis. The oxygen atoms in these layers are slightly shifted from their ideal perovskite positions, producing the puckering of the layers indicated in Fig. 13. This feature, therefore, seems to be common to all the superconductors discovered so far. The ordering of the oxygen vacancies also affects the coordination of the other metal atoms. The yttrium atom is eightfold coordinated and the shape of the polyhedron is prismatic. The barium atoms, on the other hand, are tenfold coordinated and the coordination polyhedron can still be described as a cubo-octahedron with two oxygen atoms missing.

All the powder neutron diffraction studies considered so far [44-49] were carried out on samples of composition $6.8 \leq \delta \leq 7.0$ and, as we have mentioned previously, for $\delta = 7.0$ all the O(4) sites are occupied. When $\delta < 7.0$ there must be oxygen vacancies in the structure and it has been shown [48] that these are always confined to the sites O(4) in the chains. If we assume for this compound a simple ionic model, the charge compensation is accomplished entirely by oxidation of copper from Cu^{2+} to Cu^{3+}, as was mentioned in the case of the system $La_{2-x}M_xCu\ O_4$. For $\delta = 7.0$, the average formal valence of the copper atoms is $Cu^{2.33+}$, corresponding to a mixture of $2Cu^{2+}$ + Cu^{3+}.

It has been observed that the total oxygen stoichiometry decreases smoothly with increasing temperature [50, 51] and that the vacancies associated with this decrease involve only the O(4) sites. As this process goes on, the positions O(5) at ½00 are gradually filled until the structure from orthorhombic Pmmm becomes tetragonal P4/mmm. The transition, induced by oxygen stoichiometry, is a function of the temperature and of the oxygen partial pressure in the sample container. At one atmosphere of 0_2 pressure, the orthorhombic phase exists over the range $6.5 < \delta < 7.0$; the tetragonal phase is formed at $\delta \simeq 6.5$ and exists over the range $6.0 < \delta < 6.5$. However, if at low temperatures the oxygen is removed or the tetragonal phase is annealed, the orthorhombic modification may be retained down to values of δ as low as 6.3 [52]. Studies of the superconducting properties as a function of the oxygen content show that in the orthorhombic compound the value of T_c decreases as the oxygen is removed, and it becomes zero as the transition is approached. No superconductivity has ever been reported in the tetragonal phase, which in fact is semiconducting. In Table 7 the refined parameters of the tetragonal compound $Ba_2YCu_3O_6$ are given, and the structure is schematically illustrated in Fig. 14.

It is interesting to determine the structural changes taking place in this system as the oxygen content is decreased from $\delta = 7.0$ to $\delta = 6.0$. Some relevant distances are given in Table 8 for various compositions. The coordination of the barium atoms changes from tenfold for $\delta = 7.0$ to eightfold for $\delta = 6.0$. An analysis of the data reported in Table 8 shows that the distances Ba-0(1) and Ba-0(4) (see Fig. 13 for the nomenclature of the atoms) increase with decreasing oxygen content, while the distances Ba-O(2) and Ba-O(3) decrease. This means that the Ba atoms move away from the plane of the O(2) and O(3) atoms. In other words, they move along the c-axis towards the bottom of the "cup" of oxygen atoms left from the cube-octahedron after atoms O(4) and O(5) are eliminated. The coordination of yttrium, on the contrary, does not change significantly, and the coordination polyhedron remains a near rectangular prism over the entire range $6.0 \leq \delta \leq 7.0$.

For $\delta = 7.0$, the Cu(1) atoms are four-coordinated and are located at the center of near rectangles connected by vertices and forming the already mentioned chains. For $\delta = 6.0$ these atoms are two-coordinated, suggesting for this compound the formulation $Ba_2YCu^{1+}Cu^{2+}O_6$, with Cu(1) = Cu^{1+} and Cu(2) = Cu^{2+} [53]. The distances Cu(1)-O(1) and Cu(1)-O(4) decrease continuously as the oxygen is removed. The description of the Cu(1) coordination for compositions comprised between these two extremes is rather complex and we will use here a model proposed to explain the twin laws

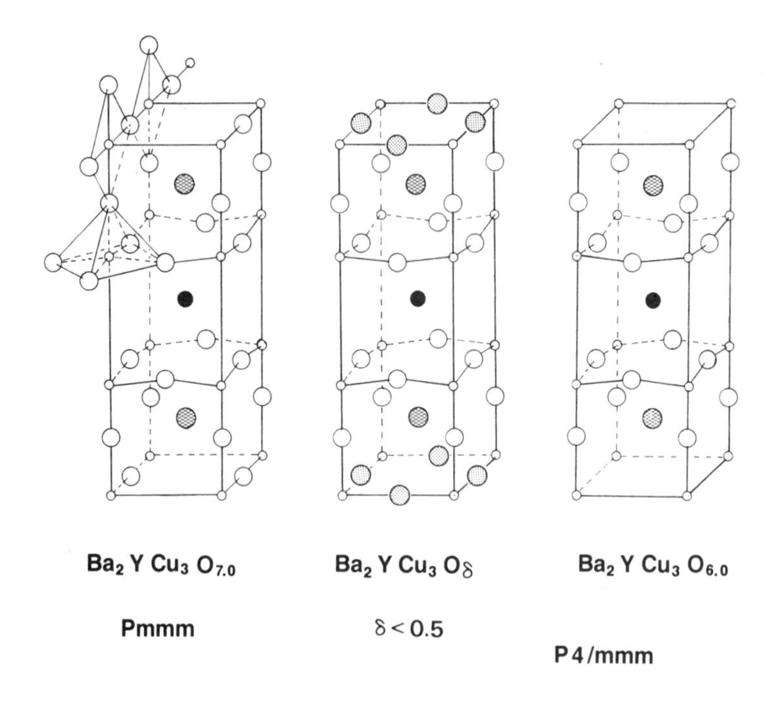

Fig. 14. Structures of $Ba_2YCu_3O_7$ (left), $Ba_2YCu_3O_{6.5}$ (center), and $Ba_2YCu_3O_6$ (right). The dotted circles indicate sites partially occupied by oxygen atoms.

observed in single crystals of the orthorhombic phase [55]. Figure 15 shows the plane with z = 0 of the tetragonal phase for a composition of $\delta \simeq 6.5$. If the oxygen atoms in excess of $\delta = 6.0$ are distributed at random over the positions at ½00 and 0½0, most of the Cu(1) atoms will have threefold coordination, which is very unusual for copper. The number of atoms with this coordination may be minimized if the occupancy of either ½00 or 0½0 takes place by blocks as shown in Fig. 15. The larger these blocks are, the less threefold coordinated are the Cu(1) atoms in the structure. The size of the blocks, however, does not have to be so large as to show the local orthorhombic symmetry of the domains. This model is supported by the type of twinning observed in the orthorhombic phase. In fact, as the oxygen content increases, the domains will become increasingly larger, and for $\delta = 7.0$ we will have a crystal composed of two identical orthorhombic individuals rotated exactly 90° about the c-axis. Each one of these individuals will then twin [twin law (110)], creating four individuals, in agreement with the splitting into four spots observed in the x-ray diagrams.

Table 7. Refined structural parameters of $Ba_2YCu_3O_{6.07}$ at room temperature [53].

Space group P4/mmm, n = 1, ρ_c = 6.14 g/cm^3

a = b = 3.8570(1) Å, c = 11.8194(3) Å , V = 175.83 Å

Atom	Position	x	y	z	B(Å^2)	Occupancy
Ba	2h 4mm	½	½	0.1952(2)	0.50(4)	1.0
Y	1d 4/mmm	½	½	½	0.73(4)	1.0
Cu(1)	la 4/mmm	0	0	0	1.00(4)	1.0
Cu(2)	2g 4mm	0	0	0.3607(1)	0.49(3)	1.0
O(1)	2g 4mm	0	0	0.1518(2)	1.25(6)	1.0
O(2)	4i mm	0	½	0.3791(1)	0.73(4)	1.0
O(3)*						
O(4)	2f mmm	0	½	0	0.9	~0.03
O(5)*						

R_N = 3.89, R_P = 6.10, R_W = 7.95, R_E = 5.64, χ = 1.41

Symmetry operations

xyz, $\bar{x}\bar{y}z$, $\bar{x}yz$, $x\bar{y}z$, $\bar{y}xz$, $y\bar{x}z$, yxz, $\bar{y}\bar{x}z$, plus center of symmetry.

*Positions 0(2) and 0(3) and positions 0(4) and 0(5) are equivalent in space group P4/mmm.

Table 8. Relevant bond distances in $Ba_2YCu_3O_\delta$

		δ = 7*	δ = 6.8*	δ = 6.5†	x - δ = 6.06‡
Ba - O(1)	x4	2.7408(4)	2.7470(8)	2.7690(8)	2.7751(5)
Ba - O(2)	x2	2.984(2)	2.972(3)	2.930(4)	
					2.905(1)
Ba - O(3)	x2	2.960(2)	2.938(6)	2.902(4)	
Ba - O(4)	x2	2.896(2)	2.922(4)	2.956(2)	
Y - O(2)	x4	2.409(1)	2.403(3)	2.404(3)	
					2.4004(8)
Y - O(3)	x4	2.386(1)	2.389(2)	2.408(3)	
Cu(1) - O(1)	x2	1.846(2)	1.843(3)	1.795(3)	1.795(2)
Cu(1) - O(4)	x2	1.9429(1)	1.9428(1)	1.9374(1)	
Cu(2) - O(1)	x1	2.295(3)	2.323(4)	2.429(4)	2.469(2)
Cu(2) - O(2)	x2	1.9299(4)	1.9305(6)	1.9366(6)	
					1.9406(3)
Cu(2) - O(3)	x2	1.9607(4)	1.9585(8)	1.9478(6)	

*From ref. [48]. The distances reported are those obtained for the samples at room temperature

†Ref. [54].

‡Ref. [53].

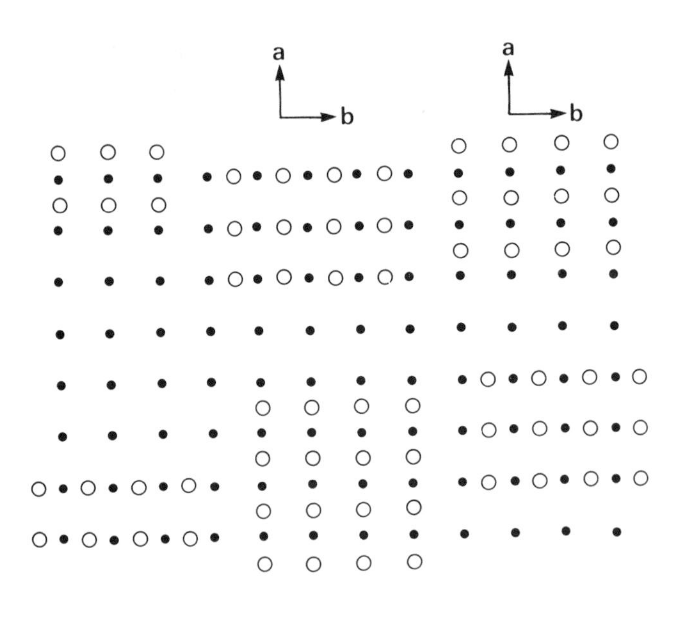

Fig. 15. Atomic arrangement of the copper and oxygen atoms on the plane (001) with z = 0, for a compound of approximate composition $Ba_2YCu_3O_{6.5}$. The oxygen atoms above and below each copper atom are not indicated. The Cu atoms at the edge of each domain have threefold coordination. Drawing taken from [55].

The environment of Cu(2) also changes significantly with composition. The distance Cu(2)-O(1) increases in going from δ = 7.0 to δ = 6.0 and, correspondingly, the distance of Cu(2) from the average plane of the atoms O(2) and O(3) decreases. This shift shows that the coordination of Cu(2) tends to be more square-planar and less pyramidal with decreasing oxygen content.

In all refinements reported in Table 8, it was found that the atoms O(4) have an unusually large temperature factor, a feature reported also by other groups [47, 44]. Low temperature experiments [48] have shown that this parameter does not decrease significantly with temperature, thus indicating the possible presence of static disorder. Anisotropic analysis revealed that O(4) is highly anisotropic in the direction of the a-axis of the orthorhombic structure. An attempt has therefore been made to refine the structure with O(4) split over the two positions 2k of Pmmm, x½0 and $\bar{x}$½0, instead of the constrained position 0½0. The agreement factors after this refinement did not differ significantly from the others, but the parameter x was found to be significantly different from zero (x = -0.053(4)) and, in addition, the thermal parameter of O(4) refined to a reasonable value. If we assume that the splitting of O(4) is real, then the near square $Cu\text{-}O_2$ units forming the chains along the b-axis are buckled much in the same way as the two-dimensional layer of copper and oxygen atoms perpendicular to the c-axis.

4.7. The System $Ba_{2-x}La_{1-x}Cu_3O_\delta$

In all compounds belonging to the system $Ba_2MCu_3O_\delta$ there is always a significant difference in the ionic radii of the cations Ba^{2+} and M^{3+}, and, because of this difference, the ordering of the two cations in the structure is complete, giving a ratio Ba/M equal to 2. Where the size of M^{3+} becomes large, as happens for La^{3+}, a complete ordering is no longer present and the ratio Ba/M may differ from two.

The compound $Ba_{2+x}La_{1+x}Cu_3O_\delta$, with x = 0.5 and δ = 7.2, has been found by x-ray powder diffraction [56] to have a tetragonal defect perovskite structure with $a \simeq a_p\sqrt{2}$ and $c \simeq 3\ c_p$, so that the unit cell volume is about double that of $Ba_2MCu_3O_\delta$. The formula has been given accordingly as $Ba_3La_3Cu_6O_{14.4}$. The postulated structure had a cation skeleton similar to that of $Ba_2MCu_3O_\delta$, but with a different ordering of the oxygen atoms, causing the doubling of the unit cell. More recently, neutron powder diffraction work [57, 58] has confirmed the tetragonal symmetry, but no evidence for doubling the unit cell has been found, so that the formula consistent with these results can be written as $Ba_{1.5}La_{1.5}Cu_3O_{7.2}$. The structure of this composition is the same as that of tetragonal $Ba_2YCu_3O_\delta$ with δ < 6.5, except for the ordering of the cations, since now the perovskite A sites are occupied by 75% Ba and 25% La. The structure determined by profile refinement using neutron powder data is shown in Fig. 16 and the refined parameters are given in Table 9.

The tetragonal symmetry in this compound rises from the addition of oxygen atoms to the system (δ > 7.0) and it requires that the sites at ½00 and 0½0 be equally occupied. Unequal filling of these two sites would preserve orthorhombic symmetry, but this configuration is apparently less favorable than equal occupation. As a consequence, there is no long range order of one-dimensional chains along the a- and b-axes.

Recent results [59] have shown that $Ba_{2-x}La_{1+x}Cu_3O_\delta$ is a solid

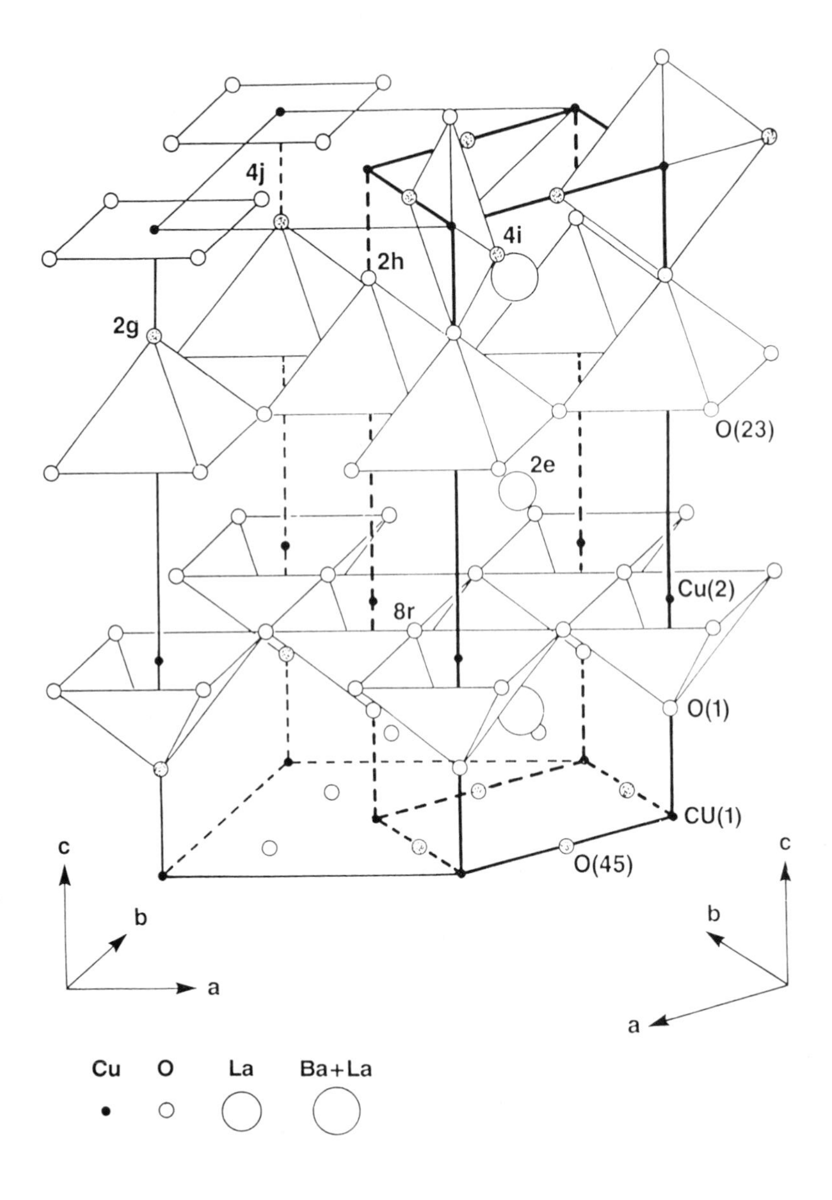

Fig. 16. The structure of $Ba_{1.5}La_{1.5}Cu_3O_{7.3}$. The large unit cell is the cell proposed in [56]. The small cell with the dark outline is the one determined by neutron powder diffraction. Atoms in the large cell are identified by means of the coding scheme used in space group notation. The other symbols refer to atoms in the small cell according to Table 9. The dotted circles indicate oxygen sites that are partially occupied. For clarity, only the cation in the small cell has been drawn.

Table 9. Refined parameters of $Ba_{1.5}La_{1.5}Cu_3O_{7.3}$ [58].
Space group P4/mmm, n = 1
a = 3.9150(1) Å, c = 11.6929(6) Å, V = 179.22 Å^3

Atom	Position	x	y	z	B(Å^2)	Occupancy
Ba	2h mm	½	½	0.1790(2)	1.14(5)	0.75
La(1)	2h mm	½	½	0.1790(2)	1.14(5)	0.25
La(2)	1d 4/mmm	½	½	½	0.65(7)	1.0
Cu(1)	1a 4/mmm	0	0	0	1.40(7)	1.0
Cu(2)	2q 4mm	0	0	0.3463(1)	0.65(4)	1.0
O(1)	2q 4mm	0	0	0.1583(3)	2.7(1)	0.96(1)
0(23)	4i mm	0	½	0.3642(2)	1.01(5)	1.0
0(45)	2f mmm	0	½	0	4.1(2)	0.66(1)

R_N = 5.63 R_P = 6.88 R_W = 8.97 R_E = 5.71 χ = 1.57

Symmetry operations are the same as in Table 7.

solution for 0.0 ≤ x ≤ 0.5. Compounds with 0.0 ≤ x ≤ 0.3 exhibit bulk superconductivity which depends on the value of the Ba/La ratio. For x = 0.2 a T_c^{onset} = 60 K was found, with a resistivity R = 0.0 at 53 K. For x < 0.1 the value of T_c is still higher, and for x = 0.0, T_c^{onset} = 89 K and R becomes zero at 78 K. For values of x larger than 0.3 the material is metallic and as x increases further it becomes semiconducting.

X-ray diffraction patterns show that all samples with 0.2 ≤ x ≤ 0.5 can be indexed in terms of a tetragonal cell similar to that of $Ba_{1.5}La_{1.5}Cu_3O_{7.2}$. For samples with x < 0.2, splitting of the diffraction lines indicates the formation of an orthorhombic structure. The amount of oxygen in the unit cell decreases as x decreases in such way that the formal valence on Cu remains +2.33, the same value observed in $Ba_2YCu_3O_7$.

Structural studies on this system are far from complete, and further work is in progress at the present time.

4.8. The System $Bi_2CaSr_2Cu_2O_\delta$

The discovery of superconductivity in the previous systems has sparked an intense search for new materials. Such effort is under way not only to find compounds with values of T_c higher than ~ 90K, but also to provide a broad materials base on which to formulate more precise theoretical models to understand the experimental results.

Recently, superconductivity has been detected in the multiphase systems Bi-Aℓ-Ca-Sr-Cu-O and Bi-Ca-Sr-Cu-O [59]. In some cases the phases present in the various materials have been identified [60] and structural determinations have been attempted by single crystal x-ray diffraction techniques. The compound $Bi_{2.2}Sr_2Ca_{0.8}Cu_2O_{8+\delta}$ with $T_c \simeq 84$ K, has been found to be orthorhombic (Fmmm), with lattice parameters a = 5.414, b = 25.79, c = 30.89 Å, and with a substructure based on a unit cell of parameters a = 5.414, b = 5.418, c = 30.89 Å [61]. The structure of the atoms in the subcell shows layers of copper and oxygen perpendicular to the c-axis and quite similar to those observed in $YBa_2Cu_3O_7$. These layers are separated by others containing calcium, strontium and oxygen, and bismuth and oxygen. The bismuth is surrounded by six oxygen atoms in octahedral coordination, while the copper has a pyramidal environment similar to that present in $YBa_2Cu_3O_7$. The structure of the true unit cell has not been determined so far, and the diffraction patterns seem to indicate the presence of modulation.

A tetragonal subcell (I4/mmm) of parameters a = 3.817 and c = 30.6 Å has been found for the composition $Bi_2Sr_{1.5}Cu_2O_8$ (T_c = 85 K) [62]. The substructure based on this cell is very similar to the one described previously, as the two subcells are related to one another in a very simple way.

The substructure of the composition $Bi_2Sr_{2.33}Ca_{0.67}Cu_2O_{8+\delta}$ ($T_c \simeq$ 95K) is based on an orthorhombic cell of parameters a = 5.399, b = 5.414, c = 30.904 Å, with space group Amaa[63]. Also in this case the sequence of the layers perpendicular to the c-axis is similar to that found for the previous two compositions. Electron microscopy shows that the superstructure is incommensurate along the a-axis.

In all the above studies the oxygen atoms are located rather imprecisely. Neutron diffraction studies are still in a preliminary stage due to the difficulty of preparing large single phase samples.

Structural determinations on the system $T\ell_2Ba_2Ca_{n-1}Cu_nO_{2n+4}$ have been carried out by x-ray [64] and neutron diffraction [65] for the compositions corresponding to n = 1,2,3. In this case the atomic positions are known quite accurately. The structures of the thallium compounds are practically the same to those of the bismuth system and they consist of sequences of layers such as:

$$(BaO)_c(T\ell O)_o(T\ell O)_c(BaO)_o(CuO_2)_c(Ca)_o(CuO_2)_c(BaO)_o(T\ell O)_c(T\ell O)_o(BaO)_c\cdots$$

(in this expression the chemical symbols give the actual composition of each layer, and the subscripts o and c indicate if the cation is located at the origin of the layer's net or at the center, respectively).

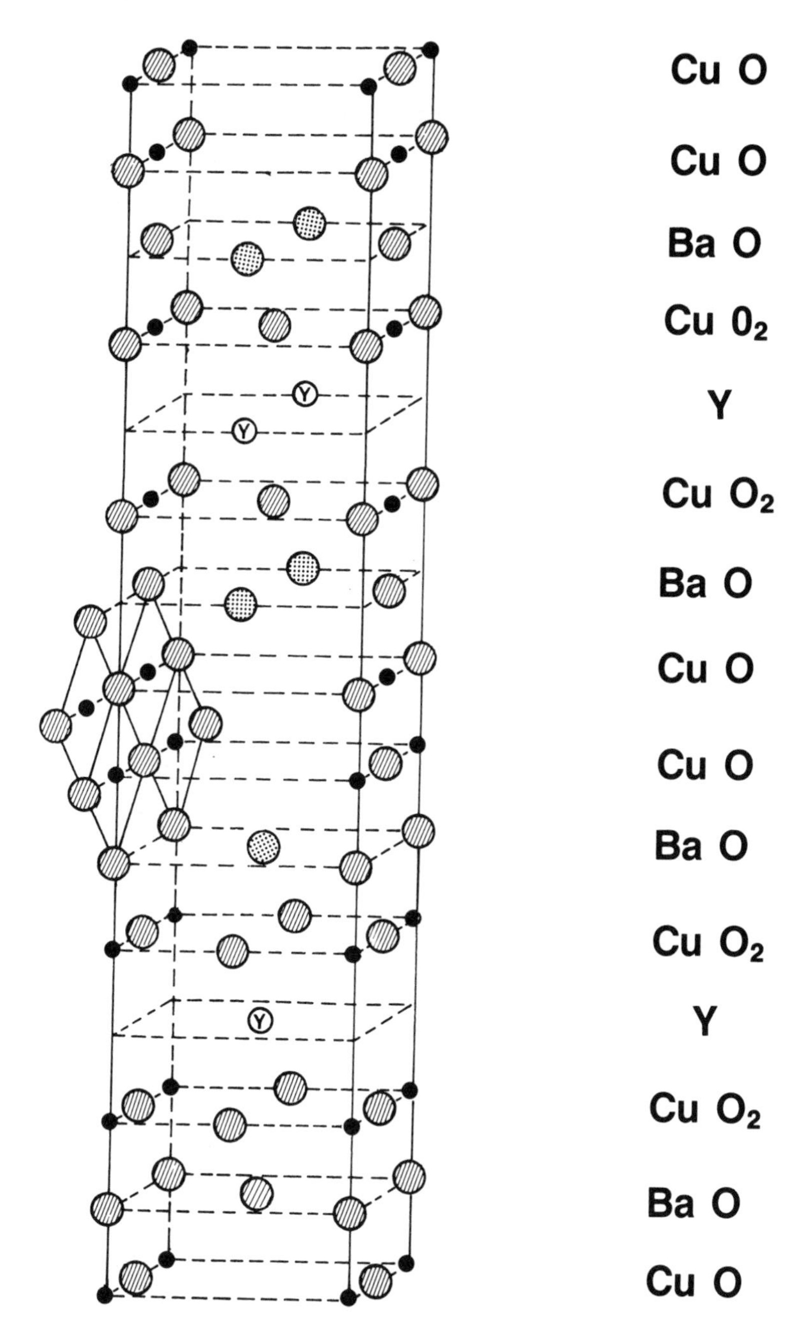

Fig. 17. Crystal structure of the compound $Ba_2YCu_4O_8$. Visible in the figure are the two consecutive CuO layers shifted one with respect to the other by ½ **a**. The effect of this shear is the formation of the double chains of edge sharing squares with oxygen atoms at the corners and copper atoms at the center.

4.9. Materials With Crystallographic Shear

Recently, two compounds have been discovered in the Ba-Y-Cu-O system, namely $Ba_2YCu_4O_8$ [66] and $Ba_4Y_2Cu_7O_{14+x}$ [67]. They exhibit superconductivity transitions at about 80 K and 40 K, respectively. The structures of these compounds are closely related to that of $Ba_2YCu_3O_{6+x}$. The main feature of these compounds is the presence of two consecutive, oxygen deficient CuO layers whose stacking is accompanied by a shift of origin of ½00. This configuration is associated with the formation of double chains of edge-sharing squares with oxygen atoms at the corners and copper atoms at the center. The geometry of the double chains is illustrated in Fig. 17 for $Ba_2YCu_4O_8$. It is clear from the figure that in this compound all the CuO layers are involved in the formation of double chains. On the other hand, the structure of $Ba_4Y_2Cu_7O_{14+x}$ is built by blocks of $Ba_2YCu_3O_7$-type alternating with blocks of $Ba_2YCu_4O_8$. Thus in this structure, in addition to the double chains, there are also the single chains characteristic of the structure of $Ba_2YCu_3O_7$, and for this reason the oxygen content of the compound may vary continuously from O_{14} to O_{15}.

References

[1] H. P. Klug and L. E. Alexander, *X-Ray Diffraction Procedures for Polycrystalline and Amorphous Materials*, 2nd Edition, J. Wiley & Sons (1974), Chapter 8.

[2] H. M. Rietveld, *J. Appl. Cryst.* **2**, 65 (1969).

[3] G. E. Bacon, *Neutron Diffraction*, Clarendon Press, Oxford (1975).

[4] E. Prince, *J. Appl. Cryst.* **14**, 157 (1981).

[5] J. M. Buerger, *Crystal Structure Analysis*, J. Wiley, p. 171 (1960).

[6] E. Prince and A. Santoro, National Bureau of Standards, U. S. Tech. Note 117, p.11, Ed. F. Shorten (1980).

[7] G. Caglioti, A. Paoletti, and F. P. Ricci, *Nucl. Instrum. Methods* **3**, 223 (1958).

[8] A. Santoro, R. J. Cava, D. W. Murphy, and R. S. Roth, *Proceedings of the Symposium on Neutron Scattering*, ANL, Argonne, Ill. August 12-14, 1981.

[9] R. D. Shannon, *Acta Cryst.* **A32**, 751 (1976).

[10] A. W. Sleight, J. L. Gillson, and P. E. Bierstedt, *Solid State Commun.* **17**, 27 (1975).

[11] D. E. Cox and A. W. Sleight, *Proceedings Conf. on Neutron Scattering*, Gatlinburg, Tenn., June 6-10, 1976; R. M. Moon, Editor.

[12] R. D. Shannon and P. E. Bierstedt, *J. Am. Ceram. Soc.* 635 (1970).
[13] G. Thornton and A. J. Jacobson, *Mat. Res. Bull.* **11**, 837 (1976).
[14] D. E. Cox and A. W. Sleight, *Solid State Comm.* **19**, 969 (1976).
[15] D. E. Cox and A. W. Sleight, *Acta Cryst.* **B35**, 1 (1979).
[16] G. Thorton and A. J. Jacobson, *Acta Cryst.* **B34**, 351 (1978).
[17] C. Chaillout, Universite Scientifique et Medical de Grenoble, Thesis, (1986).
[18] G. K. Wertheim, J. P. Remeika, and D.N.E. Buchanan, *Phys. Rev.* **B26**, 2120 (1982).
[19] L. F. Mattheiss and D. R. Hamann, *Phys. Rev.* **B28**, 4227 (1983).
[20] C. Chaillout, J. P. Remeika, A. Santoro, and M. Marezio, *Solid State Comm.* **56**, 833 (1985).
[21] *International Tables for Crystallography*, Volume A, Reidel Publishing Co. (1983).
[22] W. Fisher, H. Burzlaff, E. Hellner, and J. D. H. Donnay, *Space Groups and Lattice Complexes*, NBS Monograph 134 (1973).
[23] J. G. Bednorz and K. A. Müller, *Z. Phys.* **B64**, 189 (1986).
[24] S. Uchida, H. Takagi, K. Kitazawa, and S. Tanaka, *Jpn. J. Appl. Phys.* **26**, L1 (1987).
[25] H. Takagi, S. Uchida, K. Kitazawa, and S. Tanaka, *Jpn. J. Appl. Phys.* **26**, L123 (1987).
[26] J. M. Longo and P. M. Raccah, *J. Solid State Chem.* **6**, 526 (1973).
[27] J. D. Jorgensen, H. B. Schuettler, D. G. Hinks, D. W. Capone II, K. Zhang, and M. B. Brodsky, *Phys, Rev. Lett.* **58**, 1024 (1987).
[28] D. M. Paul, G. Balakrishnan, N. R. Bernhoeft, W. I. F. David, and W. T. A. Harrison, *Phys. Rev. Lett.* **58**, 1976 (1987).
[29] S. C. Moss, K. Forster, J. D. Axe, H. You, D. Holwein, D. E. Cox, P. H. Hor, R. L. Meng, and C. W. Chu, *Phys. Rev.* **B35**, 7195 (1987).
[30] A. Santoro, F. Beech, and S. Miraglia, unpublished results.
[31] R. J. Cava, A. Santoro, D. W. Johnson, and W. W. Rhodes, *Phys. Rev.* **B35**, 6716 (1987).
[32] M. Francois, K. Yvon, P. Fischer, and M. Decroux, *Solid State Comm.* **63**, 35 (1987).
[33] R. J. Cava, R. B. van Dover, B. Batlogg, and E. A. Rietman, *Phys. Rev. Lett.* **58**, 408 (1987).
[34] M. K. Wu, J. R. Ashburn, C. J. Torng, P. H. Hor, R. L. Meng, L. Gao, Z. J. Huang, Y. Q. Wang, and C. W. Chu, *Phys. Rev. Lett.* **58**, 908 (1987).
[35] R. J. Cava, B. Batlogg, R. B. van Dover, D. W. Murphy, S. A. Sunshine, T. Siegrist, J. P. Remeika, E. A. Rietman, S. Zahurak, and G. P. Espinosa, *Phys. Rev. Lett.* **58**, 1676 (1987).
[36] P. M. Grant, R. B. Beyers, E. M. Engler, G. Lim, S. S. P.

Parkin, M. L. Ramirez, V. Y. Lee, A. Nazzal, J. E. Vasquez, and R. J. Savoy, *Phys. Rev.* **B35**, 7242 (1987).

[37] D. G. Hinks, L. Soderholm, D. W. Capone, J. D. Jorgensen, I. K. Schuller, C. U. Segre, K. Zhang, and J. D. Grace, *Appl. Phys. Lett.* **50**, 1688 (1987).

[38] K. Kadowaki, Y. K. Huang, M. van Sprang, and A. A. Menovsky, *Physica* **B145**, 1 (1987).

[39] P. Ganguly, R. A. Mohanram, K. Sreedhar, and C. N. R. Rao, *J. Phys.* **28**, L321 (1987).

[40] T. Siegrist, S. Sunshine, D. W. Murphy, R. J. Cava, and S. M. Zahurak, *Phys. Rev.* **B35**, 7137 (1987).

[41] Y. Syono, M. Kikuchi, K. Oh-ishi, K. Hiraga, H. Arai, Y. Matsui, N. Kobayashi, T. Sasaoka, and Y. Muto, Jpn. J. Appl. Phys. **26**, L498 (1987).

[42] R. M. Hazen, L. W. Finger, R. J. Angel, C. T. Prewitt, W. L. Ross, H. K. Mao, C. S. Hadidiacos, P. H. Hor, R. L. Meng, and C. W. Chu, *Phys. Rev.* **B35**, 7238 (1987).

[43] Y. LePage, W. R. McKinnon, J. M. Tarascon, L. H. Greene, G. W. Hull, and D. M. Hwang, *Phys. Rev.* **B35**, 7245 (1987).

[44] M. A. Beno, L. Soderholm, D. W. Capone, D. G. Hinks, J. D. Jorgensen,J. D. Grace, I. K. Schuller, C. U. Segre, and K. Zhang, *Appl. Phys. Lett.* **51**, 57 (1987).

[45] J. E. Greedan, A. O'Reilly, and C. V. Stager, *Phys. Rev.* **B35**, 8770 (1987).

[46] J. J. Capponi, C. Chaillout, A. W. Hewat, P. Lejay, M. Marezio, N. Nguyon, B. Raveau, J. L. Soubeyroux, J. L. Tholence, and R. Tournier, *Europhys. Lett.* **3**, 1301 (1987).

[47] W. I. F. David, W. T. A. Harrison, J. M. F. Gunn, D. Moze, A. K. Soper, P. Day, J. D. Jorgensen, M. A. Beno, D. W. Capone, D. G. Hinks, I. K. Schuller, L. Soderholm, C. U. Segre, K. Zhang, and J. D. Grace, *Nature* **327**, 310 (1987).

[48] F. Beech, S. Miraglia, A. Santoro, and R. S. Roth, *Phys. Rev.* **B35**, 8778 (1987).

[49] S. Katano, S. Funahashi, T. Hatano, A. Matsushita, K. Nakamura, T. Matsumoto, and K. Ogawa, *Jpn. J. Appl. Phys.* **26**, L1046 (1987).

[50] J. D. Jorgensen, M. A. Beno, D. G. Hinks, L. Soderholm, K. J. Volin, R. L. Hitterman, J. D. Grace, I. K. Schuller, C. U. Segre, K. Zhang, and M. S. Kleffisch, *Phys. Rev.* B. in press.

[51] G. S. Grader and P. K. Gallagher, *Adv. Cer. Mat.* **2**, 649 (1987).

[52] R. J. Cava, B. Batlogg, C. H. Chen, E. A. Rietman, S. M. Zahurak, and D. Weder, *Phys. Rev. Lett.* submitted.

[53] A. Santoro, S. Miraglia, F. Beech, S. A. Sunshine, D. W. Murphy, L. F. Schneemeyer, and J. V. Waszczak, *Mat. Res. Bull.* **22**, 1007 (1987).

[54] S. Miraglia, F. Beech, A. Santoro, D. Tran Qui, S. A. Sunshine,

and D. W. Murphy, *Mat. Res. Bull.* **22**, 1733 (1987).

[55] J. L. Hodeau, C. Chaillout, J. J. Capponi, and M. Marezio, *Mat. Res. Bull.* submitted.

[56] L. Er-Rakho, C. Michel, J. Provost, and B. Raveau, *J. Solid State Chem.* **37**, 151 (1981).

[57] W. I. F. David, W. T. A. Harrison, R. M. Ibberson, M. T. Weller, J. R. Grasmeder, and P. Lanchester, *Nature* **328**, 328 (1987).

[58] S. Sunshine, L. F. Schneemeyer, J. V. Waszczak, D. W. Murphy, S. Miraglia, A. Santoro, and F. Beech, *J. Cryst. Growth* **85**, 632 (1987).

[59] C. W. Chu, J. Bechtold, L. Gao, P.H. Hor, Z. J. Huang, R. L. Meng, Y. Y. Sun, Y. Q. Wang, and Y. Y. Xue, *Phys. Rev. Lett.* **60**, 941 (1988).

[60] R. M. Hazen, C. T. Prewitt, R. J. Angel, N. L. Ross, L. W. Finger, C. G. Hadidiacos, D. R. Veblen, P. J. Heaney, P. H. Hor, R. L. Meng, Y. Y. Sun, Y. Q. Wang, Y. Y. Xue, Z. J. Huang, L. Gao, J. Bechtold, and C.W. Chu, *Phys. Rev. Lett.* **60**, 1174 (1988).

[61] S. A. Sunshine, T. Siegriest, L. F. Schneemeyer, D. W. Murphy, R. J. Cava, B. Batlogg, R. B. van Dover, R. M. Flemming, S. H. Glarum, S. Nakahara, R. Farrow, J. J. Krajewski, S. M. Zahurak, J. V. Waszczak, J. H. Marshall, P. Marsh, L. W. Rupp, Jr., and W. F. Peck, preprint.

[62] J. M. Tarascon, Y. LePage, P. Barboux, B. G. Bagley, L. H. Greene, W. R. McKinnon, G. W. Hull, M. Giroud, and D. M. Hwang, preprint.

[63] M. A. Subramanian, C. C. Torardi, J. C. Calabrese, J. Gopalakrishnan, K. J. Morrissey, T. R. Askew, R. B. Flippen, U. Chowdhry, and A. W. Sleight, *Science* **239**, 1016 (1988).

[64] C. C. Torardi, M. A. Subramaniam, J. C. Calabrese, J. Gopalakrishnan, K. J. Morrissey, T. R. Askew, R. B. Flippen, U. Chowdhry, and A. W. Sleight, *Science*, **240**, 631 (1988).

[65] D. E. Cox, C. C. Torardi, M. A. Subramaniam, J. Gopalakrishnan, and A. W. Sleight, *Phys. Rev. Lett.* (submitted).

[66] P. Marsh, R. M. Fleming, M. L. Mandich, A. M. DeSantolo, J. Kwo, M. Hong, and L. J. Martinez-Miranda, *Nature* **334**, 141 (1988).

[67] P. Bordet, C. Chaillout, J. Chenavas, J. L. Hodeau, M. Marezio, J. Karpinski, and E. Kaldis (preprint).

CHAPTER 5

Electronic Structure, Lattice Dynamics, and Magnetic Interactions

Ching-ping S. Wang

5.1. Introduction

Superconductivity in the copper oxide systems occurs near a structural instability (orthorhombic↔tetragonal) as well as a metal-insulator transition which can be controlled by doping or oxygen stoichiometry. At low temperatures, the metallic phase becomes superconducting, while the insulating phase shows long range antiferromagnetic order. The structure of most of the high-T_c superconductors are **layered** perovskite with large anisotropy (see Chapter 4), and the most unique feature is the presence of $Cu\text{-}O_2$ layers which seem to be crucial for the high temperature superconductivity. The only exception is the **cubic** perovskite $Ba_{1-x}K_xBiO_{3-y}$, which has an onset temperature of 30 K but has the $Cu\text{-}O_2$ layers missing [1]. Instead of antiferromagnetism (spin density wave), the superconducting phase is near a planar breathing-type displacement of the oxygen atoms away from one of the Bi neighbors (charge density wave). It will be interesting to see if this new cubic material has the same mechanism to achieve higher T_c as the cuprate superconductors.

Much of the theoretical effort has focussed on the origin of the high temperature superconductivity. The transition temperatures of the copper oxides are three to five times higher than those of intermetallics with comparable densities of states at the Fermi energy (E_F). The extremely small isotope effects [2] and many other unusual electronic properties have led to the widespread belief that the electron-phonon interaction by itself cannot be the dominant mechanism for the superconductivity. Instead, the occurrence of antiferromagnetic correlations in the copper oxides seems to support the notion that superconductivity in these compounds may be due to magnetic interactions [3,4], while the presence of the oxygen breathing mode distortion in $Ba_{1-x}K_xBiO_{3-y}$ seems to suggest an alternative pairing mechanism due to charge fluctuations [5,6]. With the exception of the electron-phonon interaction, however, precise calculation of T_c within these models is not possible at the moment. In addition, there remain many fundamental questions as to how any of these simplified theoretical models are related to the structure and chemistry of the compounds. What is the mechanism of the observed structural transformation? What is the role of ion doping and oxygen vacancies?

The purpose of this chapter is to review our current theoretical understanding of the normal state properties of the new cuprate superconductors with emphasis on their electronic properties, lattice dynamics, and magnetic interactions. The superconducting properties are discussed by Allen in Chapter 9. It will be shown that many interesting questions can be answered based on the results of band structure calculations. For example, the unusual superconducting transition phase diagram of $YBa_2Cu_3O_{7-x}$ as a function of oxygen vacancy concentration can be understood from (1) the hole count on the $Cu\text{-}O_2$ layer which modifies the electron-electron correlations, and (2) the metallic nature of the chain layers near E_F which control the interlayer coupling between the $Cu\text{-}O_2$ layers. Similar arguments can also be applied to the new bismuth- and thallium-based copper oxides, where T_c increases with increasing number of $Cu\text{-}O_2$ layers. In addition, the Hall coefficients and Fermi surfaces predicted by band theory have been confirmed by experiment. However, there are still several important discrepancies between band theory and experiment. This is evident from comparisons with the photoemission experiments, which show significant renormalization from the theoretical density of states (*DOS*). Furthermore, the observed antiferromagnetic ground state in La_2CuO_4 cannot be stabilized in the standard spin-polarized band structure calculations. These results reflect an inadequacy of the local spin density approximation in describing the intra-atomic Coulomb repulsion U. The source of the error will be analyzed based on our understanding of two related compounds: Mott insulators and heavy fermion superconductors. Finally, magnetic interactions will be discussed, based on the results of recent quantum Monte Carlo simulations of a two-dimensional extended Hubbard model.

5.2. Electronic Structure of $La_{2-x}(Ba,Sr)_xCuO_4$

$La_{2-x}(Ba,Sr)_xCuO_{4-y}$ crystallizes in the tetragonal phase when x >

0.2, and undergoes an orthorhombic distortion for $x \leq 0.2$, due to rotation of the oxygen octahedra [7]. The superconducting transition temperature depends critically on the amount of Sr or Ba doping and oxygen vacancy concentration. In particular, T_c reaches a maximum value near $x = 0.15$. For $x > 0.15$, detailed neutron powder diffraction data indicates that the oxygen vacancy concentration increases sharply and more than compensates the effects of the Ba or Sr doping so that T_c drops rapidly to zero [8]. In the limit of small x, the resistivity versus temperature curve of the orthorhombic phase shows characteristics of a doped semiconductor at low temperatures. This semiconducting nature may be associated with antiferromagnetic ordering which is observed in undoped La_2CuO_{4-y} with T_N between 240 and 290 K depending on the number of oxygen vacancies [9-12]. In the superconducting compound where Ba or Sr are substituted on the La site, no antiferromagnetism has been observed. It was proposed that doping with Ba or Sr stabilizes the metallic, tetragonal phase so that the compound becomes superconducting at lower temperatures [13,14]. Thus, it is quite surprising that superconductivity was later reported in stoichiometric La_2CuO_{4-y} near 40 K. Furthermore, Fermi surfaces were observed by Tanigawa et al. [15] in orthorhombic La_2CuO_{4-y} in positron 2D-ACAR measurements on samples which exhibit semiconducting behavior in their resistivity. They proposed that the orthorhombic phase of La_2CuO_4 is metallic with an unusual temperature dependence of its resistivity due to the nature of scattering centers or scattering mechanisms. To date, the exact nature of the orthorhombic phase of La_2CuO_4 remains an open question.

5.2.1. Band Structure of Stoichiometric La_2CuO_4

The band structure of La_2CuO_4 was first reported by Mattheiss [13] and Yu et al. [14]. The energy bands of Mattheiss for the body-centered-tetragonal (bct) phase of La_2CuO_4 are shown in Fig. 5.1 along the symmetry lines in the Brillouin zone (*BZ*). Of particular importance among the 17 bands that are centered at -3 eV below E_F are the two bands labeled as **A** and **B** in Fig 5.1. They are the antibonding and the bonding states, respectively, of the Cu_{3d} orbital with x^2-y^2 symmetry and the O_{2p} orbital of x and y symmetry along the nearest neighbor Cu-O bonds in the xy plane. Notice that the antibonding band, being exactly half filled, plays an important role in the superconducting and magnetic properties of the system. The 15 intermediate bands correspond to nonbonding Cu_{3d}-O_{2p} states and the weak pdσ bonds along the c axis where the Cu-O distance (2.40 Å) is significantly longer than that in the a-b plane (1.90

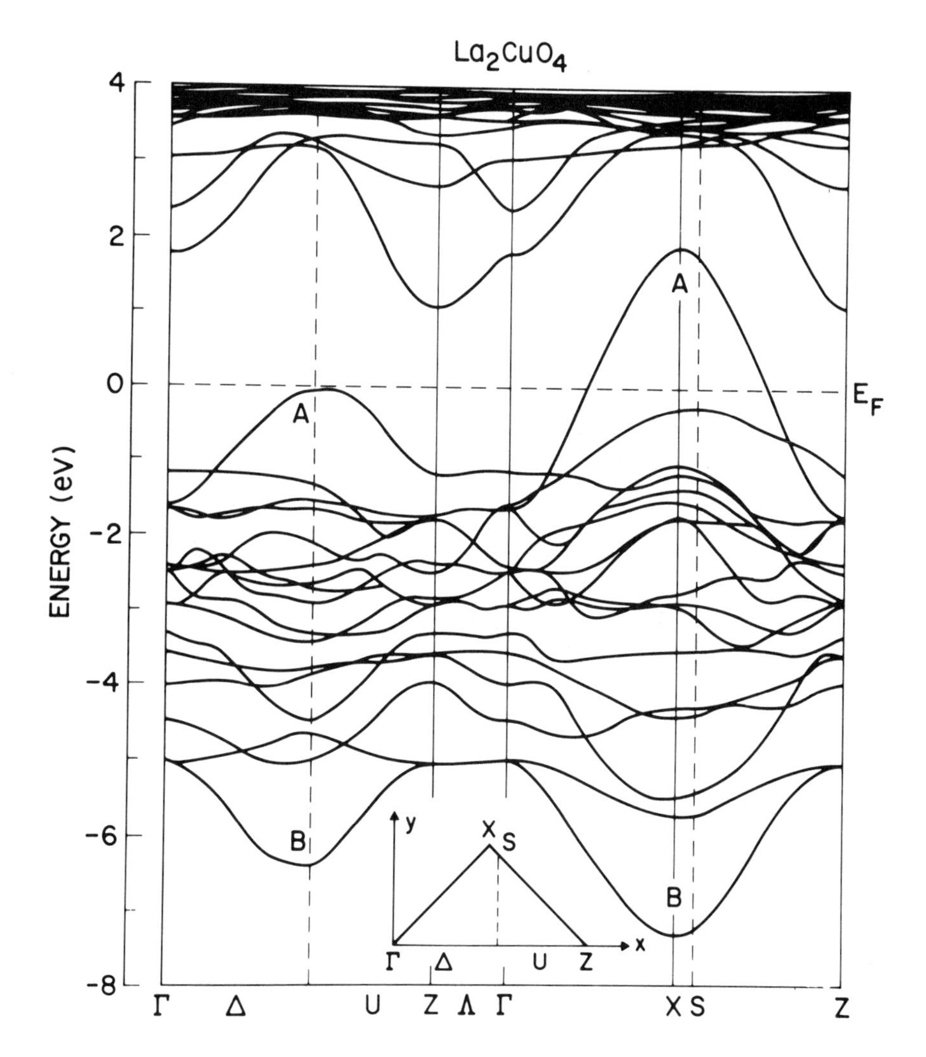

Fig. 5.1. Energy bands of La_2CuO_4 along some high-symmetry directions, from Mattheiss [13]. The Brillouin zone for the bct structure is included for clarity.

Å). The La_{5d} states starting around 1 eV merge with the La_{4f} bands at 3 eV above E_F, while the La_{5p} bands lie well below E_F at -14 eV. Thus, there is a significant charge transfer away from the La atoms with very little charge left in the occupied conduction bands.

The band structure of La_2CuO_4 is quasi-two-dimensional in character. This is evident from the fact that (1) there is very little dispersion in the energy bands along the c-axis, and (2) the charge distribution is negligibly small between the layers (see Fig. 4 of Yu et al. [14]). By contrast, the energy bands in the basal plane have large dispersions which correspond to a rather wide band width of ≥ 9 eV. Mattheiss [13] has shown that the essential feature of the bands labeled **A** and **B** in Fig. 5.1 can be described by an extremely simple two-dimensional tight-binding model of the Cu-O plane with two parameters ($E_d = E_p = -3.2$ eV, and pd$\sigma = -1.85$ eV).

5.2.2. The Effects of Sr or Ba Doping

Since the La atoms contribute very little to the occupied conduction bands, substitutional alloying for trivalent La with either a divalent (Sr or Ba) or tetravalent element may be described by the simple rigid-band model. It is assumed in this model, that the changes in the density of states (*DOS*) are negligible and E_F can be shifted rigidly to accommodate the extra charge.

To understand in more detail the effects of the Ba doping, Pickett et al. [16] have calculated the band structure of $LaBaCuO_4$. Their total and the projected *DOS* for $LaBaCuO_4$ (right panel), are compared with that of La_2CuO_4 (left panel) in Fig. 5.2. As expected, there is a substantial downward shift of E_F in $LaBaCuO_4$.

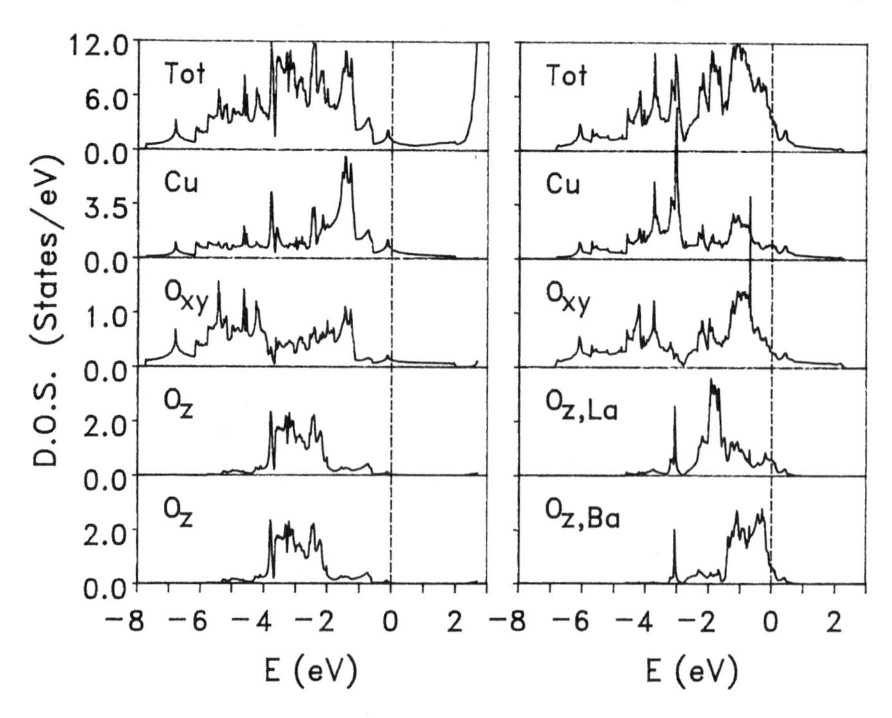

Fig. 5.2. Total and projected *DOS* per atom for La_2CuO_4 (left panel) and $LaBaCuO_4$ (right panel), from Pickett et al. [16]. The dashed line indicates E_F.

In the Cu-O plane, the $O_{x,y}$ *DOS* is changed very little while the spectral weight on the Cu ion is shifted to lower energy by the addition of the Ba atom. Significant changes were found in the *DOS* of oxygen atoms normal to the Cu-O plane, which becomes inequivalent in the 50% Ba compound. One of them (O_{La}) is nearly coplanar with the La atom, while the other (O_{Ba}) lies in a Ba layer. The divalent Ba atom is less ionic than the trivalent La atom, which accounts for the O_{Ba} *DOS* being nearer E_F than the O_{La} *DOS*. The rigid band model does appear to hold near E_F, however, where only the antibonding Cu_{3d} and O_{2p} states contribute. This applies to $La_{2-x}(Ba,Sr)_xCuO_4$ in the superconducting range of $x < 0.3$. It is interesting to note that there is a van Hove singularity in the *DOS* which occurs very near the concentration of Ba or Sr doping where T_c reaches its highest value ($x = 0.15$). Consequently, many transport and thermodynamic properties are sensitive to x in this region.

Qualitatively, these results are similar to that of Schwarz [17], who calculated the band structure of $La_{1.5}Ba_{0.5}CuO_4$ by doubling the bct unit cell of La_2CuO_4 along the z direction and replacing one La atom by a Ba atom. There are two copper atoms with different symmetry in this supercell of simple tetragonal structure. Since the O_{Ba} *DOS* was shifted closer to E_F, one of the two Copper atoms (Cu_{Ba}) effectively becomes fivefold coordinated while the other Cu_{La} atom remains fourfold coordinated. Accordingly, the *DOS* for the Cu_{Ba} atom lies lower in energy than that of the Cu_{La} and so the possibility of valence fluctuations between the two Cu atoms was proposed. This model is similar in spirit to the exciton model originally proposed by Varma et al. [15], which emphasizes the charge fluctuations between Cu and O ions.

5.2.3. Fermi Surface Nesting in $La_{2-x}(Ba,Sr)_xCuO_4$

The Fermi surface of $La_{2-x}(Ba,Sr)_xCuO_4$ has been studied in detail by Xu et al. [18] based on the rigid band model discussed above. As can be seen from Fig. 5.3a, the Fermi surface for $x = 0$ consists of hole-like cylinders along the z-axis in the *BZ* centered about the X points. At the concentration $x = 0.17$, at which E_F touches the van Hove singularity, the Fermi surface undergoes a dramatic change from hole-like cylinders centered at the corner X points to electron cylinders centered at Γ (see Fig. 5.3b). There is significant Fermi surface nesting in the [110] direction where a large portion of the Fermi surface is spanned by a common wave vector $\mathbf{q}$. As shown by Xu et al. [18], the resulting generalized susceptibility $\chi(\mathbf{q})$ (calculated from the constant matrix element approximation) is substantially enhanced near the zone boundary points X and N. For $x > 0$, the magnitude of the spanning vector $|\mathbf{q}|$ is reduced so that the

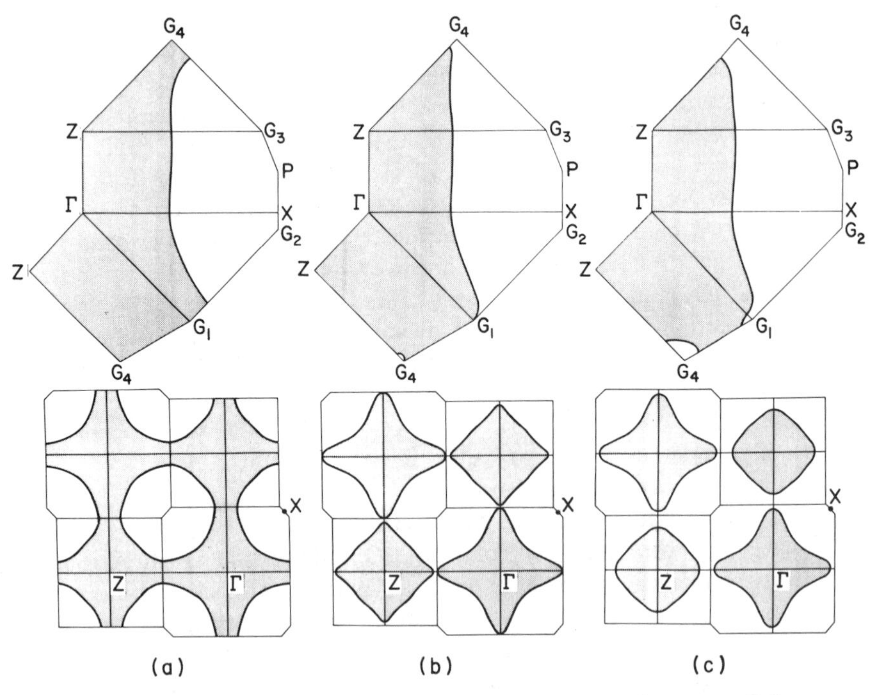

Fig. 5.3. Fermi surface of $La_{2-x}Ba_xCuO_4$ in the extended zone scheme: (a) x=0, (b) x=0.17 at the van Hove singularity and (c) x=0.2, from Xu et al. [18].

center of the peak in $\chi(q)$ is shifted away from the boundary of the *BZ*.

Fermi surface nesting is likely to drive a phonon branch soft, thereby enhancing the electron-phonon interaction. The near perfect Fermi surface nesting also suggests the possibility of charge density wave or spin density wave instabilities. In the case of pure La_2CuO_4, where $q = 2k_F=(\frac{1}{2},\frac{1}{2},0)$, the strong electron-phonon interaction may induce some kind of lattice distortion and open a semiconducting band gap at E_F. The distorted system is stabilized by the gain in the electronic energy near the gap. It is interesting to note that the intensity of the peak in $\chi(q)$ reaches its maximum value near the concentration of $x = 0.15$, where T_c is maximized. Within this Peierl's instability picture, the essential role of the Ba or Sr doping is to shift q away from the boundary of the *B*rillouin *Z*one (*BZ*), thereby stabilizing the tetragonal phase which is metallic and therefore superconducting. Mattheiss [13] and Weber [19] have considered a Fermi-surface-induced planar breathing type displacement of O atoms away from the central Cu site, which is semiconducting with orthorhombic symmetry. However, the space group is different from that of room temperature La_2CuO_4. Instead, an antiferromagnetic ordering was observed in the orthorhombic phase in which the Neel temperature depends sensitively on the amount of oxygen vacancy. Similarly, a very small amount of Ba or Sr doping ($x \simeq 0.02$) could destroy the long-range antiferromagnetic order [20].

5.2.4. Tetragonal-Orthorhombic Transition

Based on group theoretical analysis, Kasowski et al. [21] demonstrated that the two Cu atoms in the observed orthorhombic phase remain equivalent so that the energy bands must be degenerate on certain faces of the new *BZ*. This would indicate that La_2CuO_4 is metallic, contrary to the resistivity measurement. This metallic behavior is confirmed by several first principles energy band calculations of the orthorhombic phase, which is predicted to be metallic with very little changes from that of the undistorted tetragonal phase. To reconcile this with the observed semiconducting behavior requires an additional perturbation that breaks up the symmetry of the two Cu atoms, possibly by long-range antiferromagnetic order, by oxygen vacancies, or by Ba and Sr doping. Another possibility proposed by Kasowski et al. [21] is that at very low temperature La_2CuO_4 becomes a semiconductor due to an orthorhombic to monoclinic structural phase transition. A third possibility proposed by Tanigawa et al. [15], is that the orthorhombic phase of La_2CuO_4 is metallic with an unusual temperature dependence of its resistivity due to strong electron correlations. This is supported by the observation of Fermi surfaces on orthorhombic La_2CuO_{4-y} in positron annihilation experiments. The measured Fermi surfaces of the nonstoichiometric samples appear to be more complicated than that predicted by the band structure for stoichiometric La_2CuO_4. As discussed in Sec. 5.2.3, the topology of the Fermi surface is extremely sensitive to doping or oxygen stoichiometry.

5.2.5. Effects of Oxygen Vacancies

Unlike the La atom, which does not contribute much to the occupied conduction band, the oxygen atoms actively participate in the bonding and antibonding states. Therefore, the effects of oxygen vacancies cannot be understood purely on the basis of the rigid band model. Recently, Sterne and Wang [22] have studied the effects of oxygen vacancies on the electronic properties of La_2CuO_4 using a supercell geometry. The supercell is constructed by doubling the tetragonal unit cell in the two planar directions with 28 atoms in

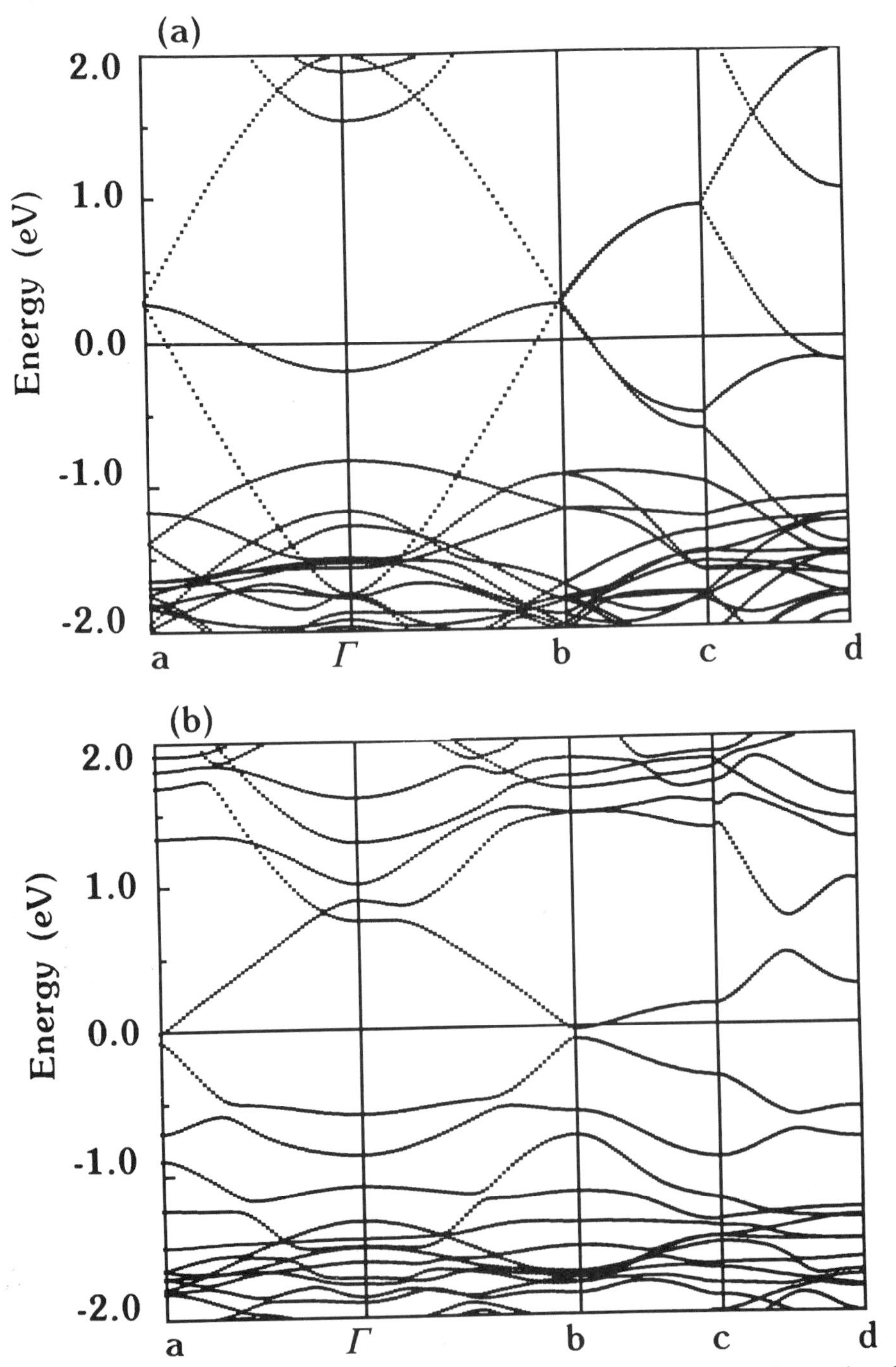

Fig. 5.4. Energy bands of (a) La_2CuO_4 folded back to the reduced zone of the 28-atom supercell, and (b) $La_2CuO_{3.75}$ with one out of 16 oxygen atoms removed from the two-dimensional Cu-O plane in the supercell, from Sterne and Wang [22].

the unit cell. One out of the 16 oxygen atoms in the supercell was removed from the two-dimensional Cu-O_2 network which formally corresponds to $La_2CuO_{3.75}$.

The effects of the oxygen vacancy are clearly indicated in Fig. 5.4, where the band structure of La_2CuO_4 in this 28 atom supercell (Fig. 5.4a) is compared with that of the $La_2CuO_{3.75}$ (Fig. 5.4b). Oxygen vacancies alter the band structure significantly near E_F with the fourfold degeneracy at **a** and **b** completely lifted. Removing an oxygen atom reduces the coordination numbers of the neighboring copper atoms, which leads to a narrower conduction band width. It is interesting to note that the system is very close to being a semiconductor. The electron pocket around **b** is compensated by a hole pocket elsewhere in the *BZ*, but the overlap is clearly very small.

Another consequence of the oxygen vacancy is to produce an upward shift of E_F, which can be seen by comparing the fourfold degenerate levels at **a** and **b** that lie **above** E_F in Fig. 5.4a with the corresponding gaps **at** E_F in Fig. 5.4b. A similar conclusion can be drawn by comparing the band structures of $YBa_2Cu_3O_7$ and that of $YBa_2Cu_3O_6$. This can be understood from the ionicity of the oxygen atom in a manner similar to the case of the Ba or Sr doping. If we assume that the oxygen atom is present as O^{2-}, then removing an oxygen atom will leave two extra electrons which are accommodated in part by an upward shift of E_F. It is assumed that removing a neutral oxygen atom with four valence p-electrons will at the same time remove six valence p-states, thereby pushing E_F upward. We found that 0.3 electrons remain in a sphere of radius 1.9 a.u. around the vacancy site, which is stabilized by the Madelung potential somewhat similar to the F-center in ionic oxides. These results indicate the importance of self-consistency to model oxygen vacancy. Recently, Papaconstantopoulos et al. [23] have used the tight-binding coherent potential approximation to understand the disorder phase of oxygen vacancies and found that deoxygenation does not raise E_F, contrary to simple doping arguments. However, the vacancy on-site energy is taken as infinite in their calculation, so that all charge is expelled from the vacancy sites. It remains interesting to see if the discrepancy reflects errors of non-self-consistent charge distribution in the CPA calculations or effects of disorder not included in the supercell model.

So far, the calculations have not included the relaxation of atoms around the vacancy site. Depending on the size of the relaxation, there may be some quantitative changes to the results, although the qualitative conclusions regarding the increase of E_F should be valid. Band gaps were found to be opening up near E_F in a superlattice of ordered oxygen vacancies. The vacancy-vacancy separations in the superlattice are reasonably far apart, so that the band gaps will probably be smeared into a deep valley in the *DOS* in samples of randomly arranged oxygen vacancies. Accordingly, the static dielectric screening should be reduced in oxygen deficient samples.

5.3. Electronic Properties of $YBa_2Cu_3O_{7-x}$

The unique feature of the structure of $YBa_2Cu_3O_{7-x}$ is the absence of 2+y oxygen atoms from the perfect triple perovskite $YCuO_3(BaCuO_3)_2$ [24]. In the limit $x = 0$, the vacancies are due to the absence of the oxygen atoms from (1) the Y plane which separates the adjoining Cu-O_2 layers, and (2) half of the oxygen sites in the Cu-O plane between the Ba-O layers which leads to the formation of the Cu-O chain. There are three weakly coupled Cu-O layers in the unit cell, consisting of two quasi-two-dimensional Cu-O_2 planes and a one-dimensional Cu-O chain. Along the c-axis, the oxygen atom in the Ba-O plane (O_{Ba}) is significantly closer to the fourfold coordinated Cu atoms in the chains (1.85 Å) than the fivefold coordinated Cu atom in the planes (2.30 Å). As x is increased from 0 to 1, the oxygen atom in the chain layer is removed and some of the oxygen atoms are moved to the empty sites, which eventually leads to the orthorhombic to tetragonal transformation. One notable change in the structure as the oxygen atoms are removed is that O_{Ba} atom moves away from the Cu-O_2 plane towards the Cu-O chain along the c-axis, so that the Cu atoms in the chain layer are more isolated and the interlayer coupling between Cu-O_2 planes are further reduced. There is a remarkable correlation between T_c and the oxygen vacancy concentration [25]. T_c is around 90 K for $0 \leq x \leq 0.2$ before decreasing sharply to 60 K, where it remains up to $x = 0.5$. Beyond $x = 0.5$, T_c drops sharply and antiferromagnetic order has been observed near $x = 0.7$, 0.85 and 1.0 [26].

5.3.1. Band Structure of $YBa_2Cu_3O_7$

The band structure of $YBa_2Cu_3O_7$ was first reported by Mattheiss and Hamann [27] and by Massidda et al. [28,29]. Since detailed neutron analysis was not available at the time, the Cu-O chains were chosen along the a-axis rather than the experimentally determined b-axis in the calculation of Mattheiss and Hamann. The resulting band structure differs somewhat from that of Massidda et al. and subsequent calculations, especially near E_F, due to a slightly different Cu-O distance in the chain. Away from E_F, they found that the electronic structure of $YBa_2Cu_3O_7$ consists of occupied O_{2s} bands centered at -15 eV, Ba_{5p} bands centered at -10 eV, as well as

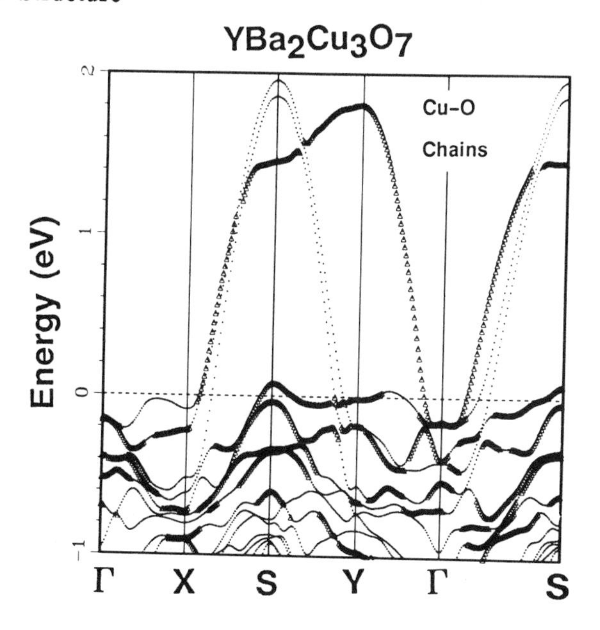

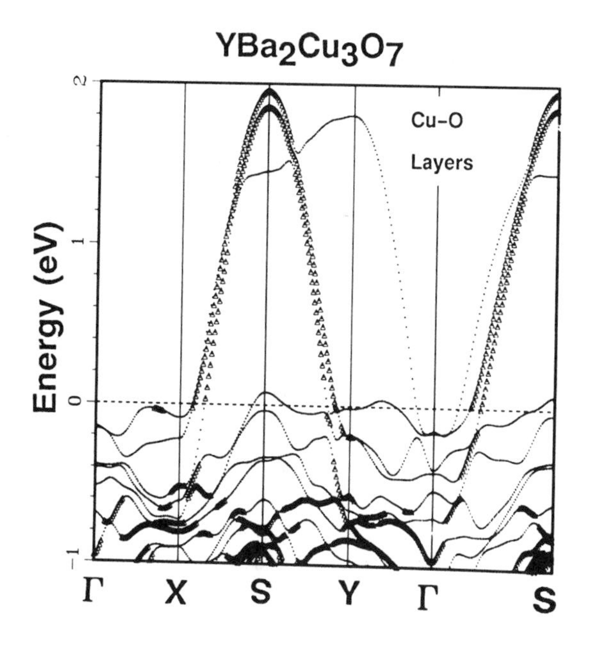

Fig. 5.5. Energy bands of $YBa_2Cu_3O_7$ along some high-symmetry directions for $k_z=0$ of the orthorhombic *BZ*, from Krakauer and Pickett [30]. (top) States with more than 60% of their charge on the Cu-O chains are emphasized with the large symbols, (bottom) those with more than 80% charge on the two Cu-O layers are emphasized.

unoccupied Ba_{5d} and Y_{4d} bands above 3.4 eV. Like La_2CuO_4, states near E_F are dominated by pdσ bands of Cu-O_2 atoms.

Figure 5.5 shows the band structure of Krakauer and Pickett [30], which is in close agreement with earlier calculations of Massidda et al. [28] and Ching et al. [31]. The states with more than 60% of their charge on the Cu-O chains are emphasized with the large symbols in Fig. 5.5a, while states with more than 80% charge on the two Cu-O layers are shown in Fig. 5.5b. Among the four bands that cross E_F, two of them have the majority of their charge on the Cu-O_2 plane corresponding to one band for each layer. These bands are similar to those of La_2CuO_4 except that they are less than half filled. The Cu-O chains are responsible for the two other bands: (1) a steep band crossing E_F, which is strongly dispersive only in the chain direction, that is, along the X-S and Γ-Y directions, and (2) a flat band just at E_F along the Y-S direction.

By examining the projected *DOS* of $YBa_2Cu_3O_7$, Massidda et al. [28] pointed out that the strong pdσ bond in the Cu-O_2 planes leads to a rather wide band width of $\geq$ 8 eV. The corresponding width for the Cu-O chains is noticeably narrower. Along the c-axis, the O_{Ba} atoms lie significantly closer to the copper atoms in the chain (Cu_{chain}) than those in the plane (Cu_{plane}), which leads to some noticeable dispersion in the two chain bands along the c-axis [28]. The reasons that O_{Ba} atoms lie closer to Cu_{chain} than Cu_{plane} have been discussed by Massidda et al. [28]. First of all, the shorter Cu_{chain}-O_{Ba} bond is stronger so that only the bonding states are filled, leaving their antibonding partners empty. By contrast, the Cu_{plane}-O_{Ba} bonding and antibonding states, which have a smaller splitting, are fully occupied. Secondly, the Cu_{plane} atoms are five-fold coordinated, with the O_Y atom on the opposite side of the Cu_{plane} missing, while the fourfold coordinated Cu_{chain} atoms are balanced by the O_{Ba} atoms above and beneath them. Together, these two effects exert a Coulombic repulsion on the O_{Ba} atom and push them away from the Cu_{plane}.

Detailed analysis of the charge density distributions confirmed the one-dimensional nature of the Cu-O chain as well as the two-dimensional nature of the Cu-O_2 planes. The ionic Y and Ba atoms were found to act as electron donors and do not otherwise participate, even if the fully substituted material ordered magnetically at very low temperature. The lack of conduction electron density near the Y site explains the stability of the high superconducting critical temperature when the isolated Y atoms are replaced by strongly magnetic rare-earth such as Gd or Er [32,33]. The absence of exchange coupling of the Gd ion with the conduction electrons, responsible for superconductivity, has recently been confirmed by Alp et al. [34] utilizing Mössbauer spectroscopy.

5.3.2. Fermi Surface of $YBa_2Cu_3O_7$

The Fermi surface and charge distribution of $YBa_2Cu_3O_7$ have been studied in detailed by Yu et al. [29]. With the exception of the flat band at E_F, a possible nesting feature can be seen in the Fermi surfaces corresponding to the three other bands. Unlike La_2CuO_4, however, the spanning vectors are not close to a reciprocal lattice vector so that **commensurate** charge density waves or spin density waves will not be possible.

Recently, Fermi surfaces have been measured by two-dimensional angular correlation of the positron annihilation radiation (2D-ACPAR) technique on a single crystal $YBa_2Cu_3O_{7-x}$ [35]. The momentum distribution shows structures associated with the higher zones, which indicates positron annihilation with extended rather than localized electrons. Three nearly cylindrical Fermi surface sheets have been observed. There is a square piece centered at the S point, which corresponds to the two bands of predominantly Cu-O_2 layer character which are too close to one another to be resolved. Qualitatively, the agreement for the other two bands of predominantly chain character appears to be reasonable considering the facts that (1) they are much more sensitive to oxygen vacancies, and (2) there are some discrepancies in different theoretical predictions for one of the chain bands with very large effective mass. Further study will be necessary to establish a more quantitative comparison with band theory.

Qualitative agreement between theory and experiments for the topology of the Fermi surfaces raises serious doubt about the resonating valence bound theory, which is probably the most innovative model that has been proposed for superconductivity in these materials. In this picture, the normal state consists of highly degenerate, short range, singlet pairs of electrons, and superconductivity arises from Bose condensation of topological disorders. In the half-filled limit, the system is an insulator. Away from the half-filled limit, there may be a "pseudo Fermi surface" in the normal state, if the band gap vanishes along certain lines or points in the Brillouin zone, but it will be very different from that of a paramagnetic band structure.

5.3.3. Band Structure of $YBa_2Cu_3O_6$

The band structure of tetragonal $YBa_2Cu_3O_6$ by Yu et al. [36] is shown in Fig. 5.6. Due to the missing oxygen atoms along the chain, the 1D chain bands in $YBa_2Cu_3O_7$ that are emphasized with large symbols in Fig. 5.5a are completely absent. In their place, there are two completely filled degenerate states (d_{yz} and d_{xz}) of Cu_{chain}

at Γ and M. These orbitals form π-bonding bands with O_{Ba} atoms. The two Cu-O_2 bands remain the same as in $YBa_2Cu_3O_7$, but there is an upward shift of E_F due to the ionicity of the missing oxygen atoms. Accordingly, there are significant changes in the Fermi surface and the hole count in the Cu-O_2 bands. In particular, the band along ΓM in Fig. 5.6 suggests possible Fermi-surface nesting along the (110) direction, which may explain the occurrence of antiferromagnetic order near $x \cong 1$. Yu et al. [36] found an enhancement of the Stoner factor $S = N(E_F)I$ in $YBa_2Cu_3O_{7-x}$, from 1.12 ($x = 0$) to 1.38 ($x = 1$), even though the *DOS* at E_F is actually reduced.

5.3.4. Oxygen Vacancies and the Phase Diagram of $YBa_2Cu_3O_{7-x}$

The phase diagram of the high-T_c superconductors is very sensitive to the filling factor of the pdσ antibonding bands, which is a measure of the hole concentration on the Cu-O_2 layers. In the

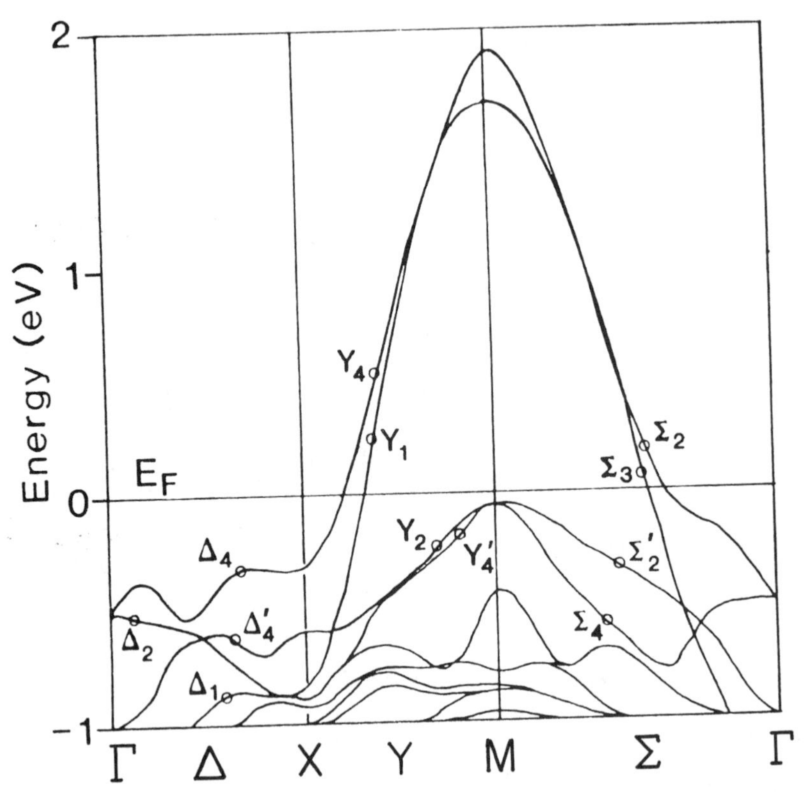

Fig. 5.6. Energy bands of $YBa_2Cu_3O_6$ along some high-symmetry directions, from Yu et al. [36].

half-filled limit, the compound is an antiferromagnetic insulator due to Fermi surface nesting. Superconductivity occurs away from the half-filled limit, where the spanning vector is no longer commensurate with the periodicity of the lattice, so that the long-range antiferromagnetic order is suppressed. In the case of $La_{2-x}(Ba,Sr)_xCuO_{4-y}$, this can be controlled by doping and oxygen vacancy. In the case of $YBa_2Cu_3O_{7-y}$, the Cu-O chain layer essentially dopes the Cu-O_2 layers even in the limit $y = 0$. As we shall discuss in Sec. 5.4.3, the Cu-O chain layer may also play an additional role of providing a metallic intervening layer that enhances the interlayer coupling among the Cu-O_2 layers and therefore raises T_c significantly.

The experimental phase diagram of $YBa_2Cu_3O_{7-x}$ as a function of oxygen-vacancy concentration is shown in the top part of Fig. 5.7. In the superconducting phase, T_c shows plateaus, which have recently been explained by Zaanen et al. [37] using a tight binding model for a unit cell doubled along the a-axis to accommodate two inequivalent chains. One of the chains is either intact ($0 \leq x \leq 0.5$) or empty ($0.5 \leq x \leq 1$) while the other chain has periodic oxygen vacancies arranged according to the nonstoichiometry. The electronic structure is described by removing the linking oxygen atoms, without changing the tight-binding parameters. Starting from the empty chain, each additional oxygen atom introduces one antibonding level and two holes. If the level is above E_F, the antibonding level will absorb two holes and there is no doping in the Cu-O_2 layer. The hole count on the Cu-O_2 layer is modified only when the level lies below E_F, which is excluded in certain ranges of x. As can be seen from the center part of Fig. 5.7, the resulting hole count on the Cu-O_2 layers shows a plateau that correlates nicely with T_c.

5.4. Electronic Structure of the Bismuth and Thallium Superconductors

The recent discovery of high temperature superconductivity in bismuth [38] and thallium [39] based copper oxides has provided valuable insight into the mechanism of superconductivity. The structures of these materials have been determined [40-42], and T_c depends on the number of Cu-O_2 layers in a very simple way: The more Cu-O_2 layers, the higher the T_c. For example, $Bi_2Sr_2CuO_6$, which has only one Cu-O_2 layer, has a comparatively low T_c of about 6-22 K [43], while the two-layer compound $Bi_2Sr_2CaCu_2O_8$ superconducts at 84 K [40]. Aside from the small orthorhombic distortion in the unit cell and the superlattice structure along the orthorhombic b direction, both structures are body-centered tetragonal. The structure of the one-layer $Bi_2Sr_3CuO_6$ ($Tl_2Ba_2CuO_6$)

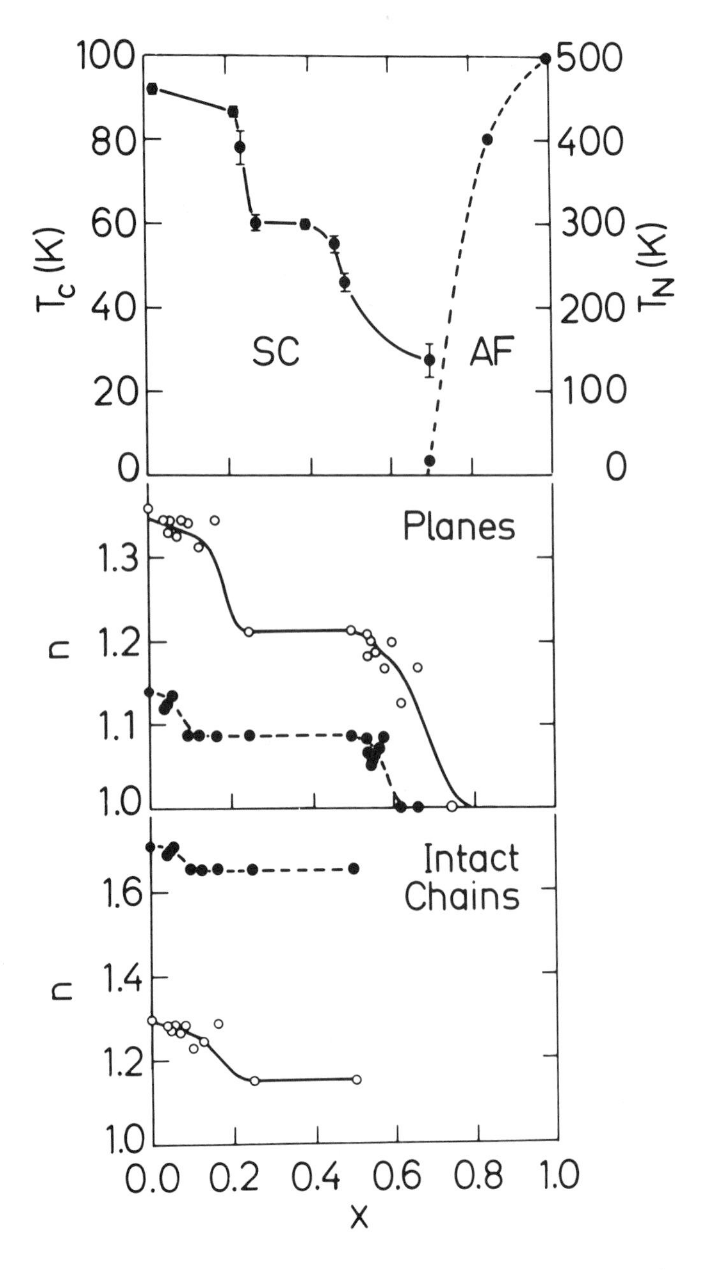

Fig. 5.7. Experimental phase diagram of $YBa_2Cu_3O_{7-x}$ (top) together with hole counts calculated for a plane (center) and an intact chain (bottom), from Zaanen et al. [37]. Two sets of parameters were considered in the theoretical calculations, with the solid circles reproducing the LDA band structure and the open circles representing adjusted parameters to line up the center of the Cu-O_2 plane band and the intact chain band.

consists of a square $Cu\text{-}O_2$ plane above which is a layer of Sr-O (Ba-O) followed by two Bi-O (Tl-O) layers and another Sr-O (Ba-O) layer before the whole structure repeats with the copper-oxygen layer shifted to the body-centered tetragonal position. The two-layer structure differs in that the copper-oxygen plane is replaced by two $Cu\text{-}O_2$ layers separated by a layer of calcium. The three-layer structure includes an additional set of $Ca\text{:}Cu\text{-}O_2$ layers to the two-layer structure.

Sterne and Wang [44] have calculated the electronic structure for the one- and two-layer bismuth compounds and compared them to see what may be responsible for the large difference in their transition temperatures. We found essentially identical $Cu\text{-}O_2$ bands in both compounds, but the Bi-O planes which lie between the $Cu\text{-}O_2$ layers are metallic in the high-T_c compound but are almost insulating in the low T_c compound. Since the metallic nature of the chain layer bands in $YBa_2Cu_3O_{7-x}$ also decreases with decreasing T_c as oxygen vacancies are introduced, we proposed that the metallic nature of the layers between the $Cu\text{-}O_2$ layers could enhance T_c significantly. This model may also apply to the new thallium compounds.

5.4.1. Band Structure of $Bi_2Sr_2CaCu_2O_8$

Figure 5.8a shows the band structure of Sterne and Wang [44] for the high-T_c $Bi_2Sr_2CaCu_2O_8$ compound which is very similar to the calculations of Hybertsen and Mattheiss [45], Krakauer and Pickett [46] and Massidda et al. [47]. There are two $Cu\text{-}O_2$ pdσ antibonding bands extending from about 1eV below E_F to 1.5 eV above E_F, similar to the earlier high-T_c compounds. The new feature is the presence of two bismuth p electron bands which dip below E_F by about 0.7 eV around the K point. These interact with the $Cu\text{-}O_2$ antibonding bands producing a complicated band structure around K. We find a shallow band of mixed Bi and $Cu\text{-}O_2$ character which barely crosses E_F and a more dispersive band which has almost pure bismuth character crossing E_F, indicating that the bismuth oxide layers are metallic. At higher energies of about 1 to 1.5 eV above E_F, the characters of the Bi p bands mix with the p orbital of the oxygen atoms on the Bi and Sr layers, indicating that these states are derived from the Bi-O antibonding bands.

5.4.2. Band Structure of $Bi_2Sr_2CuO_6$

The band structure of the low T_c $Bi_2Sr_2CuO_6$ compound is shown in Fig. 5.8b. The copper-oxygen antibonding band looks essentially

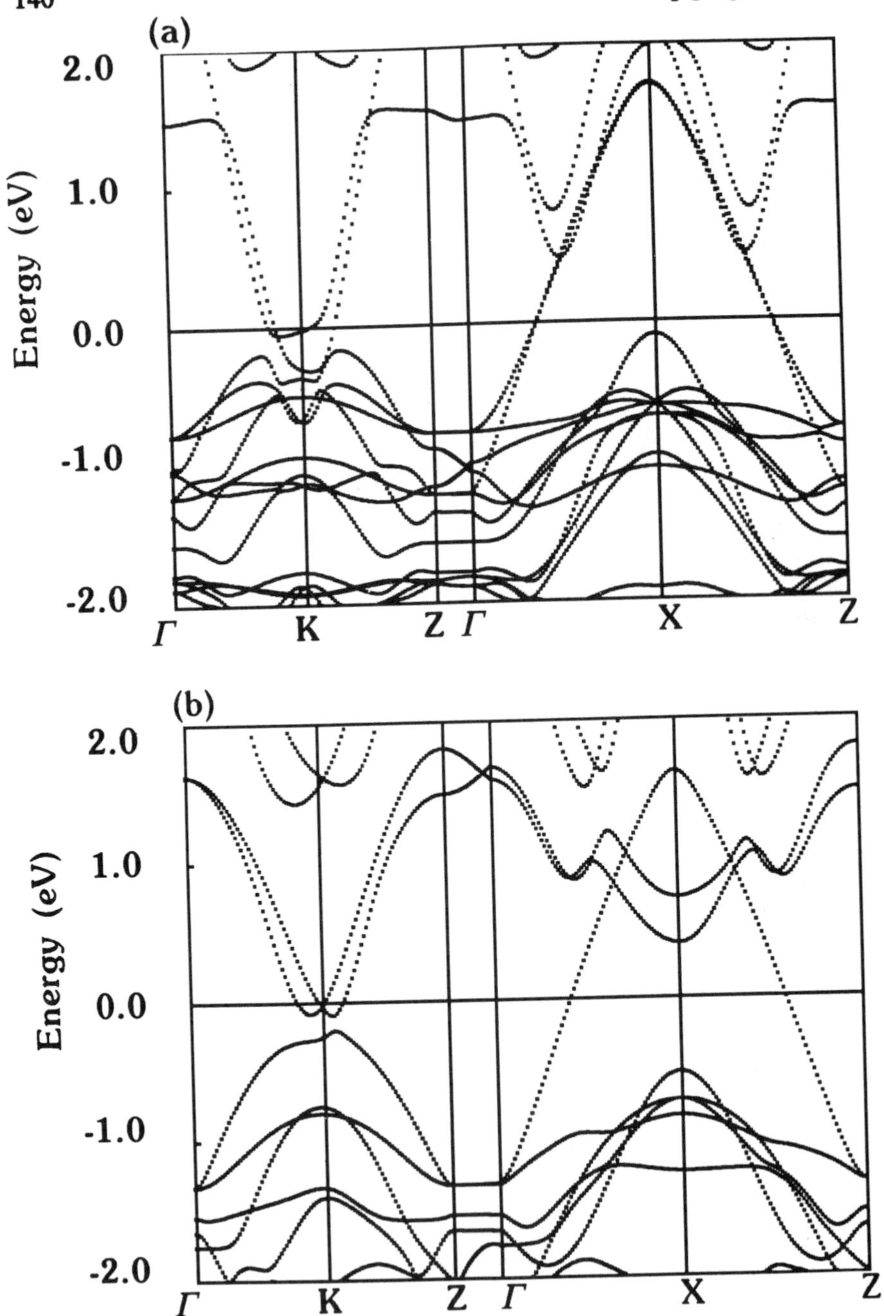

Fig. 5.8. Energy band structures around the Fermi energy for (a) the high temperature superconductor $Bi_2Sr_2CaCu_2O_8$ and (b) the low temperature superconductor $Bi_2Sr_2CuO_6$, from Sterne and Wang [44]. Note the presence of the Cu-O antibonding band in both systems and the difference in the bismuth bands around the K point.

identical to the corresponding bands in the $Bi_2Sr_2CaCu_2O_8$ compound. Around the K point, the two Bi p-bands barely cross and extend less than 0.1 eV below E_F, compared with 0.7 eV in the case of the higher-T_c system. Furthermore, the low T_c system shows a gap between the Bi-p bands and the antibonding Cu-O bands at K which is not present in the higher T_c compound so the metallic nature of these layers will be greatly reduced.

5.4.3. Interlayer Coupling

The differences in the electronic structures noted above suggest that the metallic nature of the Bi-O layers may be of importance in achieving higher transition temperatures. The situation is somewhat similar to the $YBa_2Cu_3O_{7-x}$ system in which the Cu-O chain layer is metallic. As we discussed in Sec. 5.3.4, the metallic nature of the chain layer bands begins to vanish when oxygen is removed from the chain layer. For y around 0.5 to 0.7, the chain layer becomes insulating and the material ceases to superconduct. In this picture, we assume that the superconducting transition occurs through Bose condensation of quasiparticles in the two-dimensional $Cu-O_2$ layers. According to the Mermin-Wagner theorem, this transition cannot be entirely two-dimensional, since fluctuations prevent the transition from occurring at a finite temperature.

The metallic nature of the Bi-O layers in $Bi_2Sr_2CaCu_2O_8$ or the Cu-O chains layers in $YBa_2Cu_3O_{7-x}$ are directly tied to the hole count in the $Cu-O_2$ layers, since they arise from charge transfer from one another. The former enhances interlayer coupling while the latter is an important parameter for the strength of electron correlations. To emphasize the effects of interlayer coupling, we show in Table 5.1 the experimentally measured coherence lengths normal to the $Cu-O_2$ planes, ξ_c, and the $Cu-O_2$ interlayer spacing, d_{Cu-Cu} for the $La_{2-x}Ba_xCuO_4$, $YBa_2Cu_3O_7$, and $Bi_2Sr_2CaCu_2O_8$ systems. In $La_{2-x}Sr_xCuO_4$, where there is no metallic layer between the $Cu-O_2$ planes, the coherence length is long enough to allow supercurrents to flow directly between the layers, but in $YBa_2Cu_3O_7$ and $Bi_2Sr_2CaCu_2O_8$, ξ_c is shorter than the distance between the two $Cu-O_2$ planes. In both cases, however, ξ_c is comparable to the distance d_{Cu-M} from the $Cu-O_2$ layer to the metallic chain layer or Bi-O plane, so the supercurrents could then flow between $Cu-O_2$ layers by taking advantage of the metallic states on the intervening layers, essentially hopping from copper-oxygen layer to copper-oxygen layer by tunneling through the metallic interlayer. Since the coherence length can be defined within the phenomenological Ginzburg-Landau theory, the above argument is in principle applicable to most pairing mechanisms that have been proposed for the high-T_c

Table 5.1. Interlayer separations and coherence lengths (in Å) for a number of high temperature superconductors [44]. $d_{Cu\text{-}Cu}$ is the perpendicular distance between $Cu\text{-}O_2$ planes. $d_{Cu\text{-}M}$ is the perpendicular distance from the $Cu\text{-}O_2$ layer to the intervening metallic layer, Cu-O in $YBa_2Cu_3O_7$ and Bi-O in the bismuth compounds. ξ_c is the coherence length perpendicular to the $Cu\text{-}O_2$ plane.

	$d_{Cu\text{-}Cu}$	$d_{Cu\text{-}M}$	ξ_c
$La_{2-x}Sr_xCuO_4$	6.6[a]	-	7 - 13[e]
$YBa_2Cu_3O_7$	8.2[b]	4.1[b]	4.8-7.0[f]
$Bi_2Sr_2CaCu_2O_8$	12.1[c]	4.4[c]	4.0[g]
$Bi_2Sr_2CuO_6$	12.3[d]	4.5[d]	-

References: [a][48] [b][49] [c][40] [d][43] [e][50] [f][51] [g][52]

superconductors. However, the unusually short coherence length could lead to an unusual flux-pinning picture and thus the critical field measurements may need to be reevaluated in this light [53]. In the resonating valence bond model, T_c is proportional to $\delta\, t^2/J$, where δ is the hole count in the $Cu\text{-}O_2$ layers, t the interlayer hopping matrix element, and J the interlayer exchange constant [54]. The metallic nature of the layers between the $Cu\text{-}O_2$ layers will raise T_c by enhancing t and suppressing J as electrons become delocalized. However, T_c will eventually drop in the limit of (1) large δ because the magnetic interactions among the electrons will diminish, or (2) large t, because the electron correlations become three dimensional.

5.4.4. Electronic Structure of the Thallium Compounds

T_c increases with increasing $Cu\text{-}O_2$ layers in the thallium compounds: $Tl_2Ba_2CuO_6$, $Tl_2Ba_2CaCu_2O_8$, and $Tl_2Ba_2Ca_2Cu_3O_{10}$ have T_c's of 80, 110, and 125 K, respectively. The band structures of the two- and three-layer thallium compounds have been reported by Yu et al. [55], and that of the one-layer compounds by Hamann and Mattheiss [56]. Present once again is the antibonding $Cu\text{-}O_2$ band, whose filling factor is controlled by charge transfer to the Tl-O band that crosses E_F. In general, the Tl-O bands lie higher in energy relative to the valence band complex than the corresponding Bi-O bands in the bismuth compounds, so that it is the $Tl_{6s}\text{-}O_{2p}$ state that crosses E_F. Detailed charge density contours show strong interplanar coupling between the two Tl-O layers that were absent in the Bi compounds.

Near E_F, each Ca:Cu-O_2 layer introduces one antibonding band which is half filled. However, the three antibonding Cu-O_2 bands are nearly degenerate in $Tl_2Ba_2Ca_2Cu_3O_{10}$, with filling factors very similar to that of $Tl_2Ba_2CaCu_2O_8$, which indicates that there is charge transfer from the Cu-O_2 bands to the Tl-O bands as more Ca:Cu-O_2 layers are added. Thus, T_c correlates with the metallic nature of the intervening Tl-O bands, similar to the Bi compounds.

Recently, Kasowski et al. have calculated the electronic structure of the one- to four-layer Tl compounds [57]. In addition to the antibonding Cu-O_2 bands and Tl-O bands, they found narrow Cu-O bands that lie very close to and sometimes overlaps E_F. The narrow Cu-O bands are a unique feature of their method, which is also present in all their early work on La_2CuO_4 [58], $YBa_2Cu_3O_7$ [59], and the Bi compounds [60]; but is absent from the band structures shown in previous sections. Due to the flatness and the multiplicity of these narrow bands, the hole concentration can become quite large even for a very small overlap. Therefore, this method is not particularly accurate to predict the filling factors of the Cu-O_2 and the Tl-O bands. For example, they found only one antibonding Cu-O_2 band intersecting E_F in $Tl_2Ba_2CuO_6$ that is significantly less than half filled, and the Tl-O band appears to be insulating. However, there is one electron to be shared between the antibonding Cu-O_2 and the Tl-O bands. If the Tl-O layer was indeed insulating, then the Cu-O_2 band has to be exactly half filled and the system would be an antiferromagnetic insulator as found in La_2CuO_4.

Once again we see that the intervening metallic Tl-O layers play the dual role of doping the Cu-O_2 layers and enhancing the interlayer coupling between the Cu-O_2 layers. These effects increase with increasing number of Ca:Cu-O_2 layers and therefore increasing T_c. However, the degree of charge transfer among the Tl-O and Cu-O_2 layers appears to be much weaker than that of the bismuth compounds considering the magnitude of the superconducting temperatures. This apparent discrepancy can be understood from recent neutron diffraction measurements on many different samples of $Tl_2Ba_3CuO_6$, which have shown that T_c varies from 4 to 70 K depending on the precise structural arrangement [61]. In particular, there is an orthorhombic phase with a well-ordered superstructure which is non-superconducting, while the superconducting material is pseudo-tetragonal with **disordered** oxygen within the Tl-O plane.

Recently, Lee et al. [62] have observed significant frequency shifts of thallium nuclear magnetic resonance in the superconducting state of the high temperature superconductor $Tl_2Ba_3Ca_3Cu_4O_{10+x}$, which suggest that the Tl-O layers participate directly in the superconductivity. Within the model, the superconductivity occurs in the two-dimensional Cu-O_2 layers, and interlayer tunneling is essential because the coherence length normal to the Cu-O_2 layer is so short. Thus, the observed superconductivity in the metallic Tl-O layer may be induced by the proximity effect.

Alternatively, the observed superconductivity in the Tl layers can be understood from the excitonic models proposed earlier for $YBa_2Cu_3O_{7-x}$, with the Tl-O layers substituting for the role of the chain layers in $YBa_2Cu_3O_{7-x}$. Yu et al. [55] proposes that charge-transfer excitations from occupied Cu-O pdπ bonding orbitals to their empty Cu-O pdσ antibonding orbitals in the chains leads to strong polarization which induces attractive interactions between conduction electrons in the Cu-O_2 plane. Ashkenazi and Kuper [63] suggest that the Cooper pairs come from two distinct CuO_2 planes, and the "glue" which binds them is an electronic breathing mode of charge transfer within an interplanar O_4-Cu-O_4 complex associated with the chain layer.

5.5. Electron-Phonon Interaction

The electron-phonon interaction in La_2CuO_4 was first calculated by Weber [19] in the framework of the nonorthogonal tight-binding (NTB) theory of lattice dynamics based on energy bands of Mattheiss [13]. The dominant NTB gradient for the electron-phonon calculations is obtained by fitting the results of a further band calculation involving slightly reduced (3%) lattice constants. The essential conclusion is that the electron-phonon coupling is dominated by phonon modes in a very limited region near the **X** point of the *BZ*, due to an enormous Kohn anomaly arising from Fermi surface nesting. In this region, the phonon mode is characterized by a planar breathing-type displacement of O atoms away from the central Cu site. The magnitude of the phonon frequency is depressed by almost an order of magnitude compared to that at the center of the *BZ* due to strong renormalization of the conduction electrons. From the Eliashberg function for the strong-coupling theory of superconductivity [64], this soft phonon mode leads to a fairly high-T_c of 30-40 K because of the light oxygen mass. With doping, the Kohn anomaly is weakened and shifts away from the **X** point.

The above conclusions are in agreement with the calculation of Pickett et al. [16], who evaluated the Fermi-surface-averaged electron-ion matrix elements using the rigid muffin-tin approximation of Gaspari and Gyorffy [65]. They found the electron-phonon interaction dominated by the Cu_{3d} and O_{2p} interaction, and concluded that soft phonon modes due to the vibrations of the light oxygen atoms may be sufficient to account for the high-T_c of 30-40 K in these systems.

The major obstacle is that the predicted oxygen breathing mode has not been observed experimentally. Instead, the tetragonal phase is unstable toward an orthorhombic distortion below 500 K. The orthorhombic distortion arising from the tilting of the oxygen octahedra was recently predicted by Cohen et al. [66], who applied the non-empirical Potential Induced Breathing (PIB) model to La_2CuO_4. The potential used in the PIB model is constructed from spherically averaged overlapping **ionic** charge densities (La^{++}, Cu^{++}, and O^{2-}) which are not determined self-consistently. In the tetragonal structure, the ionic model predicts several phonon branches that are harmonically **unstable.** The most unstable mode at -366 cm^{-1}, which occurs at the **X** point in the tetragonal *BZ*, is a rotational mode in which only the planar O_{xy} atoms move around the [001] axis. The most unstable mode at the Γ point is a sliding motion of the La and O_z atoms in the x-y plane. The Γ mode becomes the most unstable mode in the orthorhombic structure with a harmonic frequency of -185 cm^{-1}. Both of these predictions are in accordance with neutron diffraction studies of Jorgensen et al. [7] which indicate large amplitude thermal ellipsoids for O_{xy}, O_z and La ions.

Fu and Freeman [67] have performed a self-consistent frozen phonon calculation of the oxygen breathing distortion in a two-dimensional slab of La_2CuO_4. They found a **stable** harmonic frequency of 930 cm^{-1}, which is in reasonable agreement with the PIB estimates of 1142 cm^{-1}. The O_z motion, neglected in Fu and Freeman's calculation, was predicted to be about 25% of the O_{xy} motion for the breathing mode in the PIB model.

5.5.1. Conventional Superconductivity

The electron-phonon interaction for $YBa_2Cu_3O_7$ has been calculated by Yu et al. [29] using the rigid muffin-tin approximation of Gaspari and Gyorffy [65]. The resulting electron-phonon coupling constant, dominated by the Cu and O ions along the chain, led to a T_c of less than 32 K even when the most optimistic parameters were assumed for the phonon density of states. Thus, the conventional phonon mechanism was found to be inadequate for obtaining high-T_c in $YBa_2Cu_3O_7$.

A precise theoretical prediction of T_c depends on $<\omega^2>$, which is not known independently. Recently, Allen et al. [68] have used the measured neutron scattering function $G(\omega)$ as the shape of $\alpha^2F(\omega)$ to estimate $<\omega^2>$ for La_2CuO_4 and $YBa_2Cu_3O_7$. The resulting values of λ, 0.65 and 0.32, represent moderate to weak coupling, and would predict T_c = 8 and 0.4 K, which is much smaller than the experimental values of 40 and 95 K, respectively.

The uncertainty in the state-of-the-art calculation of

electron-phonon interaction strength remains fairly large. First of all, the ionic charge distribution should be sensitive to the interatomic separation. This important effect is completely neglected in the rigid muffin-tin approximation, and the empirical NTB model, as well as the ionic PIB model. Secondly, it will be shown in Sec. 5.6 that the electron-electron correlation in the high-T_c compounds is not well described in the local density band theory, and consequently its effects on the predicted electron-phonon interactions are not clear. Thirdly, Hardy and Flocken [69] have shown that if the oxygen ions are moving in a double-well potential, the electron-phonon interaction strength could be enhanced by an order of magnitude. Although it is unlikely that the electron-phonon coupling mechanism alone will account for the extraordinarily high temperature superconductivity, phonon renormalization is important, since phonons are within the critical energy range either for the bonding of the Cooper pairs through charge fluctuations (excitons or plasmons) or for the formation of spinons or holons within the resonating valence bond model (RVB). Assuming that the same kind of mechanism is responsible for the superconductivity in all these materials, the theoretical calculations predict that phonon renormalization may be more important in La_2CuO_4 than in $YBa_2Cu_3O_7$. This appears to be consistent with the measured isotope effect, which is much larger in La_2CuO_4 than in $YBa_2Cu_3O_7$ [2].

5.5.2. Normal State Transport Properties

Early optical measurements of Etemad et al. [70], for a series of polycrystalline samples of $La_{2-x}Ba_xCuO_{4-y}$ from $x = 0$ to $x = 0.3$, showed a relatively intense and narrow dopant-induced absorption band near 0.5 eV, whose oscillator strength correlates with the Meissner effect as well as T_c measured on the same set of samples. This was interpreted as evidence for an electronically driven pairing mechanism for superconductivity mediated by the 0.5 eV excitation. Similar conclusions were reached from experiments on polycrystalline samples of $La_{2-x}Ba_xCuO_{4-y}$ [71,72], as well as the measurement of Kamaras et al. [73] for $YBa_2Cu_3O_{7-x}$, which showed a low-energy peak near 0.37 eV in the superconducting phase that is absent in the insulating phase ($x = 0.8$). This peak is attributed to a charge transfer exciton brought on by strong electron correlation. However, more recent measurements for single crystal La_2CuO_4 and $La_{1.8}Sr_{0.2}CuO_4$ [74] and highly aligned thin films of $YBa_2Cu_3O_7$ [75,76] found the normal Drude-type behavior without any extra excitation in this region. The observed structures in the polycrystalline samples were explained by Orenstein and Rapkine [77] as a shift of the usual Drude peak away from the origin due to directional anisotropy of the sample.

The interband contribution to the optical properties of $YBa_2Cu_3O_7$ has been calculated by Zhao et al. [78]. A plasmon energy of 2.8 eV was predicted from a well-defined peak in the electron energy-loss function. This value is in close agreement with available experimental data. The optical conductivity shows strong directional anisotropy in the 0-3 eV range. The interband transitions, averaged over the three directions, yield a value of 12.9 for the static dielectric constant. This value is lower than most of the semimetals.

Assuming only electron-phonon scattering, the anisotropic transport properties of $La_{2-x}Sr_xCuO_4$ and $YBa_2Cu_3O_7$ have been reported by Allen et al. [68]. Electron-phonon parameters were estimated from the band structures using the rigid muffin-tin approximation of Gaspari and Gyorffy [65], and based on the neutron scattering spectral function. They found a surprisingly linear temperature dependence of the resistivity $\rho(T)$, much like experimentally measured results. However, the magnitudes shown in Table 5.2 are an order of magnitude too low. For example, single-crystal data for $La_{1.94}Sr_{0.06}CuO_4$ by Suzuki and Murukami [79] have $\rho(295)-\rho(0) \cong 450$ $\mu\Omega$cm, compared with 80 predicted. For $YBa_2Cu_3O_7$, twinned single-crystal data by Tozer et al. [80] give $\rho_{ab}(295)-\rho_{ab}(0) \cong 380$ $\mu\Omega$cm and oriented film data by Bozovic et al. [76] give 350 $\mu\Omega$cm , compared with $\rho_{xx} = 37$ and $\rho_{yy} = 16$ predicted. These results were interpreted by Allen et al. [68] as an indication that another scattering mechanism dominates. Magnetic fluctuations observed in neutron scattering experiments could provide such a

Table 5.2. The theoretical plasmon frequency tensor $\hbar\Omega_p$, resistivity tensor ρ at 295 K, and Hall coefficients R^H, from Allen et al. [68].

	$La_{1.85}Sr_{0.15}CuO_4$	$YBa_2Cu_3O_7$	Unit
$\hbar\Omega_{pXX}$	2.9	2.9	eV
$\hbar\Omega_{pYY}$	2.9	4.4	eV
$\hbar\Omega_{pZZ}$	0.55	1.1	eV
$\rho_{xx}(295)$	80.0	37.0	$\mu\ \Omega$ cm
$\rho_{yy}(295)$	80.0	16.0	$\mu\ \Omega$ cm
$\rho_{zz}(295)$	2200.0	260.0	$\mu\ \Omega$ cm
R^H_{xyz}	0.18	0.18	(10^{-9} m^3/c)
R^H_{yzx}	-0.94	-0.38	(10^{-9} m^3/c)
R^H_{zxy}	-0.94	-1.07	(10^{-9} m^3/c)

mechanism.

Quantities that are independent of the scattering process are calculated with more success. Table 5.2 shows the theoretical plasmon frequency tensor $\hbar\Omega_p$, resistivity tensor ρ at 295 K, and Hall coefficients R^H [68]. The plasmon frequency appears to be in good agreement with experiments, although a reduction of $\Omega_p{}^2$ by a factor of 5 to 10 could bring $\rho(T)$ into agreement with experiments. The band theory made the rather striking prediction that the Hall coefficients should change sign if the magnetic field is rotated to the *a-b* plane in a single crystal. This prediction has been verified by Tozer et al. [80] on single crystal $YBa_2Cu_3O_7$.

5.6. Electron-Electron Correlations

There is widespread belief that the electron correlations are important in the high temperature superconductors. This is evident from the comparison of band structure with photoemission data as well as from the failure of the band theory to stabilize the observed antiferromagnetic ground state. Recently, much theoretical effort has been applied to estimate the effective Hubbard Hamiltonian for the antibonding pdσ band:

$$
\begin{aligned}
H = {} & E_d \sum_i n_{d,i} + U_d \sum_i n_{d,i\uparrow} n_{d,i\downarrow} + E_p \sum_i n_{p,i} + U_p \sum_i n_{p,i\uparrow} n_{p,i\downarrow} \\
& + t_{pd} \sum_{\langle ij \rangle} (d^+_{i\sigma} p_{j\sigma} + p^+_{i\sigma} d_{j\sigma}) + t_{pp} \sum_{\langle ij \rangle} (p^+_{i\sigma} p_{j\sigma} + p^+_{i\sigma} p_{j\sigma}) \\
& + U_{pd} \sum_{\langle ij \rangle} n_{p,i} n_{d,j} \qquad , \qquad (5.1)
\end{aligned}
$$

where i, j denotes site indices and σ (=$\uparrow,\downarrow$) denote spins. The d^+ (d) and p^+ (p) create (destroy) a **hole** on the Cu_{3d} and O_{2p} sites, respectively; E is the on-site energy, and U_p, U_d are the corresponding intra-atomic Coulomb repulsions. Equation (5.1) also includes an interatomic Coulomb repulsion U_{pd} between holes on neighboring Cu and O sites, which describes the Madelung energy. The Slater-Koster parameters $t_{pd} = \pm\frac{1}{2}\sqrt{3}\, V_{pd\sigma}$ hybridize nearest neighbor Cu_{3d} and O_{2p} orbitals, and $t_{pp} = \pm \frac{1}{2} (V_{pp\sigma} - V_{pp\pi})$ allows direct O_{2p}-O_{2p} hopping. The $\pm$ signs are determined by the location of the oxygen atoms according to the Slater-Koster table for two-center integrals [81].

The effects of the strong electron correlations in the high temperature superconductors are not fully understood, but some insight can be gained from our understanding of two related systems:

(1) Mott insulators, and (2) heavy fermion superconductors. Finally, magnetic interactions will be discussed in Sec. 5.6.7, based on the results of recent quantum simulations of the extended Hubbard model, with parameters comparable with first-principles band structure calculations.

5.6.1. Antiferromagnetic Band Structure of Sc_2CuO_4 and La_2CuO_4

Leung et al. [82] have performed spin-polarized calculations on Sc_2CuO_4, with the lattice constants adjusted to mimic the band structure of the lanthanum compound so that their magnetic properties will be similar. Both ferromagnetic and antiferromagnetic ground states were investigated and found to be **unstable.** However, they have shown that an antiferromagnetic state can be stabilized if the degenerate bands at the Fermi surfaces are lifted by the antiferromagnetic moment and only the lower of these two bands are preferentially occupied. If the dispersion of these bands were smaller than the antiferromagnetic splittings, then E_F would naturally fall in the semiconducting gap. However, the resulting splitting is significantly smaller than the dispersion, so that antiferromagnetic Sc_2CuO_4 remains metallic and both of the degenerate bands should in principle be partially occupied. These results, later confirmed by Sterne et al. [83] for La_2CuO_4, are in apparent conflict with experiments where less than 1% oxygen vacancies are needed to stabilize the antiferromagnetic state. Nevertheless, this procedure assumed that the band width is overestimated while the band splitting is underestimated by the local density approximation, so that the antiferromagnetic La_2CuO_4 is a semiconductor due to antiferromagnetic order. Qualitatively, these calculations have shown that the pdσ antibonding states can in principle support an antiferromagnetic ground state if the intra-atomic Coulomb repulsion U were included correctly.

To further test the itinerant versus localized nature of the superconductors, Leung et al. [82] have calculated the magnetic form factor of antiferromagnetic Sc_2CuO_4 obtained by constraining the occupancy of the doubly degenerate states at the boundary of the BZ. The calculated moment on the Cu site was 0.136 μ_B, with each out-of-plane O_z site having a moment of about 10% of that of Cu and aligned in the direction opposite the Cu moment. The form factors have been renormalized to 1 μ_B per Cu atom in Fig. 5.9, where they are compared with the neutron measurement of Vaknin et al. [84]. The experimental data, normalized to 1.0 for the second reflection, appear to be similar to that of the band theory, which differ from the Cu^{++} ionic form factor shown as a smooth solid line in Fig. 5.9.

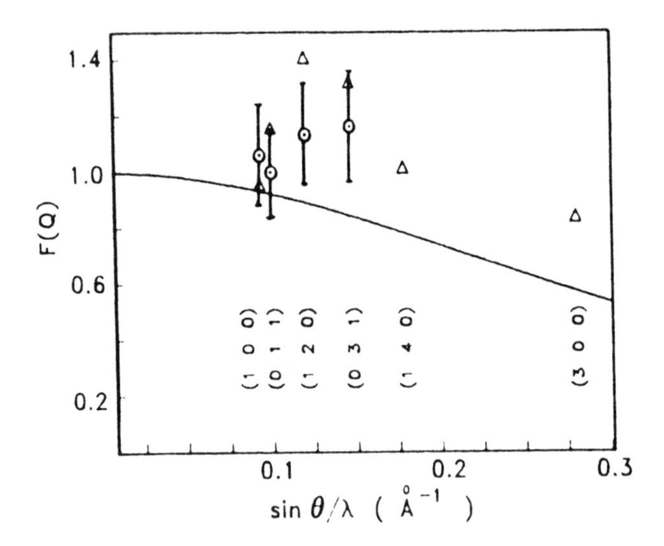

Fig. 5.9. The magnetic form factor for antiferromagnetic Sr_2CuO_4, induced by an applied staggered field corresponding to 0.5 μ_B per Cu site, from Leung et al. [82].. Both the theoretical values (Δ) and the experimental data of (∘) have been normalized to 1 μ_B for θ=0. The line is the Cu^{++} ionic form factor.

This indicates that a moment on the out-of-plane oxygen may be present. However, the measurement suggests a moment of 0.4 μ_B per Cu site, which is well out of the reach of the band theory.

5.6.2. Effects of Sr or Ba Doping and Oxygen Vacancies

In the orthorhombic $La_{2-x}Ba_xCuO_4$, the highest T_c occurs around x = 0.15, and no antiferromagnetism has been observed. Sterne and Wang [83] have performed a separate spin-polarized band structure calculation for orthorhombic La_2CuO_4 and model the effects of Ba doping by removing 0.3 electrons from the 14 atom unit cell. In good agreement with experiments, they find that antiferromagnetism cannot be sustained, even when the occupancy of the bands is constrained as described above. As we discussed in Sec. 5.2.3, there is substantial Fermi surface nesting in stoichiometric La_2CuO_4, where the antibonding pdσ band is exactly half filled. Shifting E_F away from the half-filled limit by doping will shift the spanning vector away from the boundary of the BZ, which suppresses long-range antiferromagnetic order. Recently, Chui et al. [85] have calculated the two-particle vertex function in the rigid band approximation starting from the band structure of $La_{2-x}Sr_xCuO_4$. This approach goes beyond the mean field approximation and the resulting Neel temperature as a function of doping x agrees well with the experimental observations.

The effects of oxygen vacancies on the antiferromagnetism in La_2CuO_4 were studied by Sterne and Wang [83] based on two 28-atom supercell calculations which formally correspond to $La_2CuO_{3.75}$. In one case, the 14-atom orthorhombic unit cell is doubled along the z direction so that there are two $Cu\text{-}O_2$ planes in a simple orthorhombic unit cell. An oxygen atom in one of the two $Cu\text{-}O_2$ planes is then removed to create the vacancy, while the other $Cu\text{-}O_2$ plane remains stoichiometric. As in the stoichiometric compound, self-consistent spin-polarized calculations lead to a nonmagnetic ground state. Constraining the occupancy of states near E_F to stabilize the antiferromagnetic moments leads to very different behavior on the two $Cu\text{-}O_2$ planes. The vacancy plane shows only a very **small** Cu moment (0.004 μ_B) while the stoichiometric layer retains a moment of 0.136 μ_B, very close to that found in La_2CuO_4 (0.139 μ_B).

The antiferromagnetic band structure for the 28-atom supercell is shown in Fig. 5.10. In the vacancy-free case shown in Fig. 5.10a, there are four bands crossing E_F in the reduced BZ. The bands along the U-R-T direction are split into two doubly degenerate states due to the antiferromagnetic moment, which is stabilized by occupying only the lower bands. These split bands and the corresponding highly dispersive bands in the Y-Γ-T direction remain unchanged in the stoichiometric layer of the vacancy calculation shown in Fig. 5.10b. However, significant narrowing of the band width can be seen in the vacancy layer due to the reduction of the coordination number of the Cu atoms near the oxygen vacancy. Along the U-R-T direction, the splitting in the stoichiometric layer is due to the antiferromagnetism in stoichiometric La_2CuO_4. However, the Cu atoms in the vacancy layer are essentially nonmagnetic; hence, the splitting is due entirely to the changes in the potential produced by the oxygen vacancy.

The most important effect of the oxygen defects is still an upward shift of E_F, which suppresses antiferromagnetism. This conclusion appears to contradict the observation that antiferromagnetism is enhanced as the number of oxygen vacancy is increased from 0.0 to 0.03 per unit cell. However, the reported error in the determination of the number of O vacancies is 0.02, so so that perhaps rather than creating vacancies, excess oxygen is being removed from the system, and the highest moment may actually occur for the stoichiometric compound.

In the second calculation, the supercell is constructed by

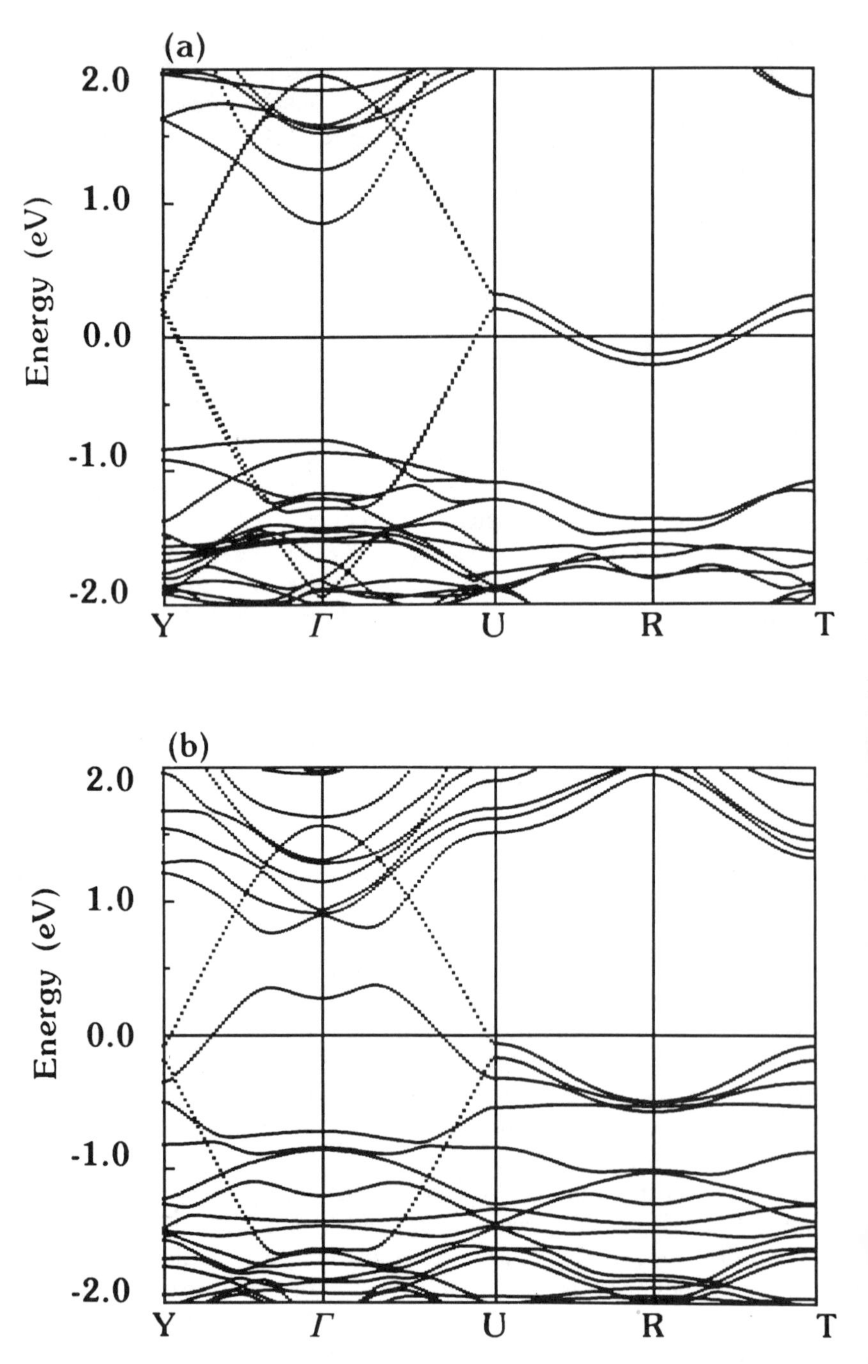

Fig. 5.10. Antiferromagnetic energy bands of (a) La_2CuO_4 folded back t the reduced zone of 28 atom supercell, and (b) $La_2CuO_{3.75}$ with one o of four oxygen atoms missing on every other Cu-O_2 plane, from Sterne a Wang [22].

doubling the tetragonal unit cell in the two planar directions, with 28 atoms in the body-centered tetragonal unit cell. There is only one Cu-O_2 plane in this unit cell from which we remove one oxygen atom. The resulting vacancy-vacancy separation is longer than that in the vacancy plane of the first supercell calculation. In this case, antiferromagnetism cannot be stabilized even when we constrain the occupancy of the states as we have done before. The clear indication is that the oxygen vacancies suppress antiferromagnetic tendencies in these compounds.

5.6.3. Photoemission

Fuggle et al. [86] have calculated from first principles the photoemission (*XPS*) and the inverse photoemission (*BIS*) spectra, which includes electron-phonon matrix elements linking the initial states and the final states, as well as a non-self-consistent step potential at the surface. The theoretical results for $La_{2-x}Sr_xCuO_4$ and $YBa_2Cu_3O_{7-x}$ are compared with experiment in Fig. 5.11. The calculated *XPS* peak of predominantly Cu_{3d} character was much too narrow and much too close to E_F. Furthermore, the *XPS* and *BIS* intensities at E_F are weaker than the theoretical estimates. Therefore, Fuggle et al. [86] concluded that the on-site correlation energies of the Cu 3d electrons (U_d) are large and not less than 4-5 eV. These conclusions are similar to (1) a configuration

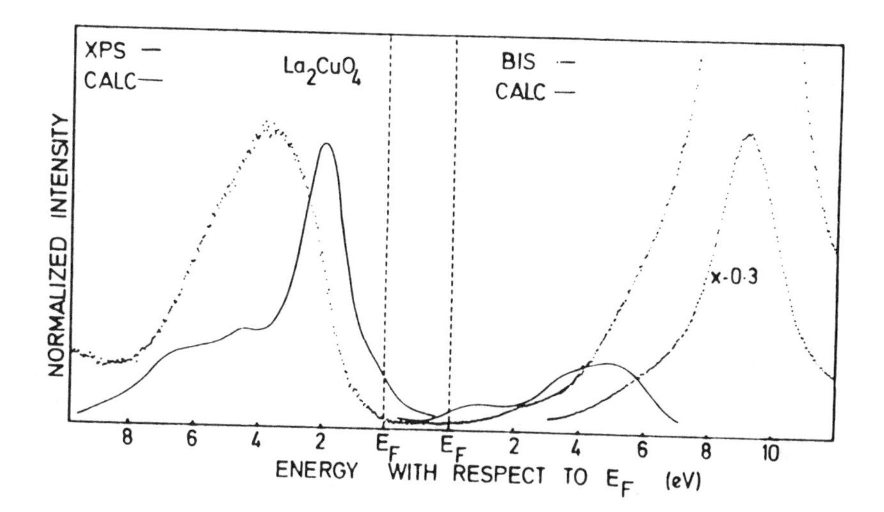

Fig. 5.11. Experimental and calculated XPS and BIS spectra at 1486.7 eV for La_2CuO_4, from Fuggle et al. [86].

interaction analysis of the photoemission data by Fujimori et al. [87], who found U_d = 5.5 eV for the Cu-O_6 cluster, (2) a similar calculation of Shen et al. [88] who reported $U_d \cong 6$ eV, the $Cu_{3d} \leftrightarrow O_{2p}$ charge transfer energy $U_{pd} \cong 0.4$ eV, and (3) a transition state calculation of Chen et al. [89] who found U_d = 5.45 eV for the $Ba_4Cu_2O_7$ cluster. Large Coulomb correlation energies have also been inferred from the satellite feature at about 12.4 eV below E_F in the valence band photoemission spectra, which corresponds to Cu-$3d^8$ final states pushed out of the valence band by U_d (Yarmoff et al. [90] and Shen et al. [88]). The fact that the $3d^8$ state occurs at such high energy indicates that charge fluctuation to $3d^8$ must be strongly suppressed in the ground state. Nücker et al. [91] studied excitations of the O_{1s} electron into the local unoccupied part of the *DOS* at the O atom by electron energy-loss spectroscopy. They proposed that the semiconducting compound is a charge-transfer semiconductor with an almost completely filled O_{2p} valence band; while for the superconducting compound, holes are created in the O_{2p} band which dominates the *DOS* at E_F. Since the cross section for the O_{2p} states are much weaker than that of the Cu_{3d} states, this model naturally explains the low *BIS* and *XPS* intensity at E_F. It also explains why very little change was found in the Cu_{2p} *XAS* spectra [92], which measures holes on the Cu site, upon variation of x and y. However, this is an important issue which is not completely resolved. Based on the resonant-photoemission results, Shen et al. [88] have emphasized that the very low *DOS* at or near E_F has a substantial contribution from Cu_{3d} states, which suggests their importance for the superconductivity.

5.6.4. Theoretical Estimates of Model Hamiltonians

The major problem of the cluster approach is that the oxygen-derived bands have a significant width independent of the Cu_{3d} state. Using the potential of La_2CuO_4, McMahan et al. [93] found a band width of 6.7 eV by explicitly decoupling the Cu_{3d} states, and a O_{2p} band width of 6.3 eV by excluding all states except those of O_{2p} character in the O sphere. Similar calculations for the Cu_{3d} states lead to a band width of 1.2-0.7 eV. The hybridization matrix element t_{pd} = -1.6 eV is comparable to t_{pp} = 0.6 eV, and the center of the Cu_{3d} hole band lies 1.2 eV higher than the center of the O_{2p} hole band. The Coulomb repulsion can be calculated by increasing or decreasing the occupancy of a single impurity site in a superlattice, and then comparing the resulting total energies. This is more difficult for the itinerant O_{2p} bands than the localized Cu_{3d} state because the Wannier functions for itinerant states are not well defined. This explains the relatively large discrepancy in

$U_p \cong$ 4-14 eV compared with $U_d \cong$ 8-10 eV, and $U_{pd} \cong$ 0-2 eV, among different calculations shown in Table 5.3 [94-96]. Making the Anderson impurity approximation and using the methods of Gunnarsson and Schönhammer [97], McMahan et al. [93] found reasonably good agreement with experimental photoemission spectra. The impurity model calculations lead to a magnetic ground state with a correlation gap E_{BIS}-E_{XPS} = 1.3 eV, which should be added to crystal field splittings. In this picture, the ground state of stoichiometric La_2CuO_4 is an insulator if the lattice broadening of the two levels is not sufficient to fill in the gap.

5.6.5. Mott Insulators

Mott insulators represent one of the most dramatic failures of the local density band theory. It has been shown by Terakura et al. [98] that the *AFM* ground states are obtained by the band theory; but in the case of FeO or CoO the observed insulating gaps were not obtained, and in the case of NiO they are an order of magnitude too small. The mechanism for the breakdown of the local-density band theory in the Mott insulators has recently been reviewed by Brandow [99]. Because of crystal field splittings of the 3d states, the different orbital exchange parameters J can be much smaller than the self-exchange parameter U. Orbital dependence of the exchange and correlation potential are neglected in the local-spin-density approximation, which amounts to averaging over these two very different quantities. It thereby missed the main point of the Mott picture, namely, that the unoccupied orbital lies above that of an occupied orbital by an amount of order U. The problem is greatly reduced in wide-band materials, where U is strongly screened so that

Table 5.3. Parameters for the two-dimensional extended Hubbard model of La_2CuO_4. Energies are given in eV.

	McMahan et al. [93]	Chen et al. [94]	Zannen et al. [96]	Schluter et al. [96]
E_d	-4.1			
E_p	-2.9			
t_{pd}	-1.6			
t_{pp}	0.65			
U_d	8.0	9.0	8.0	10 ± 1
U_p	4.1–7.3	14.3		7 ± 1
U_{pd}	≥ 0.6	1.6	0–1	2 ± 1

the mechanism for the Mott splitting is weakened. In the case of the monoxides, the O_p states lie significantly lower in energy than the transition metal 3d states, so that the overall band width W is smaller than U. In the high-T_c superconductors, $E_d \cong E_p$, which leads to a rather large band width ($W \geq 9$ eV). Nevertheless, the system is near a metal-insulator transition so that the screening is not very effective.

Recently, Svane and Gunnarsson [100] have included self-interaction corrections (SIC) to the local-spin-density (LSD) approximation of a two-dimensional Hubbard model on a Cu-O_2 lattice. The SIC approximation removes the unphysical Coulomb interaction of an electron with itself, as well as the corresponding LSD exchange-correlation potential. Unlike the LSD approximation, this potential is orbital dependent. They have used a set of realistic values for the screened Coulomb repulsion (U_d = 8 eV and U_p = 5 eV) in the LSD calculation, and subtracted "bare" interactions U_d = 25 eV and U_p = 20 eV in the SIC calculation since the other electrons do not feel the SIC potential. They found that the SIC-LSD approximation gives a moment for the system which is somewhat too large when compared with experiments, while the LSD approximation greatly underestimates the tendency towards antiferromagnetism.

The Mott-Hubbard theory was later generalized by Zaanen et al. [101] to describe a wide range of different transition compounds. They solved the Anderson impurity Hamiltonian which includes quantum charge fluctuations beyond the mean field approximation. They showed that the correlation gap can have either *d-d* or charge transfer character depending on the relative magnitudes of the parameters U_d and U_{pd}, and that the gap can go to zero even if U_d is large. If $U_d < U_{pd}$, the band gap is proportional to U_d and the compound is in the Mott-insulator region with holes on the transition metal site. On the other hand, if $U_d > U_{pd}$, the gap is proportional to U_{pd} and is of the charge-transfer nature with holes on the ligand band. There is a metallic ground state where the 3d electrons are delocalized even for U_d large, provided U_{pd} is less than half of the ligand valence band width. The latter appears to be satisfied by the high-T_c superconductors, although the overall band width of the superconductors is significantly larger. In addition, there are RKKY interactions among localized moments in a solid which are neglected in the Anderson impurity model. Spin fluctuations derived from the RKKY interaction may be the pairing mechanism for heavy fermion superconductors.

5.6.6. Heavy Fermion Superconductors

In the limit of strong electron correlation, there could be

additional corrections to the band structure due to charge and spin fluctuations which cannot be described in any kind of mean field approximation. This is best illustrated in the case of the heavy fermion superconductors, which are characterized by extremely large values of the linear coefficient of the specific heat and of the magnetic susceptibility, which is interpreted as implying fermionic quasiparticles bearing a very large effective mass m^* at low temperature. By contrast, values of m^* deduced from transport measurements are not significantly enhanced, which suggests that the energy dependence of the self-energy may dominate its momentum dependence [102]. As the temperature is raised, m^* rapidly decreases to normal values. This indicates that the self-energy corrections diminish rapidly away from E_F and explains the good agreement between the band structure and angle resolved photoemission for UPt_3 even for f-states that lie within 0.1 eV from E_F (m^* is a factor of 21 larger than the already heavy band mass) [103]. The most striking result is that the complicated Fermi surfaces predicted by band theory [104] were later confirmed by experiment [105]. The picture that emerges is that the f-electrons in UPt_3 are itinerant, with quasiparticle energies $E(k)$ are well described by the local density band structure $\varepsilon(k)$, except near the chemical potential μ where there is a strongly frequency dependent self-energy correction $\Sigma(k,\omega)$:

$$E(k) - \mu = \varepsilon(\mathrm{k}) - \mu + \Sigma(k, \mathrm{E}(k) - \mu) \quad .$$

For low energy, $\Sigma(k,\omega)$, which is dominated by its frequency dependence, can be expanded linearly to yield

$$E(k) - \mu = z(k) \; [\; \varepsilon(k) - \mu + \Sigma(k \; , \; 0) \;]$$

where

$$z(k) = \frac{1}{\left| 1 - \dfrac{\partial \; \Sigma \; (k.\omega)}{\partial \; \omega} \right|_{\omega=0}}$$

is the renormalization factor. The quasiparticle Fermi surface, which is determined by $E(k) = \mu$, will not be renormalized from the band Fermi surface if $\Sigma(k \; , \; 0) \cong 0$ ($\varepsilon(k) = \mu$) no matter how large is $z(k)$. For an interacting electron gas, the energy dependence of the self-energy vanishes at μ ($\Sigma(\mathrm{k},0) = 0$) [106]. According to Luttinger's theorem [107], the volume enclosed by the Fermi surface should be identical to the non-interacting case ($U = 0$). These are some of the reasons for the success of the band theory in spite of the enormous mass enhancement factor in UPt_3.

The situation is quite different from some other heavy fermion compounds. For example, the experimental Fermi surfaces of UPd_3 cannot be reproduced by band theory until the *f*-electrons are treated as localized core states that do not hybridize with the conduction bands [108]. In this case the entire *f*-bands are renormalized with very narrow band width and the system is not a superconductor. Experimentally there are significant differences in the thermodynamic properties of the three known heavy fermion superconductors [109]. It will be interesting to see if the predicted Fermi surface of UBe_{13} [110], which has a mass enhancement factor of 91, also turns out to be in agreement with experiment, as was the case of UPt_3.

The magnetic fluctuations responsible for the mass enhancement in the thermodynamic measurements may also be the pairing mechanism for the superconductivity. Recent quantum Monte-Carlo simulations of McQueen and Wang [111] for a two-dimensional Anderson lattice show strong *AFM* correlations above the single-impurity Kondo temperature (T_K). Below T_K, both the local moments and the *AFM* correlations are suppressed somewhat, and in some cases *FM* correlations start to develop. While *AFM* correlations favor nearest neighbor anisotropic singlet pairing, the *FM* correlations could support p-wave triplet-pairing superconductivity. The sudden enhancement of the *FM* correlations can be understood based on the temperature dependence of the mass enhancement factor within Stoner theory. In the case of high-T_c superconductors, there appears to be good qualitative agreement with experiment for the Fermi surface, with experimentally measured values for m^* ranging from a factor of 1-4 when compared with the band masses [30].

5.6.7. Magnetic Interactions

Superconductivity in the cuprate oxides occurs near an AFM phase transition, with short-range AFM correlations which persist above the Neel temperature [112]. Recently, extensive neutron scattering experiments were carried out on a series of $La_2Sr_{2-x}CuO_4$ single crystals, with a Meissner fraction less than 20% [113]. Apparently, doping with Sr does not change the magnetic moments but shortens the correlation range. Otherwise, there is no important difference in magnetic scattering in the normal and superconducting states.

Recently, Bhattacharya et al. [114] have performed finite temperature quantum Monte Carlo simulations of a two-dimensional Hubbard model on a 4x4 $Cu\text{-}O_2$ lattice, using the parameters proposed by McMahan et al.: U_d = 8 eV, ε = 1.2 eV, t_{pd} = -1.6 eV, t_{pp} = 0.65 eV. The results, presented for U_{pd} = V_p = 0 only, emphasize the spin fluctuations more than the charge fluctuations.

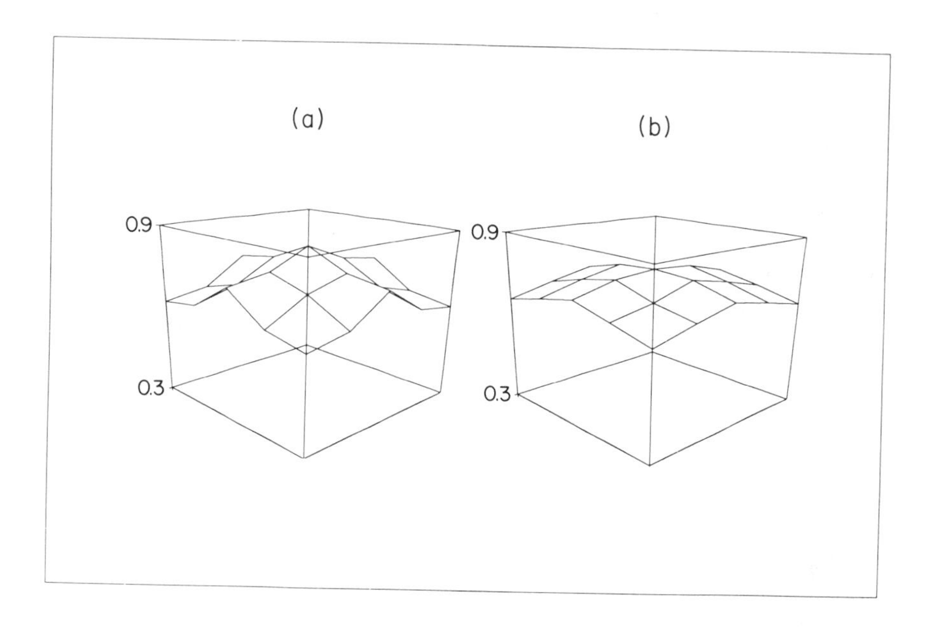

Fig. 5.12. Quantum Monte Carlo simulation of the magnetic form factors of a two-dimensional Cu-O plane: (a) n=1.0, and (b) n=1.15 (from Bhattacharya et al. [114]).

To indicate the effects of doping, Fig. 5.12 shows the magnetic structure function

$$S(\mathbf{k}) = \frac{1}{N} \sum_{ij} e^{i\mathbf{k}\cdot(\mathbf{R}_i-\mathbf{R}_j)} \left\langle (n_{d,i\uparrow} - n_{d,i\downarrow})(n_{d,j\uparrow} - n_{d,j\downarrow}) \right\rangle$$

for the Cu_{3d} on a 4x4 square lattice with (a) n = 1.0 and (b) n = 1.15, where $n = \langle n_d \rangle + 2\langle n_p \rangle$ is the average number of holes per CuO_2 unit, and N is the number of lattice sites. The temperature T = 0.5 eV is significantly higher than the Neel temperature, but the dependence on the filling factor is still very clear. The results indicate AFM correlations characterized by a peak at $\mathbf{k} = (\pi,\pi)$. The AFM correlations are strongest in the half-filled limit (n = 1.0) due to almost perfect Fermi surface nesting. As we lower the chemical potential, the maximum intensity of the (π,π) peak reduces while its width spreads out, particularly along the (1,1) direction. This indicates a reduction in the range of the AFM correlations, in agreement with the neutron scattering experiments.

Recently, Hirsch et al. [115] have studied the effects of U_p, U_d and U_{pd} in an extended Hubbard model for two-dimensional CuO_2

clusters. They calculated the binding energy for a pair of holes, by exact diagonalization of planar CuO_2 clusters with up to 16 sites, with the occupation of the Cu atoms outside the clusters taken into account in a mean field way. They find that in the presence of strong on-site Coulomb repulsion U_d, a near-neighbor Cu-O repulsion U_{pd} can give rise to an attractive interaction between holes on the O sites of the cluster. They propose that the addition of a hole on the oxygen site produces a disruption of local charge distribution which costs a significant amount of energy, and pairing results from the fact that it cost less energy to add a second hole in the vicinity of the first hole. They find that an attractive interaction is obtained for strong U_d and U_{pd} where the number of holes on the copper sites are reduced rather then increased as the cluster is doped. More specifically, they considered $E_p = E_d$. The pairing interaction is repulsive for $U_{pd} \leq 2t$ for any combination of U_p and U_d. In the region $U_{pd} \geq 4t$, the superconducting state is stable for $U_d \geq 4t$ and any U_p. Without doping (half-filled case), they find AFM correlations in the parameter range in which the doped system exhibits an attractive interaction. Comparing these results with the parameters in Table 5.3, U_{pd} appears to be unrealistically large for superconductivity to occur. It remains interesting to see how these parameters will be modified as the size of the clusters increase.

Acknowledgments

The author is indebted to the colleagues who provided figures for this chapter, and to those who sent preprints prior to publication. In particular, she wishes to thank her postdoc P. A. Sterne and her students P. G. McQueen and A. Bhattacharya for quoting some of the unpublished results. This work is supported by the National Science Foundation (Grant No. DMR-86-01708) and the Office of Naval Research (Contract No. N00014-86-K-0266). Computations were carried out under the auspices of the National Science Foundation at the Pittsburgh Supercomputer Center.

References

[1] R. J, Cava et al., *Nature* (London) **332**, 814 (1988).

[2] B. Batlogg, R. J. Cava, A. Jayaraman, R. B. van Dover, G. A. Kourouklis, S. Sunshine, D. W. Murphy, L. W. Rupp, H. S. Chen, A. White, K. T. Short, A. M. Mujsce, and E. A. Rietman, *Phys. Rev. Lett.* **58**, 2333 (1987). L. C. Bourne, M. F. Crommie, A. Zettl, H-C. zur Loye, S. W. Keller, K. L. Leary, A. M. Stacy, K. J. Chang, M. L. Cohen and D. E. Morris, *Phys. Rev. Lett.* **58**, 2337 (1987). K. J. Leary, H-C zur Loye, S. W. Keller, T. A. Faltens, W. K. Ham, J. N. Michaels, and A. M. Stacy, *Phys. Rev. Lett.* **59**, 1236 (1987). T. A. Faltens, W. K. Ham. S. W. Keller, K. J. Leary, J. N. Michaels, A. M. Stacy, H-C zur Loye, D. E. Morris, T. W. Barbee, L. C. Bourne, M. L. Cohen, S. Hoen, and A. Zettl, *Phys. Rev. Lett.* **59**, 915 (1987). B. Batlogg, G. Kourouklis, W. Weber, R. J. Cava, A. Jayaraman, A. E. White, K. T. Short, L. W. Rupp, and E. A. Rietman, *Phys. Rev. Lett.* **59**, 912 (1987). L. C. Bourne, A. Zettl, T. W. Barbee and M. L. Cohen, *Phys. Rev.* **B36**, 3990 (1987).

[3] P. W. Anderson, G. Baskaran, Z. Zou and T. Hsu, *Phys. Rev. Lett.* **58**, 2790 (1987); P. W. Anderson and Z. Zou, *ibid.* **60**, 132 (1988); S. A. Kivelson, D. S. Rokhsar and J. P. Sethna, *Phys. Rev.* B **35**, 8865 (1987).

[4] V. J. Emery, *Phys. Rev. Lett.* **58**, 2794 (1987); J. E. Hirsch, *ibid*, **59**, 228 (1987); R. H. Parmenter, *ibid* **59**, 923 (1987); J. R. Schrieffer, X.-G. Wen, S.-C. Zhang, *ibid* **60**, 944 (1988); A. Aharony, R. J. Birgeneau, A. Coniglio, M. A. Kastner, and H. E. Stanley, *ibid* **60**, 1330 (1988).

[5] C. M. Varma, S. Schmitt-Rink and E. Abrahams, *Solid State Commun.* **62**, 681 (1987).

[6] J. Ruvalds, *Phys. Rev.* **B35**, 8869 (1987).

[7] J. D. Jorgensen, H.-B Schuttler, D. G. Hinks, D. W. Capone, II,K. Zhang, M. B. Brodsky, and D. J. Scalapino, *Phys. Rev. Lett.* **58**, 1024 (1987).

[8] For structural details see Chapter 4.

[9] D. Vaknin, S. K. Sinha, D. E. Moncton, D. C. Johnston, J. M. Newsam, C. R. Safinya and H. E. King, Jr, *Phys. Rev. Lett.* **58**, 2802 (1987).

[10] T. Freltoft, J. E. Fischer, G. Shirane, D. E. Moncton, S. K. Sinha, D. Vaknin, J. P. Remeika, A. S. Cooper and D. Harshman, *Phys. Rev.*B **36**, 826 (1987).

[11] D. C. Johnston, J. P. Stokes, D. P. Goshorn and J. T. Lewandowski, *Phys. Rev.* B **36**, 4007 (1987).

[12] G. Shirane, Y. Endoh, R. J. Birgeneau, M. A. Kastner, Y. Hidaka, M. Oda, M. Suzuki, and T. Murakami, *Phys. Rev. Lett.*

59, 1613 (1987).

[13] L. F. Mattheiss, *Phys. Rev. Lett.* **58**, 1028 (1987).

[14] J. Yu, A. J. Freeman, and J.-H. Xu, *Phys. Rev. Lett.* **58** 1035 (1987).

[15] S. Tanigawa, Y. Mizuhara, Y. Hidaka, M. Oda, M. Suzuki, and T. Murakami, *MRS Symposium Proc.* **99**, 57 (1988).

[16] W. E. Pickett, H. Krakauer, D. A. Papaconstantopoulos, and L. L. Boyer, *Phys. Rev.* B**35**, 7252 (1987).

[17] K. Schwarz, *Solid State Commun.* **64**, 421 (1987).

[18] J.-H. Xu, T. J. Watson-Yang, Jaejun Yu, and A. J. Freeman, *Phys. Lett.* A**120**, 489 (1987).

[19] W. Weber, *Phys. Rev. Lett.* **58**, 1371 (1987).

[20] T. Fujita, Y, Aoki, Y. Maeno, J. Sakurai, H. Fukuba, and H. Fujii, *Jpn. J. Appl. Phys.* **26**, L368 (1987).

[21] R. V. Kasowski, W. Y. Hsu, and F. Herman, *Solid State Commun.* **63**, 1077 (1987).

[22] P. A. Sterne and C. S. Wang, *Phys. Rev.* B**37** 7472 (1988).

[23] D. A. Papaconstantopoulos, W. E. Pickett, and M. J. DeWeert, *Phys. Rev. Lett.* **61**, 211 (1988).

[24] J. D. Jorgensen, B. W. Veal, W.K. Kwok, G. W. Crabtree, A. Umezawa, L. J. Nowicki, and A. P. Paulikas, *Phys. Rev.* B**36**, 5731 (1987).

[25] R.J.Cava, B.Batlogg, C.H.Chen, E.A.Rietman, S.M.Zahurak, and D.Werder, *Nature* **329**, 423 (1987).

[26] J. M. Tranquada, D. E. Cox, W. Kunnmann, H. Moudden, G. Shirane, M. Suenaga, P. Xolliker, D. Vaknin, S. K. Sinha, M. S. Alvarez, A. J. Jacobson, and D. C. Johnston, *Phys. Rev. Lett.* **60**, 156 (1988); J. H. Brewer et al. Phys. Rev. Lett. **60**, 1073 (1988). J. W. Lynn, W. H. Li, H. A. Mook, B. C. Sales, and Z. Fisk, *Phys. Rev. Lett.* **60**, 2781 (1988).

[27] L. F. Mattheiss and D. R. Hamann, *Solid State Commun.* **63**, 395 (1987).

[28] S. Massidda, J. Yu, A. J. Freeman and D. D. Koelling, *Phys. Lett.* A**122**, 198 (1987).

[29] J. Yu, S. Massidda, A. J. Freeman and D. D. Koelling, *Phys. Lett.* A**122**, 203 (1987).

[30] H. Krakauer and W. E. Pickett, in *Novel Superconductivity*, edited by S. A. Wolf and V. Z. Kresin (Plenum, New York, 1987), p. 501.

[31] W. Y. Ching, Y. Xu, Guang-Lin Zhao, K. W. Wong, and F. Zandiehnadem, *Phys. Rev. Lett.* **59**, 1333 (1987).

[32] D. W. Murphy, S. Sunshine, R. B. van Dover, R. J. Cava, B. Batlogg, S. M. Zahurak and L. F. Schneemeyer, *Phys. Rev. Lett.* **58**, 1888 1987).

[33] P. H. Hor, R. L. Meng, Y. Q. Wang, L. Gao, Z. J. Huang, J. Bechtold, K. Foster, and C. W. Chu, *Phys. Rev. Lett.* **58**, 1891

(1987).

[34] E. E. Alp, L. Soderholm, G. K. Shenoy, D. G. Hinks, D. W. Capone II, K. Zhang, and B. D. Dunlap, *Phys. Rev.* **B36**, 8910 (1987).

[35] L. C. Semdskjaer, J. Z. Liu, R. Benedek, D. G. Legnini, D. J Lam, M. D.Stahulak, and H. Claus (to be published).

[36] J. Yu, A. J. Freeman, and S. Massidda, in *Novel Superconductivity*, edited by S. Wolf, and V. Z. Kresin (Plenum, New York 1987), p. 367.

[37] J. Zaanen, A. T. Paxton, O. Jepsen, and O. K. Andersen, *Phys. Rev. Lett.* **60**, 2685 (1988).

[38] H. Maeda, Y. Tanaka, M. Fukutomi and T. Asano, *Jpn. J. Appl. Phys. Lett.* **27**, L209 (1988); C. W. Chu et al, *Phys. Rev. Lett.* **60**, 941 (1988); C. Michel, M. Hervieu, M. M. Borel, A. Grandin, F. Deslandes, J. Povost and B. Ravean, *Z. Phys.* B **68**, 421 (1987).

[39] Z. Z. Sheng, A. M. Hermann, A. El Ali, C. Almasan, J. Estrada, T. Datta, and R. J. Matson, *Phys. Rev. Lett.* **60**, 937 (1988); S. S. P. Parkin, V. Y. Lee, E. M. Engler, A. I. Nazzal, T. C. Huang, G. Gorman, R. Savoy and R. Beyers, *Phys. Rev. Lett.* **60**, 2539 (1988); *ibid* **61**, 750 (1988).

[40] S. A. Sunshine T. Siegrist, L. F. Schneemeyer, D. W. Murphy, R. J. Cava, B. Battlog, R. B. van Dover, R. M. Fleming, S. H. Glarum, S. Nakahara, R. Farrow, J. J. Krajewski, S. M. Zahurak, J. V. Waszczak, J. H. Marshall, P. Marsh, L. W. Rupp, and W. F. Peck, *Phys. Rev.* **B38**, 893 (1988).

[41] R. M. Hazen T. C. Prewitt, R. J. Angel, N. L. Ross, L. W. Finger, C. G. Hadidiacos, D. R. Veblen, P J. Heaney, P. H. Hor, R. L. Meng, Y. Y. Sun, Y. Q. Wang, Y, Y Xue, H. J. Hung, L. Gao, J. Bechtold, and C. W. Chu, *Phys. Rev. Lett.* **60**, 1174 (1988); J. M. Tarascon, Y. Le Page, P. Barboux, B. G. Bagley, L. H. Greene, W. R. McKinnon, G. W. Hull, M. Giroud, and D. M. Hwang (preprint). M. A. Subramanian C. C. Torardi, J. C. Calabrese, J. Gopalakrishnan, K. J. Morrissey, T. R. Askew, R. B. Flippen, U. Chowdhry, A. W. Sleight, *Science* **239**, 1015 (1988).

[42] R. M. Hazen, L. W. Finger, R. J. Angel, C. T. Prewitt, N. L. Ross, C. G. Hadidiacos, P. J. Heaney, D. R. Veblen, Z. Z. Sheng, E. El Ali, and A. M. Hermann, *Phys. Rev. Lett.* **60**, 1657 (1988).

[43] J. B. Torrance, Y. Tokora, S. J. LaPlaca, T. C. Huang, R. J. Savoy and A. I. Nazzal, *Solid State Commun.* **66**, 703 (1988).

[44] P. A. Sterne and C. S. Wang, *J. Phys. C* **21**, L949 (1988).

[45] M. S. Hybertsen and L. F. Mattheiss, *Phys. Rev. Lett.* **60**, 1661 (1988).

[46] H. Krakauer and W. E. Pickett, *Phys. Rev. Lett.* **60**, 1665

(1988).

[47] A. J. Freeman, J. Yu, and S. Massidda, *Physica* **C153-155**, 1225 (1988).

[48] J. D. Jorgensen, H.-B. Schüttler, D. G. Hinks, D. W. Capone, H. K. Zhang and M. B. Brodsky, *Phys. Rev. Lett.* **58**, 1024 (1987).

[49] F. Beech, S. Miraglia, A. Santoro and R. S. Roth, *Phys. Rev.* **B35**, 8778 (1987).

[50] S. Shamoto, M. Onoda, M. Sato and S. Hosoya, *Solid State Commun.* **62**, 479 (1987).

[51] T. K. Worthington, W. J. Gallagher, and T. R. Dinger, *Phys. Rev. Lett.* **59**, 1160 (1987). A. Umezawa, G. W. Crabtree, J. Z. Liu, T. J. Moran, S. K. Malik, L. H. Nunez, W. L. Kwok, and C. H. Sowers, *Phys. Rev.* **B38**, 2843 (1988).

[52] B. Batlogg, T. T. M. Palstra, L. F. Schneemeyer, R. B. vanDover and R. J.Cava, *Physica* **C153-155**, 1062 (1988).

[53] Y. Yeshurun and A. P. Malozemoff, *Phys. Rev. Lett.* **60**, 2202 (1988).

[54] J. M. Wheatley, T. C. Hsu, and P. W. Anderson, *Phys. Rev.* **B38**, (1988); J. M. Wheatley, T. C. Hsu, and P. W. Anderson, *Nature* **333**, 121 (1988).

[55] J. Yu, S. Massidda, and A. J. Freeman, *Physica* C **152**, 273 (1988).

[56] D. R. Hamann and L. F. Mattheiss, *Phys. Rev.* **B38**, 5138 (1988).

[57] R. V. Kasowski, W. Y. Hsu, and F. Herman, *Phys. Rev.* **B38**, 6470 (1988).

[58] R. V. Kasowski, W. Y. Hsu, and F. Herman, *Phys. Rev.* **B36**, 7248 (1987).

[59] F. Herman, R. V. Kasowski, and W. Y. Hsu, *Phys. Rev.* **B36**, 2309 (1987).

[60] F. Herman, R. V. Kasowski, and W. Y. Hsu, *Phys. Rev.* **B36**, 7248 (1987).

[61] A. W. Hewat, P. Bordet, J. J. Capponi, C. Chaillout, J. Chenavas, M. Godinho, E. A. Hewat, J. L. Hodeau, M. Marezio, submitted to *Physica* **C153-155**, 369 (1988).

[62] M. Lee, Y.-Q. Song, W. P. Halperin, L. M. Tonge, T. J. Marks, H. O. Marcy, and C. R. Kannewurr, to be published.

[63] J. Ashkenazi and C. G. Kuper to be published in the Proc. of "*Superconductivity, New Models, New Applications*" Les Houches, *J. de Physique (colloque)* (1988).

[64] See, for example, D. J. Scalapino in *Superconductivity*, edited by R. D. Parks (New York: Dekker) vol. 2, chap. 10 (1969).

[65] G. D. Gaspari and B. L. Gyorffy, *Phys. Rev. Lett.* **28**, 801 (1972).

[66] R. E. Cohen, W. E. Pickett, L. L. Boyer, H. Krakauer, *Phys.*

Rev. Lett. **60**, 817 (1988).
[67] C. L. Fu and A. J. Freeman, *Phys. Rev.* B35, 8861 (1987).
[68] P. B. Allen, W. E. Pickett, and H. Krakauer, *Phys. Rev.* B37, 7482 (1988); *ibid, Phys. Rev.* B36, 3926 (1987).
[69] J. R. Hardy, and J. W. Flocken, *Phys. Rev. Lett.* **60**, 2191 (1988).
[70] S. Etemad, D. E.Aspenes, M. K. Kelly, R. Thompson, J.-M. Tarascon, and G. W. Hull, *Phys. Rev.* B37, 3396 (1987).
[71] J. Orenstein, G. A. Thomas, D. H. Rapkine, C. G. Bethea, B. F. Levine, R. J. Cava, E. A. Rietman, and D. W. Johnson, Jr., *Phys. Rev.* B36, 729 (1987).
[72] S. L. Herr, K. Karmaras, C. D. Porter, M. G. Doss, and D. B. Tanner, D. A. Bonn, J. E. Greedan, C. V. Stager, and T. Timusk, *Phys. Rev.* B36, 733 (1987).
[73] K. Kamaras, C. D. Porter, M. G. Doss, S. L. Herr, D. B. Tanner, D. A Bonn, J. E. Greedan, A. H. O'Reilly, C. V. Stager, and T. Timusk, *Phys. Rev. Lett.* **59**, 919 (1987).
[74] S. Tajima, S. Uchida, M. Ishii, H. Takaji, and S. Tanaka, *Jpn. J. Appl. Phys.* **26**, 1007 (1987).
[75] R. T. Collins, Z. Schlesinger, R. H. Koch, R. B. Laibowitz, T. S. Plaskett, P. Freitas, W. J. Gallagher, R. L. Sandstrom, and T. R. Dinger, *Phys. Rev. Lett.* **59**, 704 (1987).
[76] I. Bozovic, D. Kirillov, A. Kapitulnik, K. Char, M. R. Hahn, M. R. Beasley, T. H. Geballe, Yl H, Kim, and A. J. Heeger, *Phys. Rev. Lett.* **59**, 2219 (1987).
[77] J. Orenstein and D. H. Rapkine, *Phys. Rev. Lett.* **60**, 968 (1988).
[78] Guang-Lin Zhao, Y. Xu, W. Y. Ching, and K. W. Wong, *Phys. Rev.* B36, 7203 (1987).
[79] M. Suzuki, and T Murakami, *Jpn. J. Appl. Phys.* **26**, L524 (1987).
[80] S. W. Tozer, A. W. Kleinsasser, T. Penney, D. Kaiser, and F. Holtzberg, *Phys. Rev. Lett.* **59**, 1768 (1987).
[81] J. C. Slater and G. F. Koster, *Phys. Rev.* **94**, 1498 (1954).
[82] T. C. Leung, X. W. Wang, and B. M. Harmon, *Phys. Rev.* B37, 384 (1988).
[83] P. A. Sterne, C. S. Wang, G. M. Stocks, and W. M. Temmerman, *Materials Res. Soc. Proc.* **99**, 353 (1988).
[84] D. Vaknin, S.K. Sinha, D. E. Moncton, D. C. Johnston, J. Newsam, C. R. Safinya, and H. E. King Jr., *Phys. Rev. Lett.* **58**, 2802 (1987).
[85] S. T. Chui, Robert V. Kasowski, and William Y. Hsu, *Phys. Rev. Lett.* **61**, 207 (1988).
[86] J. C. Fuggle, P. J. W. Weijs, R. Schoorl, G. A. Sawatzky, J. Fink, N. Nücker, P. J. Durham and W. M. Temmerman, *Phys. Rev.* B37, 123 (1988).

[87] A. Fujimori, E. Takayama-Muromachi, Y. Uchida, and B. Okai, *Phys. Rev.* **B35**, 8814 (1987).

[88] Zhi-xun Shen, J. W. Allen, J. J. Yeh, J.-S. Kang, W. Ellis, W. Spicer, I. Lindau, M. B. Maple, Y. D. Dalichaouch, M. S. Torikachvili, J. Z. Sun and T. H. Geballe, *Phys. Rev.* **B36**, 8414 (1987).

[89] H. Chen, J. Callaway, and P. K. Misra, *Phys. Rev.* **B36**, 8863 (1987).

[90] J. S. Yarmoff, D. R. Clarke, W. Drube, U. O. Karlsson, A. Taleb-Ibrahimi, and F. J. Himpsel, *Phys. Rev.* **B36**, 3967 (1988).

[91] N. Nücker, J. Fink, J. C. Fuggle, P. J. Durham and W. M. Temmerman, *Phys. Rev.* **B37**, 5158 (1988).

[92] A. Bianconi, A. Congiu Castellano, M. De Santis, P Rudolf, P. Lagarde, A. M. Flank and A. Marcelli, *Solid State Commun.* **63**, 1009 (1987).

[93] A. K. McMahn, R. M. Martin, and S. Satpathy, *Phys. Rev.* **B38**, 6650 (1988).

[94] C. F. Chen, X. W. Wang, T. C. Leung, and B. N. Harmon (to be published).

[95] M. Schluter, M. S. Hybertsen, and N. E. Christensen, *Physica* **C153-155**, 1217 (1988).

[96] J. Zaanen, O. Jepsen, O. Gunnarsson, A. T. Paxton, O. K. Andersen, and A. Svane, *Physica* **C153-155**, (1988).

[97] O. Gunnarsson and K. Schönhammer, *Phys. Rev.* **B28**, 4315 (1983).

[98] K. Terakura,T. Oguchi,A. R. Williams, and J. Kubler, *Phys. Rev.* **B30** 4734 (1984).

[99] B. H. Brandow,to be published in the proceedings of the NATO Advanced Workshop, "Narrow Band Phenomena", Steverden, Netherlands, June 1-5 (1987).

[100] A. Svane and O. Gunnarsson, *Phys. Rev.* **B37**, 9919 (1988); A. Svane and O. Gunnarsson (to be published).

[101] J. Zaanen, G. A. Sawatzky, and J. W. Allen, *Phys. Rev. Lett.* **55**, 428 (1985).

[102] C. M. Varma, *Phys. Rev. Lett.* **55**, 2723 (1985).

[103] C. S. Wang, H. Krakauer, W. E. Pickett, *J. Phys.* F **16**, L287 (1986).

[104] C. S. Wang, M. R. Norman, R. C. Albers, A. M. Boring, W. E. Pickett, H. Krakauer, and N. E.Christensen, *Phys. Rev.* **B35** 7260 (1987).

[105] L. Taillefer and G. G. Lonzarich, *Phys. Rev. Lett.* **60**, 1570 (1988).

[106] L. Hedin and S. Lundqvist, *Solid State Phys.* **23**, 1 (1969).

[107] J. M. Luttinger, *Phys. Rev.* **119**, 1153 (1960).

[108] M. R. Norman, T. Oguchi, and A. J. Freeman (unpublished).

[109] G. R. Stewart, *Rev. Mod. Phys.* **56**, 755 (1984).

[110] M. R. Norman, W. E. Pickett, H. Krakauer, and C. S. Wang, *Phys. Rev.* **B36**, 4058 (1987).
[111] P. G. McQueen, and C. S. Wang (to be published).
[112] Y. Endoh, K. Yamada, R. J. Birgeneau, d. R. Gabbe, H. P. Jenssen, M. A. Kastner, C. J. Peters, P. J. Picone, T. R. Thurston, J. M. Tranquada, G. Shirane, Y. Hidaka, M. Oda, Y. Enomoto, M. Suzuki, and T. Murakami, *Phys. Rev.* **B37**, 7443 (1988).
[113] G. Shirane, Int. Conf. on Mag. (Paris, 1988), *J. de Physique (Colloque)* (to be published).
[114] A. Bhattacharya, P. G. McQueen, C. S. Wang, and T. Einstein (to be published).
[115] J. E. Hirsch, S. Tang, E. Loh, Jr., D. J. Scalapino, *Phys. Rev. Lett.* **60**, 1668 (1988).

CHAPTER 6

Synthesis and Diamagnetic Properties

Robert N. Shelton

6.1. Introduction

High transition temperature (T_c) superconductors with T_c in excess of 30 K belong to a variety of related classes of metallic oxides; for example, materials in the $La_{2-x}M_xCuO_{4-y}$ series (M = Ba, Sr, Ca) [1-5] which crystallize with either orthorhombic or tetragonal symmetry based on the K_2NiF_4-type structure, or materials in the $MBa_2Cu_3O_x$ and related series (M = rare earth element) [6-10] which form in different, but related tetragonal and orthorhombic crystallographic symmetries [see Chapter 4]. The elements comprising these multinary oxides possess diverse properties ranging from air sensitive alkaline earths to ductile copper to oxygen. Preparation techniques for these ceramic materials are decidedly different from those employed to successfully produce some of the traditional superconducting binary compounds and alloys comprised of strictly metallic constituents; for example, A-15 compounds such as Nb_3Sn and V_3Si, Laves phase (C-15) compounds such as ZrV_2 and HfV_2, or ductile intermetallic alloys such as NbTi. In contrast to these binary materials, the category of ternary and multinary superconductors includes numerous examples of compounds with at least one non-metallic constituent [11]. Chevrel phase compounds typified by $PbMo_6S_8$ utilize the chalcogen atoms in the basic crystallographic framework [12-14]. Ternary phosphides such as ZrRuP derive from a binary structure type (Fe_2P) where the non-metal is an indispensable component [15]. Some ternary oxide superconductors presaged the discovery of the current classes of high T_c materials; for example, the spinel compound $LiTi_2O_4$ [16] and the perovskite $BaPb_{1-x}Bi_xO_3$ [17].

In this chapter, we present information on sample preparation of these ceramic oxides, both in bulk polycrystalline and single crystal form. General synthesis techniques as well as specific precautions related to constituents and methods are discussed. Phase diagrams and methods of phase formation are given where available. The influence of various synthesis processes on the resulting superconducting properties is examined with reference to basic experiments on the magnetization and transport properties. We emphasize three fundamental experimental techniques suitable for studying superconductors. First, significant information is obtained by measuring the response of the superconducting material to a magnetic field. These experiments include the Meissner effect, determination of both upper (H_{c2}) and lower (H_{c1}) critical fields, the interplay between long-range magnetic order and superconductivity, and magnetization versus applied field hysteresis curves. Second, characteristic measurements of the normal state transport properties of the high T_c superconductors are presented. In addition to resistive determination of the actual superconducting transition, separate Hall effect experiments give information about the type and concentration of charge carriers present in these materials. In the superconducting state, the critical current density (J_c) is a crucial parameter for many applications. We present the current state of experiments on J_c for bulk materials. Third, the effect of compositional variation on these magnetic and transport properties is demonstrated through selected experimental results. This third category of experiments illustrates the connection between materials preparation and the ensuing properties of the superconducting compound. We present the synthesis techniques and basic superconducting properties in this one chapter in order to emphasize the strong relationship between the two.

6.2. The System $La_{2-x}M_xCuO_{4-y}$ (M = Ba, Sr, Ca)

Preparation of bulk, polycrystalline samples of compounds in the series $La_{2-x}M_xCuO_{4-y}$ (M = Ba, Sr, Ca) usually involves ceramic processing of the binary oxides La_2O_3, CuO, and BaO, or the alkaline-earth carbonate (e.g. $BaCO_3$) [18]. There is considerable variation of specific synthesis parameters in the literature, for example, sintering times and temperatures. Although the complete synthesis procedure involves a sequence with individual steps that may vary slightly, the overall goals are consistent. Details of preparation methods are described in the following paragraphs.

An initial consideration is the quality of the starting materials which affects the consistency of results. This is particularly evident in the need to use fully oxidized CuO powder and dried La_2O_3. The reaction to form the K_2NiF_4-type phase occurs via the solid state; therefore, the constituent powders must be mixed thoroughly to provide the necessary spatial uniformity. This mixing step is most frequently achieved by grinding the powders in an agate mortar for extended times (typically 60 minutes). Alternative techniques include the coprecipitation of the powders and the sol-gel process, both designed to produce a more intimate mixture of the constituents. The mechanical combination process is normally carried out under ambient conditions in air. During this step, the powder mixture absorbs moisture from the air. This moisture can be driven off in the heat treatment to minimize the influence of residual water on the reaction that forms the superconducting phase; however, some researchers are performing all of the mixing steps in a dry nitrogen atmosphere.

The mixed powders are pressed into pellets, placed into a suitable crucible, and subjected to a series of heat treatments to produce the solid state reaction that forms the K_2NiF_4-type phase. While a variety of crucible materials have been used successfully, the most common ones are platinum and alumina (Al_2O_3). Details of the initial heat treatment vary. Consistent results are obtained by reacting the powders in air at 1100°C to 1140° for 24 to 48 hours. This reaction step is sometimes repeated after regrinding the material if X-ray analysis shows the presence of secondary phases. A prefiring at a lower temperature, 750°C for example, to drive off moisture acquired during the grinding process is often used.

At this stage, the correct phase has been formed, but the oxygen content may be substoichiometric. Parameters of time and temperature for the oxygen recovery step vary widely. The kinetics of the redox equilibrium for this multicomponent system are complex, and recent research has emphasized the $YBa_2Cu_3O_x$-type compounds. Exact parameters also depend on the alkaline earth. Nevertheless, an oxygen recovery stage at temperatures below the initial reaction temperature is necessary. The range of typical parameters for heating the bulk samples in an oxygen atmosphere is: 900°C (500°C) for 12 hours (48 hours), followed by a slow (2°C/minute) furnace cool to room temperature. After such a treatment, the pellets are hard and well sintered, and the material is single phase according to X-ray diffraction standards.

Single crystal growth of superconducting oxides with the K_2NiF_4-type or related orthorhombic structure has been reported for both the pure ternary La_2CuO_4 [19] and alkaline earth substituted compositions $La_{2-x}M_xCuO_4$ (M = Ba, Sr, Ca) [20,21]. None of these compounds melts congruently; therefore, a crystal growth technique utilizes a flux to promote crystal formation below the decomposition temperature. Details of one method using a CuO flux are given by Hidaka and coworkers [20]. One significant problem with the flux growth method is the inability to attain homogeneous incorporation of the alkaline earth on the La sublattice. This nonuniform distribution leads to exceedingly broad superconducting transitions and renders any compositional dependent properties undetectable. To date, considerably less effect has been reported on crystal growth of the K_2NiF_4-type materials than on the family of superconducting oxides based on $YBa_2Cu_3O_x$.

The magnetic and transport properties of these ternary oxides with the K_2NiF_4-type structure span wide ranges, from superconductivity to antiferromagnetism and from metallic to semiconducting behavior. Only a small fraction of the total research effort can be presented, especially considering that these properties are sensitive to atomic substitution on the La sublattice and variations in the concentration of oxygen. The remainder of this section focuses on (1) magnetic field based on experiments such as the Meissner effect, magnetization data, and critical magnetic fields (H_{c1} and H_{c2}), (2) transport experiments as related to critical current densities (J_c), (3) the effect of high pressure on the superconducting state, and (4) the effects on these sets of experiments of varying the atomic composition of $La_{2-x}M_xCuO_{4-y}$.

The low temperature properties of the parent compound La_2CuO_{4-y} vary tremendously depending on oxygen content and preparation techniques that affect metal sublattice stoichiometry. It has been demonstrated [22-26] that oxygen vacancies promote an antiferromagnetic ground state with pronounced sensitivity. From an absence of antiferromagnetic order (T_{Neel} = 0 K) at stoichiometry (y=0), T_N is reported to rise to room temperature for y = 0.03 [23]. The low temperature moment is small ($<1\mu_B$/Cu atom) and resides on the copper atoms. At full oxygen occupancy, long-range magnetic order may be absent in favor of superconductivity [23]. A shift in the electrical resistivity from a semiconducting to a metallic temperature dependence correlates with the occurrence of superconductivity and the presence of vacancies on the La sites [27]. Thus, experimental evidence suggests that the superconducting state is associated with vacancies on the La sublattice in conjunction with complete occupancy of the oxygen sites, both factors influencing the oxidation state of the copper ions.

When an alkaline earth element is partially substituted for La, the tendency to antiferromagnetism is suppressed and superconductivity becomes the dominant low temperature state. Early research found that optimal superconducting transition temperatures (T_c) occurred near x = 0.15 for $La_{2-x}M_xCuO_4$ where M = Ba, Sr, or Ca [2-4,7,8,28,29]. More detailed studies later revealed a richer, more diverse behavior of T_c versus composition as presented later in this chapter. We now consider magnetic and transport properties of compositions with optimal T_c.

A common technique for determining the superconducting

transition temperature is measuring the response of the material to an external magnetic field. When the material is cooled in a constant field, magnetic flux is excluded from the interior of the sample at the onset of superconductivity. This is the Meissner effect [see Chapter 1]. In a second, related technique, the sample is cooled to below T_c in zero external field and then the measuring field is applied. In this case, a diamagnetic signal is observed due to screening supercurrents being induced on the surface of the superconducting sample. This experiment demonstrates the infinite conductivity of the material. As an example of this type of experiment, the transition into the superconducting state as determined by a Meissner effect measurement is shown in Fig. 6.1 for $La_{1.85}Sr_{0.15}CuO_4$. This particular sample shows a 60% Meissner effect which is fairly representative of these materials [30]. Reduced diamagnetic shielding occurs for applied fields greater than the lower critical field (H_{c1}), which is small, and when the penetration depth is comparable to the size of the superconducting regions. The temperature dependence of the magnetization shown in Fig. 6.1 is consistent with a material composed of high-quality superconducting grains coupled by superconducting weak links such as Josephson junctions. This picture of a granular type-II superconductor comprised of strongly superconducting grains coupled via weak superconducting links has emerged from a variety of

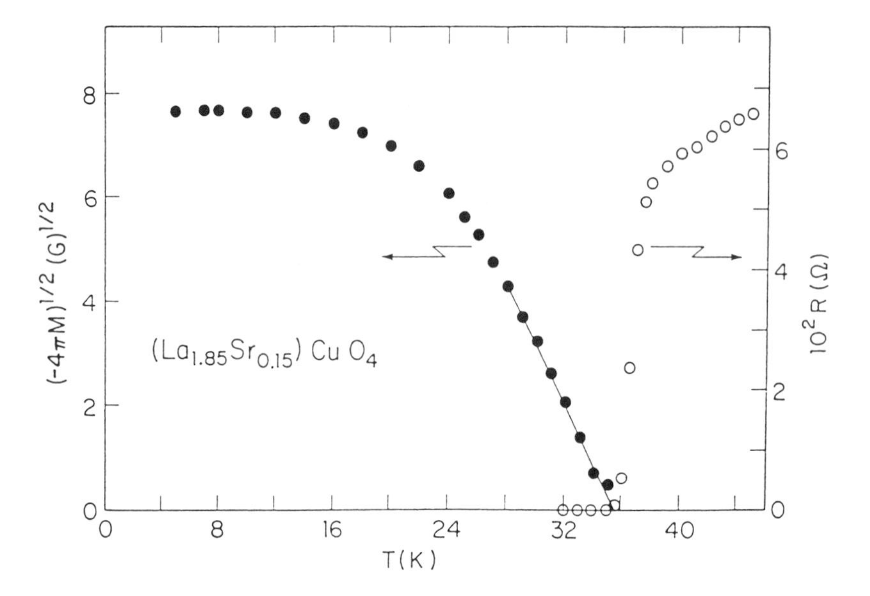

Fig. 6.1. Superconducting transition determined resistively with current of 10 mA and magnetically in an applied field of 100 Oe. See reference [31] for details.

experiments [31-33].

Critical magnetic field studies have revealed small values for the lower critical field ($H_{c1}(0) \leq$ 10 mT) and exceptionally large values for the upper critical field ($H_{c2}(0) \geq$ 50 T), thus confirming the extreme type-II nature of superconductivity in this class of compounds [34-36]. Results of resistively determined values for H_{c2} are shown in Fig. 6.2 for two samples in the series $La_{2-x}Sr_xCuO_4$. Clearly, H_{c2} is sensitive to composition. Sharp resistive transitions are important for the accurate determination of $H_{c2}(0)$

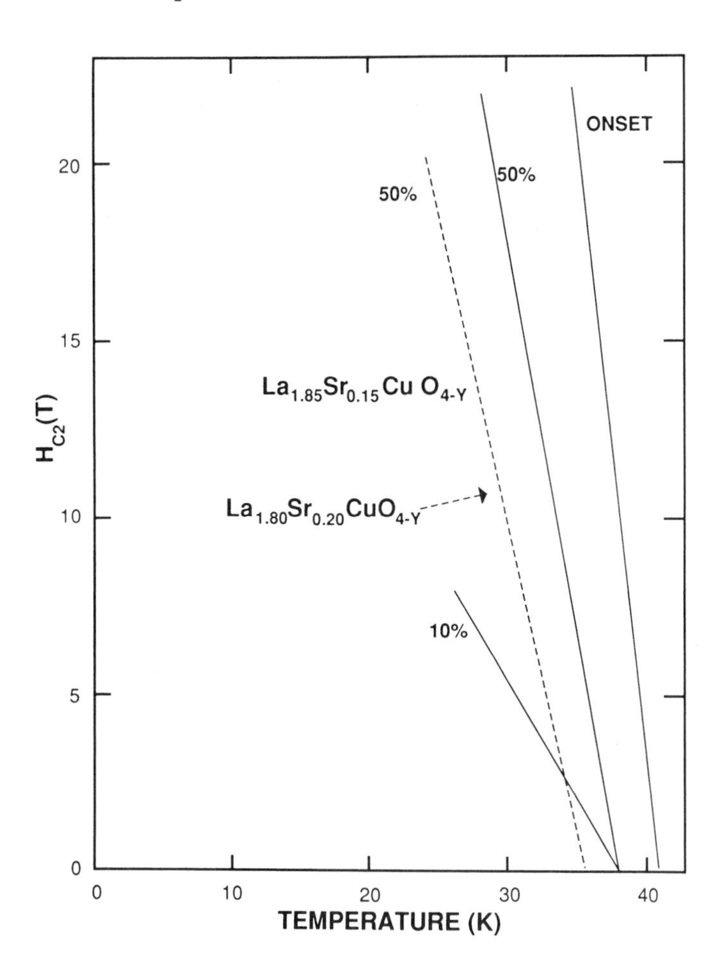

Fig. 6.2. Upper critical magnetic field H_{c2} versus temperature for two compounds in the series $La_{2-x}Sr_xCuO_{4-y}$. For x = 0.15, the solid lines show H_{c2} values determined from resistive onsets, midpoints, and 10% transitions. Only the midpoint values are shown for the x = 0.20 sample. See reference [36] for details.

and $dH_{c2}/dT)_{T=Tc}$. Experimental values for $-dH_{c2}/dT)_{T=Tc}$ based on the midpoint superconducting transition are typically 2 T/K (±10%) [34-37]. This slope is comparable to that of the A15 superconductors Nb_3Sn and V_3Si [38], but considerably smaller in magnitude than those observed in the Chevrel phase superconductor $PbMo_6S_8$ [13,39,40].

Electrical resistivity studies on compounds in the system $La_{2-x}M_xCuO_4$ have been reported by various research groups [21,41-43]. Two experimental features dominate these resistivity results; namely, the linear temperature dependence of the resistivity $\rho(T)$ which persists to the highest temperatures measured [42], and the resistivity anisotropy which indicates preferential conduction with the Cu-O planes [42]. The linearity of $\rho(T)$ is shown in Fig. 6.3, drawn from data presented in Ref. [42]. The slope $d\rho(T)/dT$ for $La_{2-x}Sr_xCuO_4$ shows no sign of decreasing up to 1100 K. This remarkable finding implies that the mean free path, l_{mfp}, is larger than the interatomic spacings over the entire temperature range. Therefore, the situation for $La_{2-x}Sr_xCuO_4$ is in sharp contrast to the behavior of resistivity saturation observed in high T_c A15 compounds such as Nb_3Sn and V_3Si [44]. This experimental evidence of large l_{mfp} at high temperatures has been used [42] in combination with band structure calculations to obtain

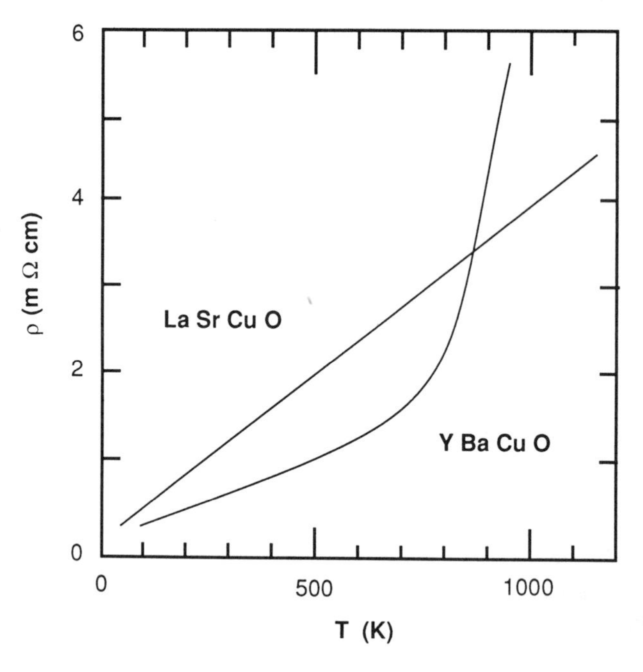

Fig. 6.3. Temperature dependence of the electrical resistivity for $La_{1.825}Sr_{0.175}CuO_4$ and $YBa_2Cu_3O_x$. See reference [42] for details.

an upper estimate for the electron-phonon coupling strength, λ_{el-ph}, in $La_{2-x}Sr_xCuO_4$ of 0.1. Using this value of λ_{el-ph} in the McMillan [45] formula to estimate T_c yields a value below 1 K. Thus, the linearity of $\rho(T)$ to the highest temperatures measured indicates that a mechanism other than, or in addition to, the electron-phonon interaction is responsible for the high superconducting transition temperatures in this class of oxides. Resistivity measurements [21] on single crystals reflect the strong anisotropy present in these materials with the K_2NiF_4-type structure. Typically, the resistivity along the c-axis is an order of magnitude larger than that measured in the a-b plane.

Another important transport property measurement is the Hall effect experiment which identifies the type and concentration of charge carriers. Various groups have reported Hall effect data for the optimal T_c concentration $La_{1.85}Sr_{0.15}CuO_4$ [43,46-48]. There is general agreement on a positive Hall constant indicating that conduction by holes dominates. The hole carrier concentration in the normal state just above T_c is low, with a value of about $(5 \pm 3) x\ 10^{21}$ cm^{-3}. Perhaps most revealing is a compositional study across the series $La_{2-x}Sr_xCuO_4$ where the hole concentration is related to the Sr doping and T_c [47,48]. In this work, the hole concentration was determined independently via wet chemistry techniques. The data taken from Ref. [48] and presented in Fig. 6.4

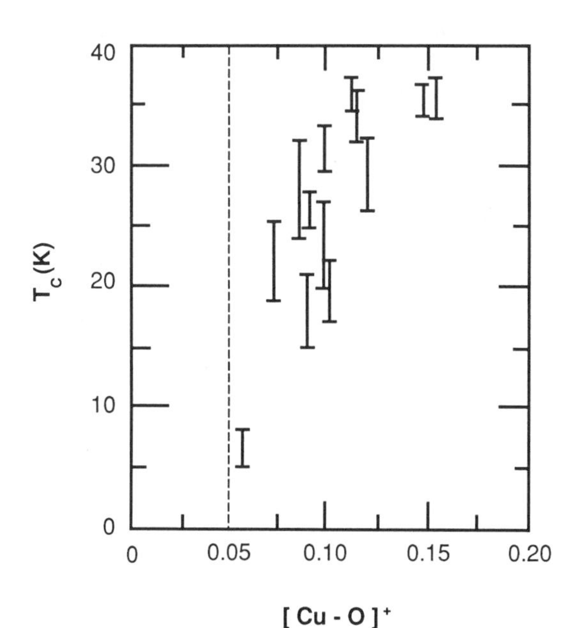

Fig. 6.4. Superconducting transition temperature (T_c) versus hole concentration $[Cu - O]^+$ for the system $La_{2-x}Sr_xCuO_4$. See reference [48] for details.

illustrate a strong correlation between T_c and hole concentration in the Cu-O layers, providing further evidence as to the importance of the low-dimensional features of these materials. The doping with alkaline earth is seen to affect T_c via a change in the hole concentration. This linear dependence of T_c on concentration of $[Cu\text{-}O]^+$ complexes is consistent with theories that incorporate a quasi-two-dimensional Bose-Einstein condensation, for example the resonant valence bond (RVB) theory [49,50].

High pressure studies stimulated early interest in the K_2NiF_4-type superconductors since the application of hydrostatic pressure enhances T_c at a rate significantly higher than that observed previously in most superconductors [51-59]. The origins of this unusually large sensitivity to pressure are not thoroughly understood. Early work including pressure measurements and determination of the compressibilities indicates that the net interaction responsible for superconductivity (electron-electron or hole-hole for example) must have a strong pressure dependence, independent of any influence of lattice stiffening under pressure. Bulk modulus measurements on the system $La_{2-x}Sr_xCuO_4$ yield values ranging from 1100 kbar to 1700 kbar depending on composition and temperature [53,60]. These values are similar to those of ternary oxides that crystallize in the perovskite structure [61]. Experimental studies on $La_{1.85}Sr_{0.15}CuO_4$ of the magnetic susceptibility under pressure reaffirm this conclusion as well as show explicitly that the rapid rise in T_c with pressure does not derive from an increase in the density of states at the Fermi level [56]. Nonlinear behavior in T_c versus pressure has also been observed for some compositions in the systems $La_{2-x}M_xCuO_4$ (M = Ca, Sr, Ba). These effects correlate well with the pressure dependence of lattice parameters. Indeed, the sensitivity of T_c to alkaline earth doping is present for dT_c/dp as well, where the slope changes sign depending on the amount of alkaline earth present. This sensitivity has been related to the details of the band structure [57]. A discussion of relevant theoretical models and their applicability to the large values of dT_c/dp concludes that the traditional BCS model is not sufficient to account for results of the pressure experiments [59].

Chemical substitution has been a major influence on the low temperature properties of the K_2NiF_4-type superconducting oxides. Early work found that doping of Ca, Sr, or Ba for La suppressed the tendency toward long range magnetic order in favor of superconductivity [62,63] and produced maximum T_c values near the composition $x = 0.15$ in the formula $La_{2-x}M_xCuO_4$ (M = Ca, Sr, Ba). As increasing experimental knowledge of phase stability and synthesis procedures became accessible, closer inspection of the composition dependence of T_c indicated a richer, more diverse behavior of T_c versus x occurs for the Ba doped materials [64,65].

Earlier in this chapter we documented the tremendous variation with oxygen concentration of the low temperature properties of the ternary compound La_2CuO_{4-y}. The role of oxygen continues to be of prime importance in the alkaline-earth doped samples. A notable demonstration of how oxygen nonstoichiometry can dominate a physical property is seen in the temperature dependence of the electrical resistivity [66]. As the oxygen deficiency (y) increases, the electrical resistivity goes from a metallic-like dependence to semiconducting-like behavior. A typical example taken from the early literature is shown in Ref. [66]. The overriding importance of oxygen content is evident in other physical properties including the crystallographic parameters [see Chapter 4]. Due to this influence of oxygen concentration, great emphasis must be placed on precise incorporation of oxygen and determination of oxygen content during the synthesis process. This prime role of oxygen will be reinforced for the system $MBa_2Cu_3O_x$ where superconducting transition temperatures in excess of 90 K are realized.

6.3. The System $MBa_2Cu_3O_x$ (M = Rare Earth)

The first occurrence of superconductivity above the boiling point of liquid nitrogen (77 K) was reported for the compound $YBa_2Cu_3O_7$ [6]. Substitution of the other rare earth elements for yttrium followed soon thereafter [7,67-71]. While attainment of T_c's above 90 K is now commonplace, the details of sample preparation and ensuing microstructure are crucial to more subtle superconducting parameters such as the critical current density (J_c) and the upper critical magnetic field (H_{c2}). These quantities are important for practical applications of the new superconducting materials. This necessity for attention to sample preparation techniques provides the starting point for synthesis of compounds in the system $MBa_2Cu_3O_x$.

As noted for the K_2NiF_4-type oxides, preparation of bulk, polycrystalline samples of the type $YBa_2Cu_3O_x$ involves ceramic processing of the oxides Y_2O_3, CuO, and a barium-containing compound. $BaCO_3$, $Ba(NO_3)_2$, and BaO are the most common choices for incorporating the barium. The quality of starting materials should be high to ensure consistent results. Considerably more quantitative information is available regarding the formation of $YBa_2Cu_3O_x$ and the associated phase diagram than for the K_2NiF_4-type compounds. In Fig. 6.5, the copper-rich part of the CuO-BaO-$YO_{1.5}$ isothermal cut at 950°C is shown [72]. Fig. 6.6 illustrates the pseudobinary phase diagram along the Y_2BaCuO_5 - to - "$Ba_3Cu_5O_y$" cut

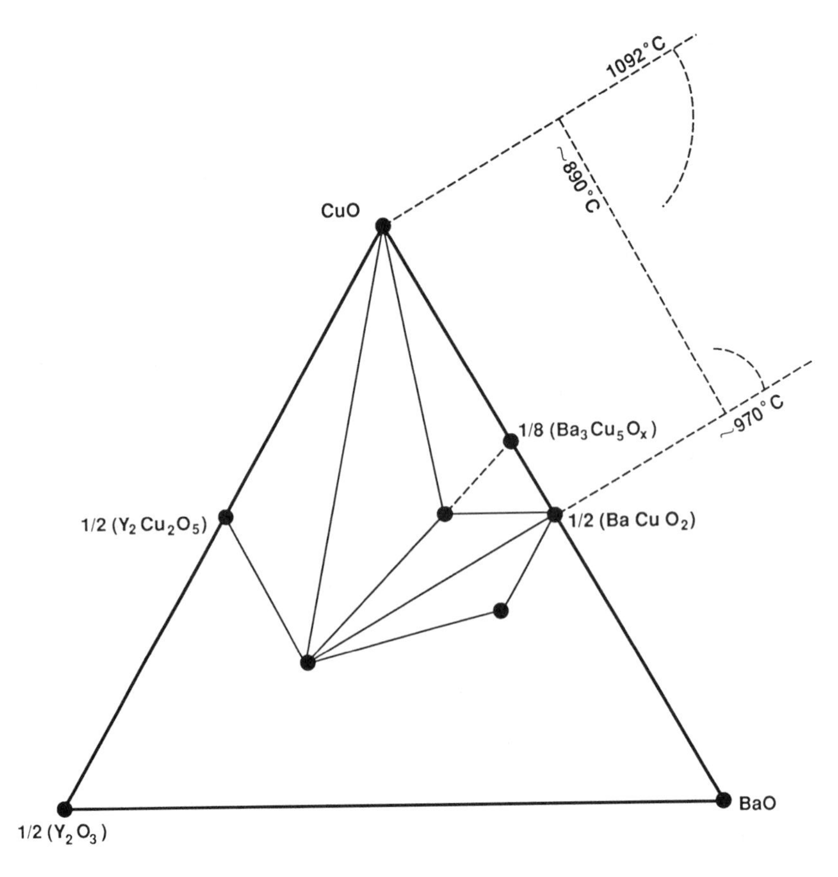

Fig. 6.5. Isothermal cut at 950°C of the copper-rich portion of the CuO-BaO-$YO_{1.5}$ phase diagram. See reference [72] for details.

of Fig. 6.5. In discussing synthesis methods, we make reference to these partial phase diagrams for this multicomponent system. There may be subtle, but important differences when other rare earth elements are substituted for yttrium.

From Fig. 6.6, we note that $YBa_2Cu_3O_7$ is not a congruently melting compound. This has significant implications for single crystal growth as discussed later. In fact, formation of this compound occurs at about 1020°C from the reaction of Y_2BaCuO_5 and a melt of approximate composition $Ba_3Cu_5O_y$. This peritectic decomposition of the superconducting phase into Y_2BaCuO_5, which melts above 1150°C, and a liquid of $BaCuO_2$ and CuO affects superconducting properties such as J_c. If this liquid phase is retained during synthesis, large grains of $YBa_2Cu_3O_x$ may be grown. However, deviation from the 1:2:3:7 stoichiometry in the Y-poor

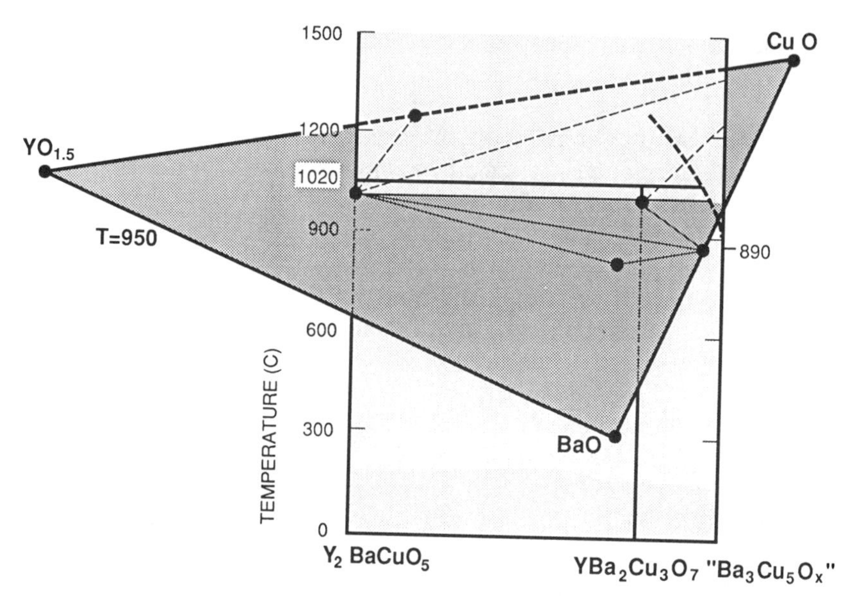

Fig. 6.6. Pseudobinary phase diagram along the Y_2BaCuO_5 - to - "$Ba_3Cu_5O_x$" cut of the isothermal plane shown in Figure 5. The 950°C isothermal plane from Figure 5 is shown in perspective. Provided courtesy of R. W. McCallum.

direction (away from $YO_{1.5}$ in Fig. 6.5) results in an excess of the eutectic liquid $Ba_3Cu_5O_y$. Upon solidification at 890°C, this phase coats the grain boundaries of the $YBa_2Cu_3O_x$ particles with detrimental effects on the conducting properties and J_c. An excess of the semiconducting phase Y_2BaCuO_5 is also not desirable under normal conditions. Therefore, stoichiometry and control of synthesis temperatures are paramount in achieving superconducting material with a microstructure optimized for J_c and H_{c2} as well as T_c.

For bulk materials, the solid state reaction to form $YBa_2Cu_3O_x$ requires a thorough mixing of the constituents to provide spatial uniformity. This combination is achieved by either mechanical grinding of the powders in an agate mortar for extended times (30-60 minutes), coprecipitation of the powders, or various sol-gel processes. Since the uniformity of particle size may be an important consideration for some applications of this superconductor, [73] the mixing technique should be matched to the desired use of the final material. After mixing, the powders are pressed into the desired shape, placed into a suitable crucible (typically platinum or alumina), and subjected to a program of heat treatments consistent with the phase diagrams in Figs. 5 and 6.

Successful formation of the "123" phase as defined by attaining a T_c above 90 K has been reported for a wide variety of heat treatments [74]. Many early reports were concerned only with achieving these high values of T_c; however, the details of sublattice order, oxygen distribution, and microstructure depend sensitively on the details of the steps of synthesis [75-79]. For the initial formation of small quantities (<500 grams) a typical heat treatment program for the mixed powders is: (1) react in air at 890°C for 36 hours, (2) regrind and repeat step 1, (3) regrind, press desired pellets, and heat in flowing O_2 according to: 950°C for 60 hours, cool to 500°C and hold for 6 hours, cool at 2.5°C/min. to room temperature. For larger amounts of material, step 2 may have to be repeated to ensure complete conversion to the $YBa_2Cu_3O_x$ phase. The specific times and temperatures of this process vary slightly depending on the research group. This method produces solid pellets of about 85% theoretical density. Typical impurities are below the 2% level and may consist of CuO, $BaCuO_2$, and Y_2BaCuO_5.

For all processing of these superconducting materials, the final oxygen atmosphere treatment is intimately connected to the superconducting properties. Deviations from ideal oxygen stoichiometry (x = 7) influence the copper valence and critically affect all physical properties, including the orthorhombic-tetragonal phase transition. The strong association of superconductivity with the copper-oxygen planes and chains connects the details of oxygen sublattice occupation with superconductivity. It is important to note that total oxygen concentration alone cannot describe adequately the variation of the properties of these materials. Rather, the relative sublattice occupancies must be considered, and these values are affected by synthesis techniques [75,77,80-84]. As an example of the importance of the synthesis process, we note that the sublattice distribution of oxygen for samples prepared in a manner similar to the method mentioned above is well documented by in situ neutron powder diffraction experiments [75]. Synthesis temperature and oxygen atmosphere both affect directly the final oxygen distribution and thus the superconductivity. Bulk samples prepared in this fashion yield a monotonically decreasing T_c as oxygen is removed. Using this high temperature quench technique, it is difficult to maintain the orthorhombic structure below about x = 6.5. In contrast, a Zr-gettering technique for removal of oxygen at lower temperatures [77,85] preserves the orthorhombic structure down to an oxygen content of about x = 6.3, and results in a step-like decrease of T_c with a plateau of T_c = 60 K near x = 6.6. This lower T_c region has been associated with an ordering of the oxygen vacancies [77,80,86]. In the final analysis, the details of sample preparation are crucial to even the most basic superconducting property, the transition temperature.

Extensive substitutions have been reported for virtually every element in the prototypical compound $YBa_2Cu_3O_x$ [87-93]. In fact, it was this substitutional chemistry that led from the original report on the Ba-La-Cu-O system [1] to the family of 90 K superconductors based on $YBa_2Cu_3O_x$ [6]. The Bi- and Tl-containing oxides with T_c's in excess of 100 K were discovered in a similar fashion [94-100]. Focusing on the $YBa_2Cu_3O_x$ material, the most successful substitution consists of replacing Y by other rare earth elements. With the exception of Sc, Ce, Pr, and Tb, all of the rare earths can completely substitute for Y and retain the high temperature superconductivity [101,102]. The radioactive rare earth element Pm has not been investigated. In the case of Sc, the Sc^{3+} ion is too small to stabilize the structure. Substitution of Tb yields a superconducting compound, but with the relatively low T_c of 35 K [68]. For Ce and Pr, no superconductivity is observed. Unlike the majority of rare earths which form the sesquioxide R_2O_3, (R = rare earth), the three elements Ce, Pr, and Tb form oxides that indicate a higher formal valence state for the rare earth; namely, CeO_2, Pr_6O_{11}, and Tb_4O_7. The valence state of the copper plays a crucial role within the framework of many theories of superconductivity in these multinary oxides, for example, the RVB model [49,50]. Thus for the Ce-, Pr-, and Tb-containing compounds, the superconductivity may be weakened or destroyed through electronic effects originating with the valence state of the rare earth ion. This important aspect of the trivalent nature of the rare earths was recognized experimentally at an early stage. Substitutions of other nominally trivalent ions such as Al and Tl and In were made, but in the $YBa_2Cu_3O_x$-type structure these elements do not raise T_c [88,103]. This work with Tl presaged the discovery of the Tl-containing compounds where T_c values exceed 100 K [97-100].

Partial substitutions have been reported for the alkaline earth Ba and the transition metal Cu [89]. These experiments result in a constant or declining transition temperature. The most common substitutions for Ba are other alkaline earths and some alkali metals [81]. Because Ba occupies one unique sublattice in the structure [see Chapter 4], most authors assume partial substitutions occur statistically on this sublattice. Successful substitution of La on the Ba sites has been reported for the system $La(Ba_{2-y}La_y)Cu_3O_x$ [84,104]. This experiment offers an additional method for altering the formal electron count in the unit cell of these materials. Numerous transition metal substitutions for Cu have been attempted, for example, Ti, V, Cr, Mn, Fe, Co, Ni, Zn [103,105,106]; however, since Cu occupies two distinct sublattices in the orthorhombic unit cell [see Chapter 4], there may be preferential site occupancy depending on the nature of the substitute element. Other metals such as Mg, Al, Ga, and Sn also have been used to replace some of the Cu atoms in $YBa_2Cu_3O_x$

[103,105,106]. In every instance, careful experimentation is necessary to determine the degree of successful substitution for Cu by elements with diverse chemical and electronic characteristics.

Exciting results were reported for elemental substitution on the oxygen sublattices. Experimenters replacing oxygen with isoelectronic sulfur atoms claim an increase in T_c up to 108 K [93]. This increase is attributed to possible effects by sulfur on the crystallographic stability. In one instance, the partial substitution of fluorine for oxygen resulted in a reported increase in T_c up to 150 K [107]. Both the sulfur and fluorine experiments await widespread confirmation. The interest in chemically adjusting the oxygen sublattices for optimal superconducting properties is based on theoretical and experimental considerations that emphasize the importance of the Cu-O planes to the occurrence of high T_c superconductivity.

The preparation of high quality single crystals is complicated by two factors; namely, the peritectic reaction that forms the compound $YBa_2Cu_3O_x$ and the strong tendency to microtwinning in the a-b plane [108-120]. Examining the latter problem first, we note that virtually every crystal growth technique successful in producing crystals of millimeter dimension yields crystals with twin planes. The pattern of microtwinning in individual grains varies with temperature. Typical spacing of twin planes is on the order of 30 nm. Whatever the origins of this twinning, the effect results in an ambiguity between the a-axis and b-axis of the crystals. With heavily twinned crystals, one cannot distinguish properties along the a-axis from those along the b-axis. This degeneracy due to synthesis deficiencies does not affect properties in the c-direction, since crystal growth selects this unique axis quite decisively.

The lack of congruent melting to form $YBa_2Cu_3O_x$ presents a problem for the formation of large single crystals [72,119,120]. Referring to the partial phase diagram in Fig. 6.5, the eutectic liquid comprised of CuO and $BaCuO_2$ (and possibly some Y_2O_3) is often used as a flux for crystal growth. Techniques based on this flux are most common in the literature [108-120] and result in crystals of millimeter linear dimensions. One basic approach is to presynthesize polycrystalline powder of the superconducting oxide, add the flux material in a ratio of 3:1, flux:compound, and subject the mixture to a series of heat treatments in air and oxygen atmospheres [121]. For this technique there is a significant problem in containing the mixture of flux plus oxide powder, due to the reactive nature of the flux. A great variety of crucible materials has been tested, none of which remains completely unreacted with the mixture at all temperatures of the synthesis process [121]. Many authors recommend using gold for the immediate contact with the charge. Successful crystal growth also has been reported in platinum and alumina crucibles [108-120], the latter representing a considerable financial savings. Alternative flux materials have been reported depending on the rare earth constituent of the superconducting compound. Two examples of other fluxes used in successful crystal growth are: (1) CuO [112] and (2) a mix of bismuth, barium and copper oxides [122]. Independent of the technique employed, the resulting crystals must be subjected to a final recovery stage of oxygen treatment in order to achieve transition temperatures in excess of 90 K.

Synthesis of high quality single crystals has played an integral role in the determination of the anisotropy of superconducting and normal state properties of bulk high T_c oxides [118,123-127]. Measurements of magnetization versus applied field demonstrate clearly the importance of crystal or grain orientation to field-dependent superconducting properties. Upper critical field (H_{c2}) data are sensitive to this anisotropy [126,127]. Experiments on oriented thin films also have quantified the degree of anisotropy present in these materials, especially with respect to critical current densities [128-130]. The degree of anisotropy present in the $YBa_2Cu_3O_x$-type superconductors is greater than that found for the K_2NiF_4-type materials. Superconducting oxides containing Bi or Tl with T_c in excess of 100 K show an even greater degree of anisotropy as determined from lattice parameters and single crystal experiments [96,98,131,132]. The rise in T_c with increased crystallographic anisotropy focuses attention on a possible connection between anisotropy as represented by Cu-O layers and the magnitude of the transition temperature.

As observed for superconducting oxides with the K_2NiF_4-type structure, the family of superconductors based on $YBa_2Cu_3O_x$ displays a diverse range of magnetic and transport properties. In this chapter, we focus on magnetic field based experiments including the Meissner effect, magnetization data, magnetic susceptibility, and critical magnetic fields. When possible, data on single crystal samples are discussed in order to illustrate the anisotropic nature of superconductivity in these oxides. Background material is also provided on the transport property of critical current densities. The effects on the magnetic and superconducting properties of varying external pressure is represented by a few selected experimental results. Although this family of compounds has been discovered very recently, the volume of research on them grows at such a rapid pace that only a small fraction of the experimental results can be presented directly.

The response of a collection of single crystals of $YBa_2Cu_3O_x$ to an external magnetic field is shown in Fig. 6.7. These data, which show an onset T_c of approximately 90 K, are a typical example of both the Meissner effect (+) and screening supercurrents (o) that results in a diamagnetic signal in the superconducting state. When the sample is cooled in zero applied field (o), then measured with increasing temperature, the diamagnetic signal is essentially 100%, even after accounting for demagnetization effects. The Meissner flux expulsion expressed as a percentage of this diamagnetic shielding is smaller, but still significant at about 15% for these crystals. This reduced effect is common among the high T_c oxides. It is associated with flux trapping within the sample which originates from defects such as twin planes, inclusions, or slight variations in oxygen concentration. The superconducting oxides are particularly susceptible to flux penetration since values of the lower critical field H_{c1} are on the order of 10 mT.

Although H_{c1} is small (≤ 10 mT), the upper critical field H_{c2} of $YBa_2Cu_3O_x$ exceeds that of all previous superconductors, with estimates of $H_{c2}(0)$ as high as 340 T [133,134]. An example of upper critical field data as a function of temperature taken from Ref.

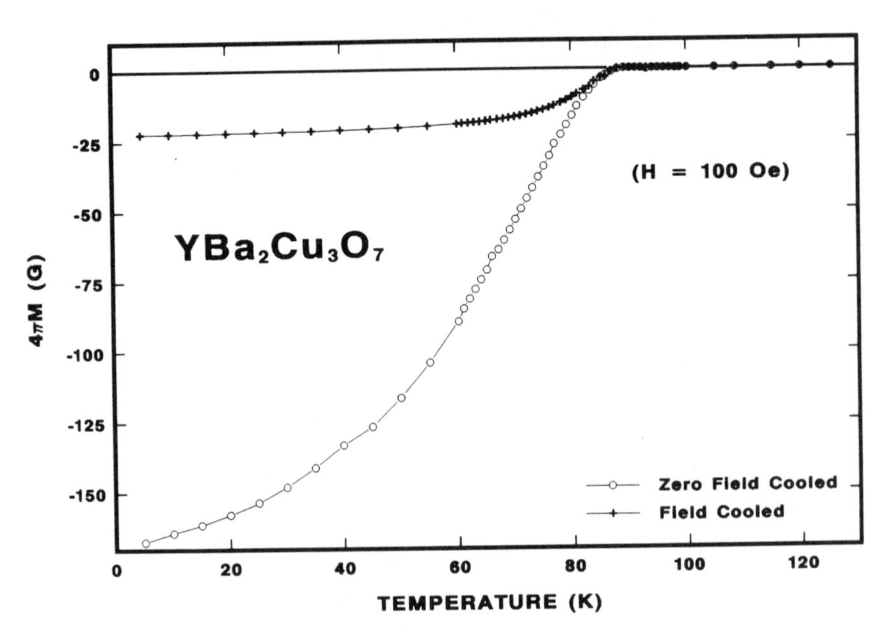

Fig. 6.7. Magnetization versus temperature for single crystals of $YBa_2Cu_3O_x$. For an applied field of 100 Oe, both dc screening (open circles) and Meissner effect (crosses) curves are shown. See reference [124] for details.

[133] is shown in Fig. 6.8, where H_{c2} is determined from the onset and midpoint of the curve of the resistive transition into the superconducting state. Superconducting transitions measured resistively broaden significantly when large external fields are applied. (See Fig. 6.9 below) Therefore, the determination of H_{c2} depends sensitively on the choice of T_c using either the onset or midpoint of the transition. For either choice, values of H_{c2} are sufficiently large that only the initial slope of $-dH_{c2}/dT$ near T_c can be measured even at the highest field facilities in the world. A typical value of $-dH_{c2}/dT)_{T=Tc}$ is 3 T/K. Measurements of H_{c2} on single crystals [135,136] reveal a high degree of anisotropy and are similar to the magnetization data of Fig. 6.8. When the applied field is parallel to the crystallographic c-axis, the slope

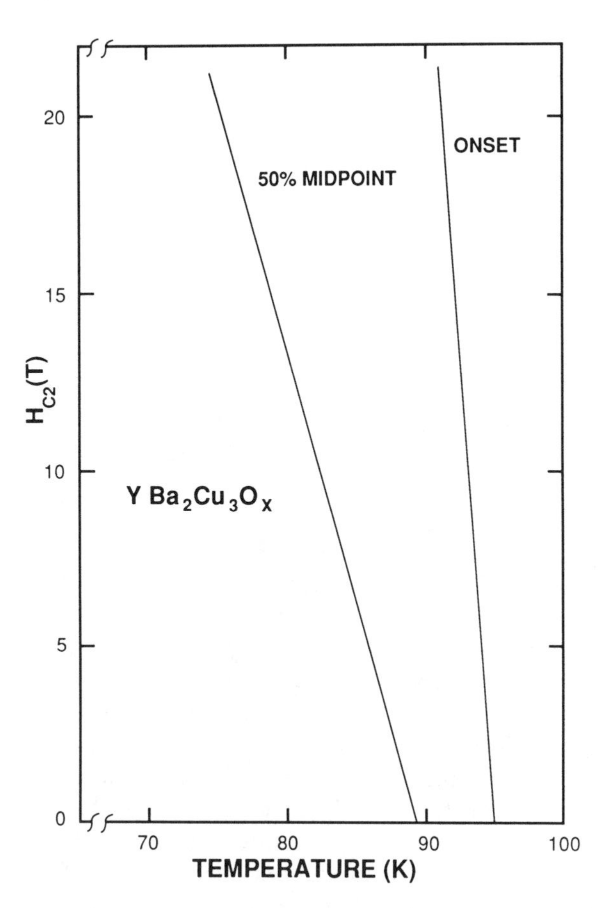

Fig. 6.8. Upper critical field H_{c2} versus temperature for superconducting $YBa_2Cu_3O_x$. Values for H_{c2} are determined from both onsets and midpoints of the resistive transition. See reference [133] for details.

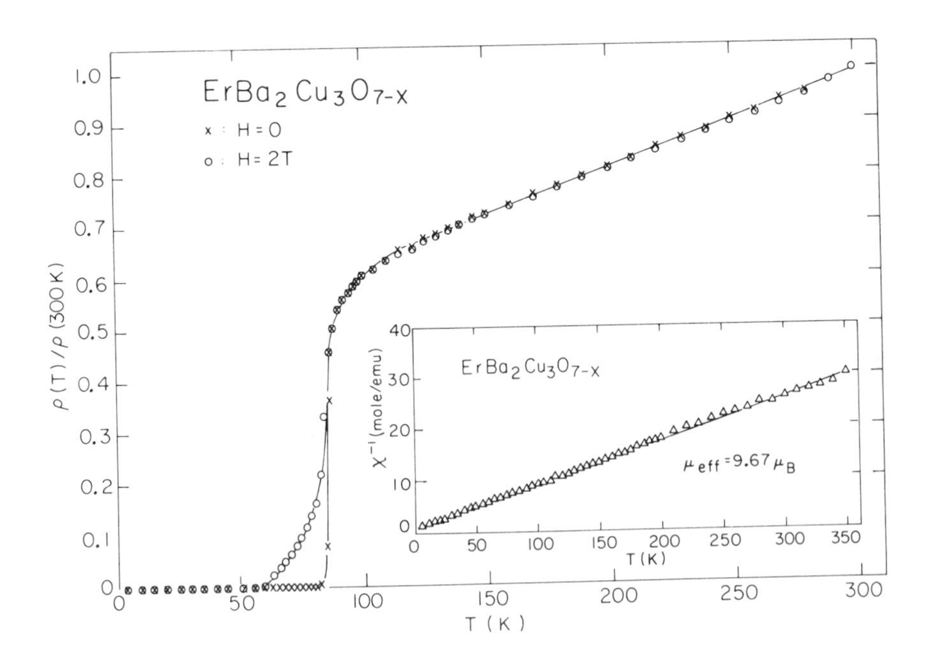

Fig. 6.9. Normalized resistivity versus temperature measured at zero and 2 Tesla applied external fields. Inverse molar magnetic susceptibility versus temperature determined in a 2 Tesla magnetic field is plotted in the inset. See reference [141] for details.

$-dH_{c2}/dT)_{T=Tc}$ is reduced to about 0.6 T/K. With the external field applied in the basal plane, the slope is comparable to that measured for polycrystalline specimens, namely 3 T/K. These experiments yield an estimate of the coherence length in the c-direction of ξ = 0.5 nm and in the basal plane of ξ = 3 nm. Coherence length estimates on other compounds in the series $MBa_2Cu_3O_x$ yield similar values [126]. This difference in the two ξ values is yet another manifestation of the anisotropic nature of these materials.

Using experimental data for the slope and the standard theory [137-139] with no Pauli-spin pair breaking, estimates of $H_{c2}(0)$ can be made [140]. For a series of eight compounds $MBa_2Cu_3O_x$ (R = Nd, Eu, Gd, Dy, Ho, Er, Tm, and Y), extrapolated values of $H_{c2}(0)$ based on the midpoint transitions are found to lie in the range (160 ± 20) T [134]. This is an enormous number. Even at the boiling point of liquid nitrogen of 77 K, which is an optimal temperature for large scale applications, H_{c2} exceeds 20 T. The constancy of these results across the rare earth series indicates that the magnetic ions have minimal interaction with the conduction electrons in this family of materials. This decoupling of the strength of the superconductive interactions and the effects of the magnetic rare earth ions is evidenced by the relatively constant T_c values across the rare earth series for $MBa_2Cu_3O_x$ compounds. Magnetic susceptibility data on $ErBa_2Cu_3O_x$ discussed below also reflect this fact. This independent behavior is possible since the rare earth ions are far from the Cu-O planes and chains (see Chapter 4), which are thought to be responsible for the high transition temperatures.

Further experimental evidence separating any magnetic tendencies of the rare earth ion from the occurrence of superconductivity in the $MBa_2Cu_3O_x$ compounds is provided by temperature-dependent magnetic susceptibility measurements. Using the Er-containing compound as an example, the insert to Fig. 6.9 shows the inverse molar magnetic susceptibility between 5 K and 350 K in a constant measuring field of 2 T. This compound becomes superconducting at 84 K as observed by resistivity measurements in Fig. 6.9 and low field dc magnetization experiments [141]. Nevertheless, the susceptibility data may be fitted to a Curie-Weiss law, $\chi_m = \chi_o + C/(T + \Theta)$, over the entire temperature range. The corresponding effective magnetic moment $\mu_{eff} = 9.67\ \mu_B$ is in good agreement with the Hund's rule ground state for the free ion Er^{3+}. The small, negative value of θ = -10.8 K is consistent with the low temperature antiferromagnetism of this compound [142]. Most interesting is the smooth continuation of the Curie-Weiss behavior through the superconducting transition. In general, when the applied field (2 T) is much smaller than the upper critical field H_{c2} (10^2 T), a diamagnetic susceptibility is observed below T_c. This holds true at low applied fields; however, for $ErBa_2Cu_3O_x$ and other isostructural rare-earth containing oxides, at high fields the induced paramagnetism from the magnetic rare-earth ions masks the superconducting onset [143-148]. This observation implies that the occurrence of superconductivity and the magnetic field induced paramagnetism in the Er sublattice are independent. Therefore, when using the magnetization versus field curves to determine the critical field, one must be cautious in defining T_c since compounds such as $ErBa_2Cu_3O_x$ possess a significant paramagnetic signal at high fields. With respect to upper critical field determination, the resistivity data of Fig. 6.9 also illustrate the importance of stating clearly how T_c is defined in a large applied field. The onset and midpoint of the transition temperature are affected very little by an applied magnetic field of 2 T, whereas, the zero resistance temperature shifts from 84 K to 60 K. Calculation of $-dH_{c2}/dT)_{T=Tc}$ depends sensitively on the choice of the transition temperatures using the criterion of 90%, 50%, or 10% of a full resistive transition.

A strong contribution to the magnetic susceptibility from the rare-earth element is eliminated for $YBa_2Cu_3O_x$ since yttrium has no localized 4f electrons. However, even for $YBa_2Cu_3O_x$, Curie-Weiss contributions to the susceptibility are observed with effective moments ranging up to 1 μ_B. This paramagnetic behavior is attributed to the copper atoms. The local moment formation occurs in oxygen-deficient materials and may be associated with particular arrangements of oxygen vacancies [80,86,145]. This relationship in $YBa_2Cu_3O_x$ between oxygen vacancies and the magnetic behavior of the copper atoms is analogous to the one described for La_2CuO_{4-y} earlier in this chapter. The common factor is the oxidation state of the copper ions. Thus synthesis techniques and thorough characterization of samples are crucial to complete understanding of the magnetic behavior of these oxides. A large (0.5 μ_B/Cu atom) moment assigned to the copper sites based on a normal state Curie Weiss law may actually be due to the presence of secondary phases. Indeed, well-characterized, single phase samples yield Curie constants that are reduced by an order of magnitude over early results [145,149,150].

While superconducting transition temperatures above the boiling point of liquid nitrogen (77 K) and upper critical magnetic fields of useful strength (>20 T) at 77 K are readily achieved in the $MBa_2Cu_3O_x$ compounds, raising the critical current density (J_c) at 77 K to practical levels poses metallurgical and microstructural challenges [151-153]. Some valuable expertise comes from working with traditional superconducting compounds such as Nb_3Sn that are brittle. Experience in powder metallurgical processing of Chevrel-phase superconducting wires provides further guidance for the ceramic oxides [154,155]. Nevertheless, the unique kinetics of phase formation, the granular nature of the resulting material, the crucial role of oxygen content, and the necessity of powder processing all combine to make the production of high J_c wires a formidable challenge for chemists, materials scientists, and physicists.

Critical current densities in excess of 10^3 A/cm^2 have been reported for bulk samples of $YBa_2Cu_3O_x$ [151-153]. This is far below values achieved in oriented thin films where J_c up to 10^7 A/cm^2 have been measured at 4.2 K [156]. A key experiment in understanding the origins of this difference is an inductive measuring technique that determines the critical current density within a single grain of a bulk powder sample [151,157,158]. This intragrain J_c has been determined for bulk powder samples to be in excess of 10^6 A/cm^2 at 4 K and low applied fields [159]. The field dependence of J_c, both intra- and intergrain, shows an exponential sensitivity to applied field at low fields [159]. This exponential behavior is not observed in single crystals [123]. Two factors could account for this dependence and the difference between inter- and intra-granular J_c values. First, the presence of secondary impurity phases at the grain boundaries would create intergranular weak links so the high J_c grains are Josephson coupled. Critical current densities in these Josephson-coupled grains may be suppressed by fields on the order of a few mT [160]. Second, a mismatch between two grains with different crystallographic orientations would lead to a strong applied field dependence through the extreme anisotropy of H_{c2}. Thus, the problem of attaining large bulk (intergranular) J_c values focuses on intergranular weak links, and grain orientation and alignment. In all likelihood, a combination of these two factors governs the ultimate attainable value of J_c. The role of twin boundaries serving as pinning centers is also important. In fact, since single crystals show high J_c values with no voids, grain boundaries, or normal-state inclusions, twin boundaries may provide the dominant mechanism for pinning.

High pressure studies on $YBa_2Cu_3O_x$ and isostructural superconductors have been done over a wide range of pressures using a variety of pressure transmitting media [28,51,161-169]. Early work on dT_c/dp of multiphase samples showed small values for the pressure derivative, with both positive and negative signs reported [162,164]. This ambiguity was resolved to give a positive value on the order of dT_c/dp = 0.1 K/kbar for single phase material with oxygen content sufficiently high to produce an ambient pressure T_c above 90 K. This quantitative value of dT_c/dp is consistent with lower pressure (to 8 kbar) experiments using He gas as the medium [165], intermediate range pressures (to 20 kbar) obtained with liquid media [163,166], and higher pressure studies (to 149 kbar) using a Bridgman anvil device [169]. Recall from earlier material in this chapter, that dT_c/dp values for the K_2NiF_4-type superconducting oxides also have positive pressure derivatives, but dT_c/dp is an order of magnitude larger than that observed for $YBa_2Cu_3O_x$. Pressure enhanced superconductivity has been observed in traditional materials such as V-based A15 compounds [170], Chevrel phases, [171] and Th_3P_4-type chalcogenides [172,173]. For the latter two classes of compounds, dT_c/dp values can exceed those observed for $YBa_2Cu_3O_x$; however, they remain less than the pressure sensitivity of the K_2NiF_4-type compounds. These large positive pressure derivatives suggest a similar mechanism for superconductivity in these high T_c oxides. An in-depth discussion of early pressure studies within the framework of various theoretical models favors three approaches (two-dimensional BCS, RVB, and bipolaron models) for accounting for the strong positive pressure effect on T_c [59]. High pressure studies of the oxide superconductors provide one further experimental characterization to test model descriptions of these high T_c materials.

6.4. Superconductivity Above 100 K

Since the discovery in 1986 and 1987 of the two families of superconducting oxides typified by $La_{1.85}Sr_{0.15}CuO_4$ and $YBa_2Cu_3O_7$, respectively, the level of activity in superconductivity research has been extraordinary. Nevertheless, there were no reproducible measurements of superconductivity at temperatures above 100 K until early in 1988 when two new classes of multinary oxides were discovered [94-98,131,132,174-185]. In each case, chemical substitution of the rare element played a key role in the discovery. A second important experimental step was that researchers extended their searches to include systems with four metallic elements plus oxygen. This added complexity proved crucial, since earlier work on both the Bi-containing [94] and Tl-containing [174] systems found maximum transition temperatures of about 20 K when only three metals were present in the structure. In the closing section of this chapter, we describe briefly the synthesis and bulk superconducting properties of two classes of superconductors with T_c values above 100 K.

In the system Bi-Sr-Ca-Cu-O, two superconducting transitions have been reported [95,96,98,131,132,177-180,184]. The atomic ratios yield a chemical formula of $Bi_2Sr_2CaCu_2O_y$ where $y = 8$ [96,98]. The relative proportion of the two alkaline earths may be varied, so the general formula is often written as $Bi_2Sr_{3-x}Ca_xCu_2O_y$. While specific values for T_c vary slightly depending on laboratory and measurement technique, the two values of 84 K and 110 K are representative. The phase with the lower T_c is dominant in most samples synthesized to date. The higher T_c phase is more difficult to stabilize and so far occurs only in the presence of the phase with the 84 K transition. There is some speculation that the 110 K phase may form in grain boundaries between the lower T_c material. It may be possible to relate the two phases by a modulation of the structure along one crystallographic axis [see Chapter 4]. This modulation could be either commensurate or incommensurate in nature.

Synthesis techniques center around powder metallurgical processing of the binary oxides, carbonates, and/or nitrates, for example, Bi_2O_3, CuO, $CaCO_3$, $SrCO_3$, $Sr(NO_3)_2$. The maximum synthesis temperature seems to be crucial to successful formation of the superconducting phase with the 110 K transition. Depending on crucible environment and ratios of the starting materials, a critical temperature near 860 °C exists. Since the phase decomposition temperature lies slightly above this critical value, precise and reproducible temperature control is necessary. Single crystal growth has been achieved [96,98,179,180,184]. Flux techniques are most commonly employed with success reported for oxide fluxes [96,180] and alkali chloride fluxes [184].

In the system of elements Tl-Ba-Ca-Cu-O, at least three distinct superconducting phases with T_c above 80 K have been reported [97,98,175,176,185]. Crystallographically these materials are related to the Bi-containing compounds, utilizing the basic perovskite building unit [see Chapter 4]. While the extent of oxygen vacancies and sublattice disorder are still being determined, the general formulae may be represented as $(TlO)_2Ba_2Ca_{n-1}Cu_nO_{2+2n}$, where n = 1,2,3. Reported transition temperatures vary slightly depending on the method of measurement; however, as the integer n increases, T_c increases. Typical values are: (n = 1), $Tl_2Ba_2CuO_6$, T_c = 85 K; (n = 2), $Tl_2Ba_2CaCu_2O_8$, T_c = 110 K; (n = 3), $Tl_2Ba_2Ca_2Cu_3O_{10}$, T_c = 125 K. Interestingly, as the value of n grows, so does the number of layers stacked along the c-direction in the unit cell. Thus anisotropy is expected to be a major factor in all normal state and superconducting properties of these materials.

Early synthesis techniques center around powder metallurgical processing of the binary oxides, carbonates, and/or nitrates. The most common starting constituents are: Tl_2O_3, CuO, $CaCO_3$, $BaCO_3$, or $Ba(NO_3)_2$. The relatively low melting point of Tl_2O_3 (717 C) and the toxic nature of Tl pose added difficulties in working with these materials. To date, powder metallurgical methods have been employed most successfully to fabricate polycrystalline specimens [97,98]. Single crystal growth occurs with an excess of CuO serving as a flux. In one growth technique, the crystals are grown in a sealed gold tube to control the relative composition [176,185].

The significance of the discoveries of superconductivity above 100 K in the two systems described above cannot be overstated. Most straightforwardly, the increase in the maximum T_c is important to potential applications of superconducting materials. Superconducting devices should be operated at a temperature well below T_c for example, at $\frac{1}{2}T_c$ to $\frac{3}{4}T_c$. The reality of transition temperatures between 1.4 to 1.6 times the boiling point of liquid nitrogen (77 K) makes this possible. Perhaps even more noteworthy, although more subtle, is the simple fact that an additional, unique family of superconducting oxides with T_c's in excess of the liquid nitrogen temperature exists. To be sure, these Bi- and Tl-containing multinary oxides share some common features with the $YBa_2Cu_3O_x$ family, such as the Cu-O planar structure which is often linked to the occurrence of high transition temperatures and significant anisotropy. Nevertheless, the latest oxides with Bi and Tl are different in their constituents, crystallographic stacking, crystal growth, and of course maximum T_c values. The existence of more than one family of high T_c multinary oxides offers indirect evidence and hope for the discovery of additional families of high transitional temperature superconductors.

References

[1] J.G. Bednorz and K.A. Müller, *Z. Phys.* B **64** (1986) 189.

[2] C.W. Chu, P.H. Hor, R.L. Meng, L. Gao, Z.J. Huang, and Y.Q. Wang, *Phys. Rev. Lett.* **58** (1987) 405.

[3] R.J. Cava, R.B. van Dover, B. Batlogg, and E.A. Rietman, *Phys. Rev. Lett.* **58** (1987) 408.

[4] S. Uchida, H. Takagi, K. Kitazawa, and S. Tanaka, *Japan. J. Appl. Phys.* **26** (1987) L1.

[5] Z.X. Zhao, L.Q. Chen, C.G. Cui, Y.Z. Huang, J.X. Liu, G.H. Chen, S.L. Li, S.Q. Guo, and Y.Y. He, *Kexue Tongbao* **32** (1987) 522.

[6] M.K. Wu, J.R. Ashburn, C.J. Torng, P.H. Hor, R.L. Meng, L. Gao, Z.J. Huang, Y.Q. Wang, and C.W. Chu, *Phys. Rev. Lett.* **58** (1987) 908.

[7] A.R. Moodenbaugh, M. Suenaga, T. Asano, R.N. Shelton, H.C. Ku, R.W. McCallum, and P. Klavins, *Phys. Rev. Lett.* **58** (1987) 1885.

[8] L.C. Bourne, M.L. Cohen, W.N. Creager, M.F. Crommie, A.M. Stacy, and A. Zettl, *Phys. Lett.* A**120** (1987) 494.

[9] Z.X. Zhao, L.Q. Chen, Q.S. Yang, Y.Z. Huang, G.H. Chen, R.M. Tang, G.R. Liu, G.C. Cui, L. Chen, L.Z. Wang, S.Q. Guo, S.L. Li, and J.Z. Bi, *Kexue Tongbao* **33** (1987) 661.

[10] S. Hikami, T. Hirai, and S. Kagoshima, *Japan. J. Appl. Phys.* **26** (1987) L314.

[11] R.N. Shelton, in *Superconductivity in d- and f-Band Metals*, edited by W. Buckel and W. Weber (Kernforschungszentrum, Karlsruhe, FRG 1982) pp. 123-131.

[12] B.T. Matthias, M. Marezio, E. Corenzwit, A.S. Cooper, and H.E. Barz, *Science* **175** (1972) 1465.

[13] O. Fischer, *Appl. Phys.* **16** (1978) 1.

[14] K. Yvon, in *Current Topics in Materials Science*, Vol. 3, edited by E. Kaldis (North-Holland, Amsterdam, 1979) p. 53.

[15] H. Barz, H.C. Ku, G.P. Meisner, Z. Fisk, and B.T. Matthias, *Proc. Natl. Acadm. Sci.* USA **77** (1980) 3132.

[16] D.C. Johnston, H. Prakash, W.H. Zachariasen, and R. Viswanathan, *Mater. Res. Bull.* **8** (1973) 777.

[17] A.W. Sleight, J.L. Gibson, and F.E. Bieldstedt, *Solid State Commun.* **17** (1975) 27.

[18] J.M. Tarascon, L.H. Green, W.R. McKinnon, G.W. Hull, and T.H. Geballe, *Science* **235** (1987) 1373.

[19] P.J. Picone, H.P. Jenssen, and D.R. Gabbe, *J. Cryst. Growth*

85 (1987) 576.

[20] Y. Hidaka, Y. Enomoto, M. Suzuki, M. Oda, and T. Murakami, *J. Cryst. Growth* **85** (1987) 581.

[21] Y. Hidaka, Y. Enomoto, M. Suzuki, M. Osa, and T. Murakami, Japan. J. Appl. Phys. **26** (1987) L377.

[22] P.M. Grant, S.S.P. Parkin, V.Y. Lee, E.M. Engler, M.L. Ramirez, J.E. Vazquez, G. Lim, R.D. Jacowitz, and R.L. Greene, *Phys. Rev. Lett.* **58** (1987) 2482.

[23] D. Vaknin, S.K. Sinha, D.E. Moncton, D.C. Johnston, J.M. Newsam, C.R. Safinya, and H.E. King, Jr., *Phys. Rev. Lett.* **58** (1987) 2802.

[24] J.I. Budnick, A. Golnik, C. Niedermayer, E. Recknagel, M. Rossmanith, A. Weidinger, B. Chamberland, M. Filipkowski, and D.P. Yang, *Phys. Lett.* **A124** (1987) 103.

[25] S. Uchida, H. Takagi, H. Yanagisawa, K. Kishio, K. Kitazawa, K. Fueki, and S. Tanaka, *Japan. J. Appl. Phys.* **26** (1987) L445.

[26] Y. Yamaguchi, H. Yamauchi, M. Ohashi, H. Yamamoto, N. Shimoda, M. Kikuchi, and Y. Syono, *Japan. J. Appl. Phys.* **26** (1987) L447.

[27] J.M. Tarascon, P. Barboux, B.G. Bagley, L.H. Greene, W.R. McKinnon, and G.W. Hull, in *Chemistry of High-temperature Superconductors*, edited by D.L. Nelson, M.S. Whittingham, and T.F. George (American Chemical Society Symposium Series **351**, 1987) pp. 198-210.

[28] K. Kishio, K. Kitazawa, S. Kanbe, I. Yasuda, N. Sugii, H. Takagi, S. Uchida, K. Fueki, and S. Tanaka, *Chem. Lett.* **22** (1987) 429.

[29] M.B. Maple, K.N. Yang, M.S. Torikachvili, J.M. Ferreira, J.J. Neumeier, H. Zhou, Y. Dalichaouch, and B.W. Lee, *Solid State Commun.* **63** (1987) 635.

[30] H. Maletta, A.P. Malozemoff, D.C. Cronemeyer, C.C. Tsuei, R.L. Greene, J.G. Bednorz, and K.A. Mueller, *Solid State Commun.* **62** (1987) 323.

[31] D.K. Finnemore, R.N. Shelton, J.R. Clem, R.W. McCallum, H.C. Ku, R.E. McCarley, S.C. Chen, P. Klavins, and V. Kogan, *Phys. Rev.* **B35** (1987) 5319.

[32] S. Senoussi, M. Oussena, M. Rigault, and G. Collin, *Phys. Rev.* **B36** (1987) 4003.

[33] K.A. Müller, M. Tahashige, and J.G. Bednorz, *Phys. Rev. Lett.* **58** (1987) 1143.

[34] S. Uchida, H. Takagi, S. Tanaka, K. Nakao, N. Miura, K. Kishio, K. Kitazawa, and K. Fueki, *Japan. J. Appl. Phys.* **26** (1987) L196.

[35] B. Batlogg, A.P. Ramirez, R.J. Cava, R. B. van Dover, and E. A. Rietman, *Phys. Rev.* **B35** (1987) 5340. 5340.

[36] T.P. Orlando, K.A. Delin, S. Foner, E.J. McNiff, J., J.M. Tarascon, L.H. Green, W.R. McKinnon, and G.W. Hull, *Phys. Rev.* **B35** (1987) 5347.

[37] M.S. Osofsky, W.W. Fuller, L.E. Toth, S.B. Qadri, S.H. Lawrence, R.A. Hein, D.U. Gubser, S.A. Wolf, C.S. Pande, A.K. Singh, E.F. Skelton, and B.A. Bender, in *Novel Superconductivity*, edited by S.A. Wolf and V.Z. Kresin (Plenum, New York, 1987) pp. 807-815.

[38] T.P. Orlando, E.J. McNiff, Jr., S. Foner, and M.R. Beasley, *Phys. Rev.* **B19** (1979) 4545.

[39] R. Odermatt, O. Fischer, H. Jones, and G. Bongi, *J. Phys.* C: Solid State Phys. **7** (1974) L13.

[40] S. Foner, E.J. McNiff, Jr., and E.J. Alexander, *Phys. Lett.* **A49** (1974) 269.

[41] J.M. Tarascon, L.H. Greene, B.G. Bagley, W.R. McKinnon, P. Barboux, and G.W. Hull, in *Novel Superconductivity*, edited by S.A. Wolf and V.Z. Kresin (Plenum, New York, 1987) pp. 705-724.

[42] M. Gurvitch and A.T. Fiory, *Phys. Rev. Lett.* **59** (1987) 1137.

[43] M.F. Hundley, A. Zettl, A. Stacy, and M.L. Cohen, *Phys. Rev.* **B35** (1987) 8800.

[44] Z. Fisk and G.W. Webb, *Phys. Rev. Lett.* **36** (1976) 1084.

[45] W.L. McMillan, *Phys. Rev.* **167** (1968) 331.

[46] N.P. Ong, Z.Z. Wang, J. Clayhold, J.M. Tarascon, L.H. Greene, and W.R. McKinnon, *Phys. Rev.* **B35** (1987) 8807.

[47] M.W. Shafter, T. Penney, and B. Olson, *Phys. Rev.* **B36** (1987) 4047.

[48] M.W. Shafer, T. Penney, and B. Olson, in *Novel Superconductivity*, edited by S.A. Wolf and V.Z. Kresin (Plenum, New York, 1987) pp. 771-779.

[49] P.W. Anderson, *Science* **235** (1987) 1196.

[50] P.W. Anderson, G. Baskaran, Z. Zhou, and T. Hsu, *Phys. Rev. Lett.* **58** (1987) 2790.

[51] C.W. Chu, P.H. Hor, R.L. Meng, L. Gao, and Z.J. Huang, *Science* **235** (1987) 567.

[52] M. Kurisu, H. Kadomatsu, H. Fujiwara, Y. Maeno, and T. Fugita, *Japan. J. Appl. Phys.* **26** (1987) L361.

[53] M.R. Dietrich, W.H. Fietz, J. Ecke, B. Obst, and C. Politis, *Z. Phys.* **B66** (1987) 283.

[54] R.N. Shelton, T.J. Folkerts, P. Klavins, and H.C. Ku, in *High Temperature Superconductors*, edited by D.U. Gubser and M. Schluter (Materials Research Society, Pittsburgh, 1987) pp. 59-52.

[55] R. Moret, A.I. Goldman, and A. Moodenbaugh, *Phys. Rev.* B (in press).

[56] C. Allgeier, J.S. Schilling, H.C. Ku, P. Klavins, and R.N.

Shelton, *Solid State Commun.* **64** (1987) 227.

[57] J.E. Schirber, E.L. Venturini, J.F. Kwak, D.S. Ginley, and B. Morosin, *J. Mater. Res.* **2** (1987) 421.

[58] R. Moret, J.P. Pouget, and G. Collin, Europhys. Lett. **4** (1987) 365.

[59] R. Griessen, *Phys. Rev.* **B36** (1987) 5284.

[60] N. Terada, H. Ihara, M. Hirabayashi, K. Senzaki, Y. Kimura, K. Murata, M. Tokumoto, O. Shimomura, and T. Kikegawa, *Japan. J. Appl. Phys.* **26** (1987) L510.

[61] Landolt-Boernstein, New Series, Group III, edited by K.-H. Hellwege and A.M. Hellwege (Springer-Verlag, Berlin 1966), Vol. 1.

[62] D.U. Gubser, R.A. Hein, S.H. Lawrence, M.S. Osofsky, D.J. Schrodt, L.E. Toth, and S.A. Wolf, *Phys. Rev.* **B35** (1987) 5350.

[63] R.M. Fleming, B. Batlogg, R.J. Cava, and E.A. Rietman, *Phys. Rev.* **B35** (1987) 7191.

[64] A.R. Moodenbaugh, Y. Xu, and M. Suenaga, MRS Fall Meeting, Boston, MA, Nov. 30 - Dec. 4, 1987.

[65] A.R. Moodenbaugh, Y. Xu, M. Suenaga, T.J. Folkerts, and R.N. Shelton, *Phys. Rev.* B (in press).

[66] H. Takagi, S. Uchida, K. Kitazawa, and S. Tanaka, *Japan. J. Appl. Phys.* **26** (1987) L218.

[67] E.M. Engler, V.Y. Lee, A.I. Nazzal, R.B. Beyers, G. Lim, P.M. Grant, S.S.P. Parkin, M.L. Ramirez, J.E. Vasquez, and R.J. Savoy, *J. Am. Chem. Soc.* **109** (1987) 2848.

[68] Z. Fisk, J.D. Thompson, E. Zirngiebl, J.L. Smith and S.-W. Cheong, *Solid State Commun.* **62** (1987) 743.

[69] P.H. Hor, R.L. Meng, Y.Q. Wang, L. Gao, Z.J. Huang, J. Bechtold, K. Forster, and C.W. Chu, *Phys. Rev. Lett.* **58** (1987) 1891.

[70] K.N. Yang, Y. Dalichaouch, J.M. Ferreira, B.W. Lee, J.J. Neumeier, M.S. Torikachvili, H. Zhou, M.B. Maple, and R.R. Hake, *Solid State Commun.* **63** (1987) 515.

[71] S. Hosoya, S. Shamoto, M. Onoda, and M. Sato, *Japan. J. Appl. Phys.* **26** (1987) L325.

[72] R.W. McCallum, R.N. Shelton, M.A. Noack, J.D. Verhoeven, C.A. Swenson, M.A. Damento, K.A. Gschneidner, Jr., E.D. Gibson, A.R. Moodenbaugh, in *Novel Superconductivity*, edited by S.A. Wolf and V.Z. Kresin (Plenum, New York, 1987) pp. 633-645.

[73] H. Kuepfer, I. Apfelstedt, W. Schauer, R. Fluekiger, R. Meier-Hirmer, and H. Wuehl, *Z. Phys.* **B69** (1987) 159.

[74] See for example, articles in: *Chemistry of High-Temperature Superconductors*, edited by D.L. Nelson, M.S. Whittingham, and T.F. George (American Chemical Society Symposium Series 351, 1987) and *Novel Superconductivity*, edited by S.A. Wolf and

V.Z. Kresin (Plenum, New York, 1987).

[75] J.D. Jorgensen, M.A. Beno, D.G. Hinks, L. Soderholm, K.J. Volin, R.L. Hitterman, J.D. Grace, I.K. Schuller, C.U. Segre, K. Zhang, and M.S. Kleefisch, *Phys. Rev.* **B36** (1987) 3608, and J.D. Jorgensen, B.W. Veal, W.K. Kwok, G.W. Crabtree, A. Umezawa, L.J. Nowicki, and A.P. Paulikas, *Phys. Rev.* **B36** (1987) 5731.

[76] A.C. Pastor and R.C. Pastor, *J. Cryst. Growth* **85** (1987) 652.

[77] R.J. Cava, B. Batlogg, C.H. Chen, E.A. Rietman, S.M. Zahurak, and D. Werder, *Nature* **329** (1987) 423.

[78] S. Nakahara, G.J. Fisanick, M.F. Yan, R.B. van Dover, T. Boone, and R. Moore, *J. Cryst. Growth* **85** (1987) 639.

[79] C.A. Costa, M. Ferretti, C.L. Olcese, M.R. Cimberle, C. Ferdeghini, G.L. Nicchiotti, A.S. Siri, and C. Rizzuto, *J. Cryst. Growth* **85** (1987) 623.

[80] B. Batlogg and R.J. Cava, in *Superconductivity in Highly Correlated Fermion Systems*, edited by M. Tachiki, Y. Muto, and S. Maekawa (North-Holland, Amsterdam, 1987) pp. 173-176.

[81] T. Wada, S. Adachi, O. Inoue, S. Kawashima, and T. Mirhara, *Japan. J. Appl. Phys.* **26** (1987) L1475.

[82] P. Strobel, J.J. Capponi, C. Chaillout, M. Marezio, and J.L. Tholence, *Nature* **327** (1987) 306.

[83] M. Tokumoto, H. Ihara, T. Matsubara, M. Hirabayashi, N. Terada, H. Oyanagi, K. Murata, and Y. Kimura, *Japan. J. Appl. Phys.* **26** (1987) L1565.

[84] C.U. Segre, B. Dabrowski, D.G. Hinks, K. Zhang, J.D. Jorgensen, M.A. Beno, and I.K. Schuller, *Nature* **329** (1987) 227.

[85] C. Chaillout and J.P. Remeika, *Solid State Commun.* **56** (1985) 833.

[86] R.J. Cava, B. Batlogg, C.H. Chen, E.A. Rietman, S.M. Zahurak, and D. Werder, *Phys. Rev.* B36 (1987) 5719.

[87] G. Xiao, F.H. Streitz, A. Gavrin, Y.W. Du, and C.L. Chien, *Phys. Rev.* B36 (1987) 8782.

[88] J.P. Franck, J. Jung, and M.A.-K. Mohamed, *Phys. Rev.* B36 (1987) 2308.

[89] Y. Maeno, T. Tomita, M. Kyogoku, S. Awaji, Y. Aoki, K. Hoshino, A. Minami, and T. Fujita, *Nature* **328** (1987) 512.

[90] D.N. Matthews, A. Bailey, R.A. Vaile, G.J. Russell, and K.N.R. Taylor, *Nature* **328** (1987) 786.

[91] C.Y. Huang, L.J. Dries, P.H. Hor, R.L. Meng, C.W. Chu, and R.B. Frankel, *Nature* **328** (1987) 403.

[92] I. Felner, I. Nowik, and Y. Yeshurun, *Phys. Rev.* **B36** (1987) 3923.

[93] K.N.R. Taylor, D.N. Matthews, and G.J. Russell, *J. Cryst. Growth* **85** (1987) 628.

[94] C. Michel, M. Hervieu, M.M. Borel, A. Grandin, F. Deslandes, J. Provost, and B. Raveau, *Z. Phys.* **B68** (1987) 421.
[95] H. Maeda, Y. Tanaka, M. Fukutomi, and T. Asano, *Japan. J. Appl. Phys.* **27** (1988) L209.
[96] M.A. Subramanian, C.C. Torardi, J.C. Calabrese, J. Gopalakrishnan, K.J. Morrissey, T.R. Askew, R.B. Flippen, U. Chowdhry, and A.W. Sleight, *Science* **239** (1988) 1015.
[97] Z.Z. Sheng, A.M. Hermann, A. El Ali, C. Almason, J. Estrada, T. Datta, and R.J. Matson, *Phys. Rev. Lett.* 60 (1988) 937, Z.Z. Sheng and A.M. Hermann, Nature 332 (1988) 55, and Z.Z. Sheng and A.M. Hermann, *Nature* **332** (1988) 138.
[98] R.M. Hazen, L.W. Finger, R.J. Angel, C.T. Prewitt, N.L. Ross, C.G. Hadidiacos, P.J. Heaney, D.R. Veblen, Z.Z. Sheng, A. El Ali, and A.M. Hermann, *Phys. Rev. Lett.* **60** (1988) 1657, and R.M. Hazen, C.T. Prewitt, R.J. Angel, N.L. Ross, L.W. Finger, C. G. Hadidiacos, D.R. Veblen, P.J. Heaney, P.H. Hor, R.L. Meng, Y.Y. Sun, Y.Q. Wang, Y.Y. Xue, Z.J. Huang, L. Gao, J. Bechtold, and C.W. Chu, *Phys. Rev. Lett.* **60** (1988) 1174.
[99] S.S.P. Parkin, V.Y. Lee, E.M. Engler, A.I. Nazzal, T.C. Huang, G. Gorman, R. Savoy, and R. Beyers, Phys. Rev. Lett., (in press).
[100] H. Kuepfer, S.M. Green, C. Jiang, Y. Mei, H.L. Luo, R. Meier-Hirmer, and C. Politis, *Z. Phys.* B (in press).
[101] H.C. Ku, M.F. Tai, S.W. Hsu, K.H. Lii, H.D. Yang, and R.N. Shelton, *Chinese J. Phys.* **26** (1988) S99.
[102] M.B. Maple, Y. Dalichaouch, J.M. Ferreira, R.R. Hake, B.W. Lee, J.J. Neumeier, M.S. Torikachvili, K.N. Yang, H. Zhou, R.P. Guertin, and M.V. Kuric, in *Superconductivity in Highly Correlated Fermion Systems*, edited by M. Tachiki, Y. Muto, and S. Maekawa (North-Holland, Amsterdam, 1987) pp. 155-162.
[103] Y. Saito, T. Noji, A. Endo, N. Higuchi, K. Fujimoto, T. Oikawa, A. Hattori, and K. Furuse, in *Superconductivity in Highly Correlated Fermion Systems*, edited by M. Tachiki, Y. Muto, and S. Maekawa (North-Holland, Amsterdam, 1987) pp. 336-338.
[104] D.B. Mitzi, P.T. Feffer, J.M. Newsam, D.J. Webb, P. Klavins, A.J. Jacobson, and A. Kapitulnik, *Phys. Rev.* B (in press).
[105] J.L. Tholence, in *Superconductivity in Highly Correlated Fermion Systems*, edited by M. Tachiki, Y. Muto, and S. Maekawa (North-Holland, Amsterdam, 1987) pp. 353-356.
[106] Y Maeno, M. Kato, Y. Aoki, T. Nojima, and T. Fujita, in *Superconductivity in Highly Correlated Fermion Systems*, edited by M. Tachiki, Y. Muto, and S. Maekawa (North-Holland, Amsterdam, 1987) pp. 357-359.
[107] S.R. Ovshinsky, R.T. Young, D.D. Allred, G. DeMaggio, and G.A. Van der Leeden, *Phys. Rev. Lett.* **58** (1987) 1028.

[108] R.A. Camps, J.E. Evetts, B.A. Glowacki, S.B. Newcomb, R.E. Somekh, and W.M. Stobbs, *Nature* **329** (1987) 229.

[109] Y. Le Page, T. Siegrist, S.A. Sunshine, L.F. Schneemeyer, D.W. Murphy, S.M. Zahurak, J.V. Waszczak, W.R. McKinnon, J.M. Tarascon, G.W. Hull, and L.H. Greene, *Phys. Rev.* **B36** (1987) 3617.

[110] M. Hervieu, B. Domenges, C. Michel, G. Heger, J. Provost, and B. Raveau, *Phys. Rev.* **B36** (1987) 3920.

[111] D.E. Farrell, B.S. Chandrasekhar, M.R. DeGuire, M.M. Fang, V.G. Kogan, J.R. Clem, and D.K. Finnemore, *Phys. Rev.* **B36** (1987) 4025.

[112] M.A. Damento, K.A. Gschneidner, Jr., and R.W. McCallum, *Appl. Phys. Lett.* **51** (1987) 690.

[113] S. Iijima, T. Ichihashi, Y. Kubo, and J. Tabuchi, *Japan. J. Appl. Phys.* **26** (1987) L1478.

[114] A.J. Melmed, R.D. Shull, C.K. Chiang, and H.A. Fowler, *Science* **239** (1987) 176.

[115] D.L. Kaiser, F. Holtzberg, M.F. Chisholm and T.K. Worthington, *J. Cryst. Growth* 85 (1987) 593.

[116] G. Balestrino, S. Barbanera, and P. Paroli, *J. Cryst. Growth* **85** (1987) 585.

[117] B.N. Das, L.E. Toth, A.K. Singh, B. Bender, M. Osofsky, C.S. Pande, N.C. Koon, and S. Wolf, *J. Cryst. Growth* **85**, 1987) 588.

[118] Y. Tajima, M. Hikita, A. Katsui, Y. Hidada, T. Iwata, and S. Tsurumi, *J. Cryst. Growth* **85** (1987) 665.

[119] H.J. Scheel and F. Licci, *J. Cryst. Growth* **85** (1987) 607.

[120] R.A. Laudise, L.F. Schneemeyer, and R.L. Barns, *J. Cryst. Growth* **85**, (1987) 569.

[121] F. Holtzberg, D.L. Kaiser, B.A. Scott, T.R. McGuire, T.N. Jackson, A. Kleinsasser, and S. Tozer, in *Chemistry of High-Temperature Superconductors*, edited by D.L. Nelson, M.S. Whittingham, and T.F. George (American Chemical Society Symposium Series **351**, 1987) pp. 79-84.

[122] M.S. Thompson, M. Dixon, G.F. Holland, and A.M. Stacy, in *High Temperature Superconductors II*, edited by D.W. Capone II, W.H. Butler, B. Batlogg and C.W. Chu (Materials Research Society, Pittsburgh, 1988) p. 339.

[123] T.R. Dinger, T.K. Worthington, W.J. Gallagher, and R.L. Sandstrom, *Phys. Rev. Lett.* **58** (1987) 2687.

[124] R.N. Shelton, R.W. McCallum, M.A. Damento, and K.A. Gschneidner, Jr., *Int. J. Mod. Phys.* **1** (1987) 401.

[125] R.N. Shelton, R.W. McCallum, M.A. Damento, K.A. Gschneidner, Jr., H.C. Ku, H.D. Yang, J.W. Lynn, W.H. Li, and Q. Li, *Physica* **148B** (1987) 285.

[126] H. Noel, P. Gougeon, J. Padiou, J.C. Levet, M. Potel, O.

Laorde, and P. Monceau, *Solid State Commun.* **63** (1987) 915.

[127] K. Takita, H. Akinaga, H. Katoh, and K. Masuda, *Japan. J. Appl. Phys.* **26** (1987) L1552.

[128] P. Chaudhari, R.H. Koch, R.B. Laibowitz, T.R. McGuire, and R.J. Gambino, *Phys. Rev. Lett.* **58** (1987) 2684.

[129] M. Naito, R.H. Hammond, B. Oh, M.R. Hahn, J.W.P. Hsu, P. Rosenthal, A.F. Marshall, M.R. Beasley, T.H. Geballe, and A. Kapitulnik, *J. Mater. Res.* **2** (1987) 713.

[130] Y. Enomoto, T. Murakami, M. Suzuki, and K. Moriwaki, *Japan. J. Appl. Phys.* **26** (1987) L1248.

[131] T.T.M. Palstra, B. Batlogg, L.F. Schneemeyer, and R.J. Cava, *Phys. Rev.* B (in press).

[132] Y. Koike, T. Nakanomyo, and T. Fukase, *Japan. J. Appl. Phys.* (in press).

[133] T.P. Orlando, K.A. Delin, S. Foner, E.J. McNiff, Jr., J.M. Tarascon, L.H. Greene, W.R. McKinnon, and G.W. Hull, *Phys. Rev.* **B35** (1987) 7249.

[134] T.P. Orlando, K.A. Delin, S. Foner, E.J. McNiff, Jr., J.M. Tarascon, L.H. Greene, W.R. McKinnon, and G.W. Hull, *Phys. Rev.* **B36** (1987) 2394.

[135] Y. Hidaka, M. Oda, M. Suzuki, A. Katsui, T. Murakami, N. Kobayashi, and Y. Muto, in *Superconductivity in Highly Correlated Fermion Systems*, edited by M. Tachiki, Y. Muto, and S. Maekawa (North-Holland, Amsterdam, 1987) pp. 329-331.

[136] W.J. Gallagher, T.K. Worthington, T.R. Dinger, F. Holtzberg, D.L. Kaiser, and R.L. Sandstrom, in *Superconductivity in Highly Correlated Fermion Systems*, edited by M. Tachiki, Y. Muto, and S. Maekawa (North-Holland, Amsterdam, 1987) pp. 228-232.

[137] E. Helfand and N.R. Werthamer, *Phys. Rev.* **147** (1966) 288.

[138] N.R. Werthamer, E. Helfand, and P.C. Hohenberg, *Phys. Rev.* **148** (1966) 295.

[139] K. Maki, *Phys. Rev.* **148** (1966) 362.

[140] T.P. Orlando, E.J. McNiff, Jr., S. Foner, and M.R. Beasley, *Phys. Rev.* **B19** (1979) 4545.

[141] H.D. Yang, H.C. Ku, P. Klavins, and R.N. Shelton, *Phys. Rev.* **B36** (1987) 8791.

[142] J.W. Lynn, W.-H. Li, Q. Li, H.C. Ku, H.D. Yang, and R.N. Shelton, *Phys. Rev.* **B36** (1987) 2374.

[143] S.-W. Cheong, S.E. Brown, Z. Fisk, R.S. Kwok, J.D. Thompson, E. Zirngiebl, G. Gruner, D.E. Peterson, G.L. Wells, R.B. Schwarz, and J.R. Cooper, *Phys. Rev.* **B36** (1987) 3913.

[144] J.P. Golben, S.-I. Lee, S.Y. Lee, Y. Song, T.W. Noh, X.-D. Chen, J.R. Gaines, and R.T. Tettenhorst, *Phys. Rev.* **B35** (1987) 8705.

[145] F. Zuo, B.R. Patton, D.L. Cox, S.I. Lee, Y. Song, J.P.

Golben, X.D. Chen, S.Y. Lee, Y. Cao, Y. Lu, J.R. Gaines, J.C. Garland, and A.J. Epstein, *Phys. Rev.* **B36** (1987) 3603.

[146] A. Oota, Y. Kiyoshima, A. Shimono, K. Koyama, N. Kamegashira, and S. Noguchi, *Japan. J. Appl. Phys.* **26** (1987) L1543.

[147] A. Freimuth, S. Blumenroeder, G. Jackel, H. Kierspel, J. Langen, G. Buth, A. Nowack, H. Schmidt, W. Schlabitz, E. Zirngiebl, and E. Moersen, *Z. Phys.* **B68** (1987) 433.

[148] J.R. Thompson, S.T. Sekula, D.K. Christen, B.C. Sales, L.A. Boatner, and Y.C. Kim, *Phys. Rev.* **B36** (1987) 718.

[149] J.Z. Sun, D.J. Webb, M. Naito, K. Char, M.R. Hahn, J.W.P. Hsu, A.D. Kent, D.B. Mitzi, B. Oh, M.R. Beasley, T.H. Geballe, R.H. Hammond, and A. Kapitulnik, *Phys. Rev. Lett.* **58** (1987) 1574.

[150] H. Riesemeier, E.W. Scheidt, I. Stang, K. Lueders, V. Mueller, H. Eickenbusch, and R. Schoellhorn, in *Superconductivity in Highly Correlated Fermion Systems*, edited by M. Tachiki, Y. Muto, and S. Maekawa (North-Holland, Amsterdam, 1987) pp. 312-314.

[151] R. Fluekiger, T. Mueller, T. Wolf, I. Apfelstedt, E. Seibt, H. Kuepfer, and W. Schauer, *Physica* C, (Interlaken Conference).

[152] S. Jin, R.C. Sherwood, R.B. van Dover, T.H. Tiefel, and D.W. Johnson, Jr., *Appl. Phys. Lett.* **51** (1987) 203.

[153] O. Kohno, Y. Ireno, N. Sadakata, S. Aoki, M. Sugimoto, and M. Nakagawa, *Japan. J. Appl. Phys.* **26** (1987) 1653.

[154] B. Seeber, C. Rossel, O. Fischer, and W. Glaetzle, *IEEE Trans. Magn. MAG-9* (1983) 402.

[155] W. Goldacker, S. Miraglia, Y. Hariharan, T. Wolf, and R. Fluekiger, *Adv. Cryo. Eng.*, **34** (1987) 655.

[156] I. Bozovic, D. Kirillov, A. Kapitulnik, K. Char, M.R. Hahn, M.R. Beasley, T.H. Geballe, Y.H. Kim, and A.J. Heeger, *Phys. Rev. Lett.* **59** (1987) 2219.

[157] A.M. Campbell, *J. Phys. C*: Solid State Phys. **2** (1969) 1492.

[158] R.W. Rollins, H. Kuepfer, and W. Gey, *J. Appl. Phys.* **45** (1974) 5392.

[159] H. Kuepfer, I. Apfelstedt, R. Fluekiger, R. Meier-Hirmer, W. Schauer, T. Wolf, and H. Wuehl, *Physica C* (Interlaken Conference).

[160] J.R. Clem and V.G. Kogan, *Japan. J. Appl. Phys.* **26** (1987) 1161.

[161] T. Kaneko, H. Yoshida, S. Abe, H. Morita, K. Noto, and H. Fujimori, *Japan. J. Appl. Phys.* **26** (1987) L1374.

[162] P.H. Hor, L. Gao, R.L. Meng, Z.J. Huang, Y.Q. Wang, K. Forster, J. Vassilious, C.W. Chu, M.K. Wu, J.R. Ashburn, and C.J. Torng, *Phys. Rev. Lett.* **58** (1987) 911.

[163] H.A. Borges, R. Kwok, J.D. Thompson, G.L. Wells, J.L. Smith,

Z. Fisk, and D.E. Peterson, *Phys. Rev.* **B36** (1987) 2404.

[164] K. Murata, H. Ihara, M. Tokumoto, M. Hirabayashi, N. Terada, K. Senzaki, and Y. Kimura, *Japan. J. Appl. Phys.* **26** (1987) L471.

[165] J.E. Schirber, D.S. Ginley, E.L. Venturini, and B. Morosin, *Phys. Rev.* **B35** (1987) 8709.

[166] M.R. Dietrich, W.H. Fietz, J. Ecke, and C. Politis, *Japan. J. Appl. Phys.* **26** (1987) 1113.

[167] M. Lang, T. Lechner, S. Riegel, F. Steglich, G. Weber, T.J. Kim, B. Leuthi, B. Wolf, H. Rietschel, and M. Wilhelm, *Z. Phys.* **B69** (1988) 459.

[168] P. Przyslupski, T. Skoskiewicz, J. Igalson, and J. Rauluszkiewicz, in *Superconductivity in Highly Correlated Fermion Systems*, edited by M. Tachiki, Y. Muto, and S. Maekawa (North-Holland, Amsterdam, 1987) pp. 289-291.

[169] M.B. Maple, Y. Dalichaouch, J.M. Ferreira, R.R. Hake, S.E. Lambert, B.W. Lee, J.J. Neumeier, M.S. Torikachvili, K.N. Yang, H. Zhou, Z. Fisk, M.W. McElfresh, and J.L. Smith, in *Novel Superconductivity*, edited by S.A. Wolf and V.Z. Kresin (Plenum, New York, 1987) pp. 839-853.

[170] T.F. Smith, *J. Low Temp. Phys.* **6** (1972) 171.

[171] R.N. Shelton, in *Superconductivity in d- and f-Band Metals*, edited by D.H. Douglass (Plenum, New York, 1977) pp. 137-160.

[172] R.N. Shelton, A.R. Moodenbaugh, P.D. Dernier, and B.T. Matthias, *Mater. Res. Bull.* **10** (1975) 1111.

[173] A. Eiling, J.S. Schilling, and H. Bach, in *Physics of Solids Under High Pressure*, edited by J.S. Schilling and R. N. Shelton (North-Holland, Amsterdam, 1981) pp. 385-396.

[174] S. Kondoh, Y. Ando, M. Onoda, M. Sato, and J. Akimitsu, *Solid State Commun.* **65** (1988) 1329.

[175] L. Gao, Z.J. Huang, R.L. Meng, P.H. Hor, J. Bechtold, Y.Y. Sun, C.W. Chu, Z.Z. Sheng, and A.M. Hermann, *Nature* **332** (1988) 623.

[176] M.A. Subramanian, J.C. Calabrese, C.C. Torardi, J. Gopalakrishnan, T.R. Askew, R.B. Flippen, K.J. Morrissey, U. Chowdhry, and A.W. Sleight, *Nature* **332** (1988) 420.

[177] C.W. Chu, J. Bechtold, L. Gao, P.H. Hor, Z.J. Huang, R.L. Meng, Y.Y. Sun, Y.Q. Wang, and Y.Y. Xue, *Phys. Rev. Lett.* **60** (1988) 941.

[178] B.C. Yang, H.-C. Li, X.X. Xi, M. Dietrich, G. Linker, and J. Geerk, *Z. Phys.* **B70** (1988) 275.

[179] D.R. Veblen, P.J. Heaney, R.J. Angel, L.W. Finger, R.M. Hazen, C.T. Prewitt, N.L. Ross, C.W. Chu, P.H. Hor, and R.L. Meng, *Nature* **332** (1988) 334.

[180] H. Takagi, H. Eisaki, S. Uchida, A. Maeda, S. Tajima, K. Uchinokura, and S. Tanaka, *Nature* **332** (1988) 236.

[181] M.S. Hybertsen and L.F. Mattheis, *Phys. Rev. Lett.* **60** (1988) 1661.

[182] H.W. Zandbergen, Y.K. Huang, M.J.V. Menken, J.N. Li, K. Kadowaki, A.A. Menovsky, G. van Tendeloo, and S. Amelinckx, *Nature* **332** (1988) 620.

[183] H. Krakauer and W.E. Pickett, *Phys. Rev. Lett.* **60** (1988) 1665.

[184] L.F. Schneemeyer, R.B. van Dover, S.H. Glarum, S.A. Sunshine, R.M. Fleming, B. Batlogg, T. Siegrist, J.H. Markshall, J.V. Waszczak, and L.W. Rupp, *Nature* **332** (1988) 422.

[185] C.C. Torardi, M.A. Subramanian, J.C. Calabrese, J. Gopalakrishnan, K.J. Morrissey, T.R. Askew, R.B. Flippen, U. Chowdhry, and A.W. Sleight, *Science* **240** (1988) 631.

CHAPTER 7

Thermal and Transport Properties

Jack E. Crow and Nai-Phuan Ong

7.1. Introduction

The unusual superconducting properties of the recently discovered high T_c oxides, in particular, the high transition temperature T_c, drove the early stages of the revival of interest in superconductivity, but it may be the understanding of the normal state which represents the largest challenge. Their strange normal state properties, which are not characteristic of other materials, is most vividly displayed in their transport properties. The conductivity is extremely anisotropic, with metallic-like behavior along one direction and semiconducting-like or insulating behavior along the other. Many standard theoretical approaches to understanding these properties have failed to properly account for this normal state behavior, i.e., band theories have predicted ground states which are inconsistent with the experimentally established ground state. Even though research may confirm that the superconducting properties may be described by a Bardeen-Cooper-Schrieffer (BCS) theory, the nature of the pairing mechanism will undoubtedly remain a mystery until more is known about the normal state.

The discussion presented in this chapter is focused on the thermal and transport properties of these new oxide systems. First, the transport properties, including Hall effect and electrical conductivity, will be presented and discussed. Following this, the thermal properties, namely the specific heat, will be discussed. The final section of this chapter is directed towards examining the role played by phonons within these systems. Since this chapter is pedagogical in nature, the review of the relevant experimental data is selective and not intended to be complete. Also, additional material or discussion has been included to provide the reader with a broader background on relevant issues than is often available in review discussions so that hopefully the literature can be read with greater understanding.[1]

7.2. Normal-State Transport Properties of the High-T_c Oxides

The normal-state transport properties of the high T_c oxides [1-3] have attracted almost as much attention as the superconducting properties. The interest stems from the belief that if the charge carriers in these unusual systems are strongly correlated (i.e. interact very strongly through the Coulomb interaction) the normal state properties should show highly unusual properties [4]. A central issue is whether Normal Fermi Liquid (NFL) theory and conventional Bloch-Boltzmann models can account for the observed transport properties. Anderson and coworkers [4,5] have discussed evidence pointing to the failure of NFL theory. Other investigators, however, adopt a less extreme viewpoint. The question of the validity of NFL theory is currently a very active area of research. We wish to discuss here the Hall and resistivity studies, especially how they relate to these issues.

One of the first puzzling features was the finding that La_2CuO_4 is an insulator at low temperatures T [6,7]. Simple valence counting implies that the Cu ions are in the $3d^9$ state which corresponds to one hole in the d-shell at each Cu site. Band structure calculations [8,9] show that in La_2CuO_4 the anti-bonding state formed from the hybridization of the $Cu(d_{x^2-y^2})$ and O(2p) orbitals forms a band of width $\sim$ 2 eV which is intersected at mid-band by the Fermi level E_F (see Chap. 5). The calculations also show that the La(4s) orbitals are all too far from E_F to influence the valence counting. Thus, La is ionized in the 3+ state. A direct interpretation would predict that La_2CuO_4 should be a good metal with a carrier concentration close to 1×10^{22} cm^{-3} (one hole per Cu site). This conflicts with the experimental results.

7.2.1. Charge and Spin Density Wave Models

Several early proposals were put forward to account for the observed

[1]Throughout the chapter "214" refers to La_2Cu0_4 and "123" to $YBa_2Cu_3O_7$ systems. The newer systems Bi-Ca-Sr-Cu-0 and Tl-Ca-Ba-Cu-0 will be referred to as Bi 2122 [or Bi 2223] and Tl 2122 [or Tl 2223] respectively, where the numbers represent the stoichiometry of the compounds with 2 or 3 adjacent $Cu0_2$ planes.

insulating state. The most popular was the charge density wave (CDW) instability [9,10]. In metals with (quasi) two-dimensional (2D) electronic dispersion, the Fermi surface (FS) possesses a "nesting" property; large portions can be brought into coincidence by rigidly translating one side of the FS through a "spanning" vector Q. This implies that a single distortion with wave vector Q interacts with a very substantial fraction of the electrons on the FS. At low temperatures the system is unstable to the spontaneous softening of the phonon mode with wave vector Q and the simultaneous development of a gap over the nested portions of the FS [11,12]. The gap which develops at the FS stabilizes the distortion through the gain in electronic energy. The system becomes insulating if the whole FS is gapped. Many examples of such CDW instabilities occur in the dichalcogenides (1T-TaS_2,2H-$NbSe_2$), which have quasi 2D electronic structures [11]. They also occur in quasi-one dimensional conductors ($K_{0.3}MoO_3$ or "blue bronze", TaS_3, and $NbSe_3$) where they are also known as Peierls instabilities [12].

For La_2CuO_4, the FS computed from band structure is very nearly a square since the $3d_{x^2-y^2}$-2p band is close to being half-filled and the dispersion is very flat in the direction perpendicular to the CuO_2 planes. The spanning vector is then along the [110] or equivalent direction. The development of a CDW would appear to account for the insulating state. It would also arise quite naturally from the 2D electronic dispersion. Doping with divalent ions such as Ba or Sr would lower the Fermi level. Eventually the nesting property would be spoiled, and the CDW phase destabilized. The destruction of the CDW would convert the system to a metal with a possible high T_c because of the enhanced electron-phonon coupling.

A related theory is the spin-density-wave (SDW) model [13]. A CDW instability may not be favored if the electron-electron interaction energy U is substantial. In this case (provided U is not excessively large) the FS nesting favors the spin-density-wave state, in which the spin density of the electrons is modulated with a period $2\pi/Q$ but the charge density remains uniform. As in the CDW case a gap develops over the nested portions of the FS, resulting in an insulating state. The SDW instability is observed in Cr [13] and in the organic conductors $(TMTSF)_2X(X=PF_6,ClO_4)$ [14]. In this system the application of pressure destroys the SDW state, replacing it with a superconducting state. It was natural to argue, as many did, that a similar phase diagram applied to the $La_{2-x}Sr_xCuO_4$ system (with the doping level x as the driving variable, instead of pressure).

7.2.2. Hall Effect Studies in $La_{2-x}Cu0_4$

In both CDW and SDW models the metallic state in $La_{2-x}Sr_xCuO_4$ arises from the destabilization of an insulating state which resulted from a Fermi surface driven instability. There are two scenarios for the destabilization. If the density wave is weakly pinned to the underlying lattice a lowering of the Fermi level by doping would alter the optimum nesting wave vector Q so that the CDW or SDW would shift its wave vector to some incommensurate value to remain stable. Eventually the instability would vanish for sufficiently large doping and the system then becomes metallic. In the second scenario the CDW or SDW is strongly pinned to the commensurate value corresponding to exact half-filling. The commensurability energy may be sufficiently strong for the wave vector to be pinned at the half-filling value despite a significant lowering of the Fermi level. An early test [15,16] of these ideas came from the Hall effect in $La_{2-x}Sr_xCuO_4$. For x in the range $0.05 < x < 0.14$, the Hall coefficient R_H is found to be independent of T. Because the band structure indicates that only one band intersects the Fermi level, and the Fermi surface is relatively simple, one may argue that R_H is equal to 1/ne, where n is the hole concentration in the sample and e the electronic charge. A plot of R_H vs. x (Fig. 7.1) showed that for this range of x, R_H is proportional to 1/x, i.e. $n = 2x/(abc)$ where abc is the volume of the unit cell (which contains two formula units) [16]. As x increases beyond 0.14, R_H deviates strongly from this simple relationship, dropping by a factor of 20 in the narrow range $0.15 < x < 0.2$. (note the T_c attains a maximum of 40 K near $x = 0.17$). The observed variation of the itinerant carrier population with x is inconsistent with either the CDW or SDW model, unless one modifies the theories in directions which are quite extreme.

The linear relationship between n and x for $x < 0.14$ implies that each Sr ion introduces one itinerant hole, i.e. the volume of the whole FS scales linearly with x. (In 214 the filling factor δ defined as the number of itinerant holes per Cu site equals x.) If the Q of the CDW or SDW shifts to optimize the nesting condition (the first scenario) we would not observe the linear n vs. x relationship. The system would either remain insulating or at best have a remnant FS pocket which need not be equal to 2x/abc in volume. The Hall data are consistent with the existence of a density wave in the undoped system only if Q is pinned at the commensurate value for all x below 0.14 (the second scenario). In this scenario Q remains pinned at the value in the insulating state independent of x (provided it does not exceed 0.14). Lowering the Fermi level with doping creates a pocket of holes below the gap which equals 2x/abc in volume. When x exceeds 0.15 the sharp decrease in R_H is interpreted as the destruction of the pinned CDW or SDW. In this model

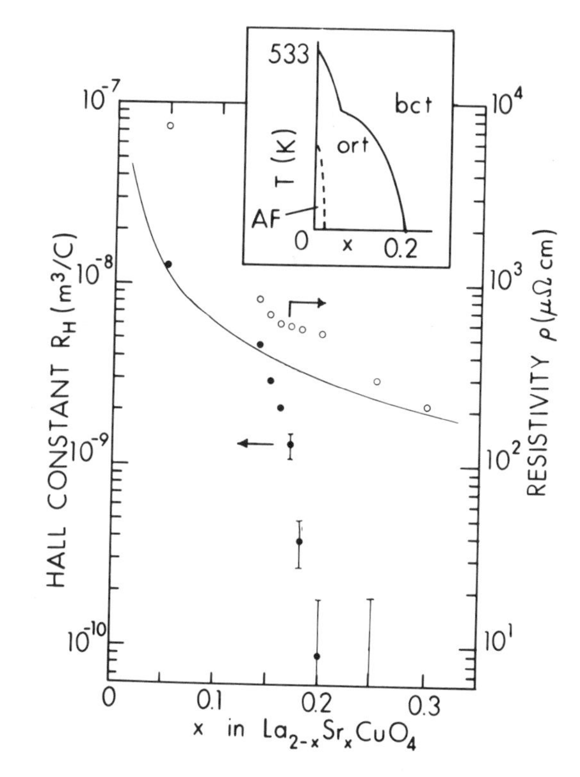

Fig. 1. Hall constant R_H and resistivity ρ versus x for $La_{2-x}Sr_xCuO_4$. The solid curve corresponds to $R_N \propto 1/x$. Inset: Magnetic and structural phase boundary versus x, with AF = antiferromagnetic, ort = orthorhombic, bct = tetragonal.

one needs to introduce a mechanism which stabilizes the density wave at the half-filling value despite the loss of a substantial fraction of occupied states near the FS.

7.2.3. The Mott-Hubbard Insulator, and Large U Models

In the large-U model first proposed by Anderson [17], the linear relationship between n and x is a direct consequence of the starting assumptions, although one has to abandon familiar concepts from NFL theory. If U/t is large (where t is the hopping integral) the proper starting ground state of the undoped system is the Mott-Hubbard insulator. For the situation where all Cu ions are in the $3d^9$ state the system is insulating because there exists a large gap (of a few eV) for excitations in which two holes (with spins opposed) occupy the same site. This is called the Mott-Hubbard gap [18]. Thus, although we have one hole per site, the system is insulating because of the large gap for charge excitations. In the insulating state the system can further lower its energy by assuming an antiferromagnetic (AF) order [19]. Physically, each hole can lower its kinetic energy by making virtual hops to its nearest neighbors and back. The Pauli principle then requires nearest neighbors to have spins opposed. The AF state widely encountered in the transition metal compounds is due to this mechanism. AF ordering occurs in the insulating members of both 214 and 123 families [20].

Doping the system by introducing additional holes results in a population of itinerant excitations which can carry current. Thus n equals 2x/abc, as observed in $La_{2-x}Sr_xCuO_4$ for $x < 0.14$ [16]. The itinerant holes move in a two-dimensional lattice of localized spin-½ objects. Much work is in progress to investigate how the holes modify the spin background, as well as the nature of the mutual interactions of the holes. In the RVB model, the charge carriers correspond to holons with zero spin and charge one. The holons interact very strongly with the spin excitations. Experimentally the 3D AF order in 214 is highly sensitive to doping: increasing x in $La_{2-x}Sr_xCuO_4$ from 0 to 0.02 suppresses the Neel temperature T_N from 240 K to 0 K. What remains is a population of

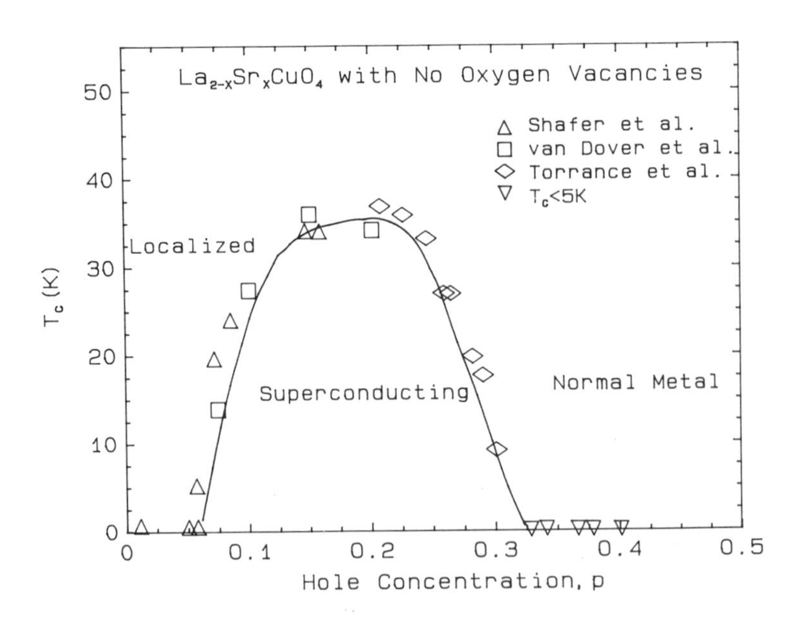

Fig. 2. Superconducting transition temperature versus hole concentration for $La_{2-x}Sr_xCuO_4$. From [22].

itinerant charge carriers (positively charged) moving in a background of spin-½ localized spins which either exist as a spin liquid state or an AF state with a very short correlation length. There is strong evidence that the holes move in orbitals with mostly oxygen 2p character. The state is written d^9L where L represents a ligand state. This is equivalent formally to a Cu^{3+} state.

7.2.4. A Normal Fermi Liquid?

It should be noted that in the large-U models the rapid fall of R_H as x exceeds 0.15 is not understood. There is a fairly sharp boundary near x = 0.2. The Hall data show that R_H is a *T*-independent for $0.05 < x < 0.14$, but decreases with *T* for $x > 0.16$. Shafer et. al. [21] and Torrance et. al. [22] have suggested that oxygen deficiency, in the region $0.15 < x < 0.4$, may strongly influence R_H. Both groups have used chemical titration techniques to measure the amount of Cu^{3+} ions in the sample. The collapse of R_H near 0.2 is ascribed by Shafer et. al. [21] to compensation of the holes by electrons introduced by oxygen vacancies. For a two-band Drude model we have

$$R_H = \frac{(n^h \nu^2 - n^e \mu^2)}{(n^h \nu + n^e \mu)^2} \tag{7.1}$$

where n^e (n^h) is the electron (hole) density and μ (ν) the electron (hole) mobility. Assuming that the electrons occupy a separate band from the original holes, we see the R_H can be greatly reduced if the numerator in Eq. (7.1) is close to zero. However, the weak *T* dependence of R_H, and the degree to which the electron and hole populations have to be balanced to reduce R_G by ~ 20 pose serious difficulties for such models. Since R_H never becomes negative for $x > 0.2$ each sample must have just enough oxygen vacancies to achieve cancellation. In Eq. (7.1), $n^h\nu^2$ and $n^e\mu^2$ have to be closely matched over the temperature range 50-300 K and over the range $0.17 < x < 0.30$ for this mechanism to be effective.

Is $La_{2-x}Sr_xCuO_4$ a normal fermi liquid in the region $x > 0.20$? In some models the sharp drop in R_H is interpreted as dividing the region in which NFL theory is invalid (low x) from the region (high x) in which it becomes valid. In the high-x region R_H measures the full FS volume ($1 \cdot 10^{22}$ cm^{-3}) whereas in the low-x region dc probes fail (by a wide margin) to measure anywhere close to the FS volume calculated from one hole per Cu ion. The decrease in T_c in this region (reaching 0 K near x = 0.32) is also not understood [22] (Fig. 7.2). Whether, in fact, the sharp drop in R_H and the disappearance of high T_c behavior are intimately related is a key question that will likely receive a good deal of future attention.

7.2.5. Temperature Dependent Hall Effect in the 123, 2122 and 2223 Systems

The Hall results are more complicated in $YBa_2Cu_3O_7$. Early data on ceramic samples [23-25] showed that the Hall density n_H (defined as $1/eR_H$) is linear in T over the range 100-300 K. Measurements on single crystals [26] showed that whereas R_H is roughly T independent (and negative) when H lies within the CuO_2 plane (i.e. normal to the c-axis), the Hall density n_H is again linear in T and positive when H is normal to the $Cu0_2$ planes. Results on ceramic samples apparently provide a good quantitative measurement of the in-plane transport because of the extreme anisotropy of the electronic conductivity. The interesting but awkward T dependence of R_H obviously precludes a direct identification of n_H with the carrier density. Several groups have proposed *ad hoc* multiband Drude models to explain the $1/T$ dependence of R_H. For example, if one assumes [27] in Eq. (7.1) that $n^h = n^e$, and μ and ν vary as $(1\text{-}a/T)/T$ and $(1+a/T)/T$, respectively, one can show that n_H is linear in T. Bloch-Boltzmann calculations [28] based on detailed band structures have also been performed to predict the difference is sign in 123 between R_H measured with H aligned parallel to c and normal to it. These machine calculations do not reproduce the $1/T$ dependence in R_H in 123, or the $n_H \sim x$ relationship in 214. The latter difficulty reflects the basic failure of band theories to reproduce the insulating state in $La_{2-x}Sr_xCuO_4$ when $x = 0$.

Recent work in doped 123 ceramics have shown that the multiband Drude models are all too specific and almost certainly invalid [29-31]. The temperature dependence of R_H has been shown to be part of a general pattern that occurs in all the high T_c oxides based on CuO_2 planes [29]. In both Bi 2122 (ceramics and single crystals) and Tl 2223, R_H is positive and decreases with increasing T. In one study [31], n_H is found to be linear in T in mixed phase ceramic Bi 2122-2223. In $La_{2-x}Sr_xCuO_4$, recent work has shown that for $x > 0.15$ R_H also decreases appreciably with increasing T. In Ref. 29, it is argued that the T dependence of R_H is, in fact, a generic property of the high T_c oxides, rather than the result of accidental cancellations in a specific band structure. An important clue is obtained by studying how chemical doping alters the T dependence of R_H. A recent study [29,30] has shown that in both 123 (doped with Ni or Co) and $La_{2-x}Sr_xCuO_4$ (doped with Ni), T_c is suppressed and this suppression of T_c invariably leads to a suppression of the slope dn_H/dT (Fig. 7.3). Apparently, the T dependence of R_H in all the high T_c oxides is due to an asymmetric scattering mechanism which reflects an anomalous property of the normal state. When the conditions favoring high T_c behavior are removed (either by reducing n or

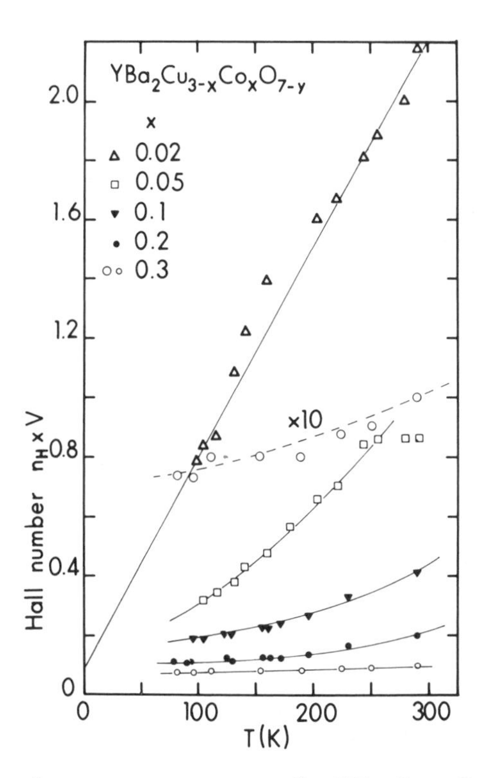

Fig. 3. Hall number versus temperature for $YBa_2Cu_{3-x}Co_xO_{7-y}$.

by creating site disorder in the CuO_2 planes) the strong T dependence of R_H is also suppressed. The T dependence of R_H recalls skew scattering in magnetic systems, mixed valent systems and in the heavy electron systems. However, the weakness of the spin-orbit scattering in the CuO_2 planes and the absence of $1/T$ susceptibility in the high T_c systems argue against such a conventional skew scattering mechanism. The generic nature of this behavior poses a serious challenge to theory. It suggests an intriguing pattern which may underlie and unify many transport properties in all the high T_c oxides based on CuO_2 planes.

The doping studies also provide direct evidence against multiband Drude models for the Hall effect in 123. It may be shown from Eq. (7.1) that R_H decreases towards negative values as E_F is raised by chemical doping. (Raising E_F increases the volume of the electron FS pocket while decreasing the hole pocket. Hence, the Hall field from the electron pocket dominates.) The Hall results from Co-doped 123 contradict this expectation. Thus, in 123 (and most likely in 214 as well) there is strong evidence against competing Hall currents: Only one band is in evidence in the transport.

7.2.6. Conductivity Anisotropy in Single Crystals

An important aspect of the normal state transport in 123 is the unexpectedly large conductivity anisotropy. Because high quality crystals of 123 can be grown, using a flux technique based on a mixture of BaO-CuO flux [3,32-34], most single crystal studies have been carried out on this system, although several studies on Bi 2122 crystals are being reported now.

Tozer et. al. [26] first reported measurements on the conductivity anisotropy in 123. Whereas the in-plane resistivity ρ_{ab} is similar to that observed in typical good sintered samples (i.e. metallic), the out-of-plane resistivity ρ_c is quite unusual. (The subscript *ab* represents an average of the resistivities along a and b.) For T above 150 K, ρ_c is weakly T-dependent. Below 150 K it increases with decreasing T even though ρ_{ab} is decreasing. The anisotropy ρ_c/ρ_{ab} increases from 30 at 295 K to 60 just above T_c. Subsequent work by a number of groups confirmed some aspects of this work, but also disagreed in other aspects. Hagen et. al. [35] studied seven samples in which the aspect ratio differed by a factor of 10 (Fig. 7.4). They obtained a much higher anisotropy (50 to 120 at 290 K and increasing to 200 to 300 at 100 K) than Tozer et. al. [26], primarily because ρ_{ab} was found to be much lower (50 $\mu\Omega$-cm at 100 K) than previously reported. ρ_c was of the same magnitude (10-15 mΩ cm) as in Tozer et al. Hagen et al. also found a shallow minimum in ρ_c at 150 K in all their samples. Buranov et. al. [36] showed that in samples which are highly oxygen deficient ρ_c can

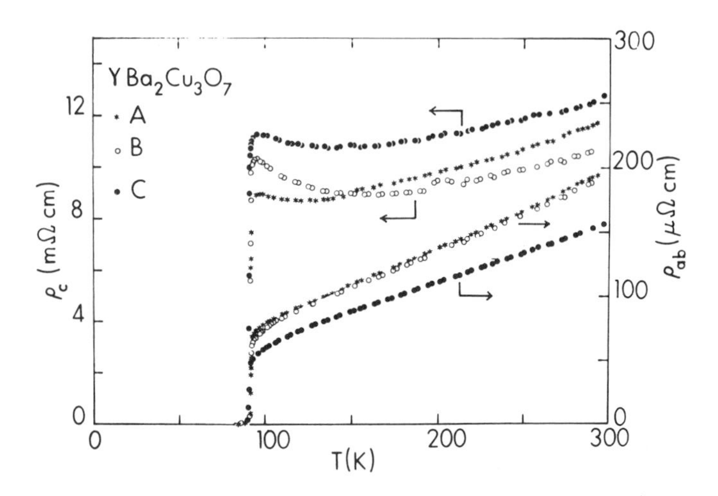

Fig. 4. Resistivity versus temperature for $YBa_2Cu_3O_7$. ρ_c is measured parallel to the c-axis, while ρ_{ab} is an average in the a-b plane.

increase steeply with decreasing T, ρ_c also goes to zero at a T significantly lower that the T_c obtained from ρ_{ab}. With extended oxygen annealing both ρ_c and ρ_{ab} approach the values reported by Tozer et. al. Iye et. al. [37] obtain a different behavior for ρ_c in their best samples. They find that ρ_c is metallic at all T. At the time of writing it is not clear if the shallow minimum observed in ρ_c is an artifact of oxygen deficiency or an intrinsic property of the fully oxygenated 123 crystals, and more work is needed. Measurements on the crystals with the best figures of merit (lowest ρ_{ab} and narrowest width of the superconducting transition) appear to support the existence of a minimum. In a thin (1 μm) single crystal of Bi 2122 Martin et. al. [38] report that ρ_c is metallic at high T (> 300 K) but increases with decreasing T below ~200 K. They derive an unusually large anisotropy in this crystal (~10^5). However, Wang et. al. [39] obtain smaller anisotropies in the same system (~10^3).

Why is the disparity (if it exists) between the signs of $d\rho_c/dT$ and $d\rho_{ab}/dT$ near T_c important? Many highly anisotropic conductors (some superconducting) are known. The most extensively studied are those that display transitions to CDW or SDW states. Among the layered dichalcogenides [11] such as 2H-$NbSe_2$ and 1T-TaS_2 the anisotropy (in-plane vs. out-of-plane conductivity) is moderately large (10 to 100). However, in all cases (with the possible exception of the last example) the $d\rho/dT$ is either positive in both directions or negative in both directions. In the quasi-one-dimensional systems [12] e.g., $NbSe_3$, $K_{0.3}MoO_3$ ("blue bronze"), and the Bechgaard salts $(TMTSF)_2X$ (where X = PF_6, ClO_4) [41], the sign of $d\rho/dT$ along the highly conducting (chain) direction is the same as that perpendicular to it (either both positive or both negative). The anisotropy in $K_{0.3}MoO_3$ and $(TMTSF)_2X$ is of the order of 10^3 and 10^4, respectively. One situation where a difference in sign is known to occur is in graphite strongly intercalated with K and $SbCl_5$. Phan et. al. [41] attribute the semiconducting ρ_c to the creation of carrier-depleted regions due to screening of the intercalants.

In 123, Anderson and Zou [5] have suggested that ρ_c should vary as $1/T$ because interplane charge transport requires a physical hole (which is a "holon" + a spinon). Thus, the tunneling amplitude depends on the spinon density, which decreases linearly with T. Further studies, particularly in crystals with much lower T_c's, will be required to test these provocative ideas.

7.2.7. The Linear Resistance vs. Temperature Behavior

The resistivity ρ of all the high T_c oxides based on CuO_2 displays a remarkable linear temperature dependence extending over a large range of T. Because of the fact that ρ in metals is linear in T (at high T) is a well-known textbook result, there has been some confusion over the issues relating to the ρ vs. T profile in the oxides. We first recall some features of the Bloch theory [42].

In the absence of Umklapp scattering, the Bloch theory for the lattice-scattering contribution ρ_L to the resistivity agrees with the data in a wide range of "well-behaved" metals. We will disregard impurity scattering for a moment. In the limit $T \to 0$, ρ_L goes at T^5 (Bloch-Gruneisen law). In the opposite limit ($T > \theta_D$, the Debye temperature) ρ_L is linear in T (with zero intercept for the extrapolation). The crossover temperature T_B between these two regimes, defined as the temperature at which the extrapolated T^5 curve intersects the linear curve, is given by $T_B/\theta_D = (4J^5(\infty))^{-1/4} = 0.21$, where J^5 is a variant of the Debye integral [42]. (For systems with very low carrier densities T_B is reduced by the factor k_Fa, where a is the lattice spacing [42]. This is because the maximum phonon wave vector is determined by the FS diameter rather than the Debye cut-off.) However, T_B is not sharply defined. In practice, strong deviation from the linear curve starts at a temperature ($\sim 0.7\theta_D$) that is quite a bit above T_B. At the low-T end, the ρ_L vs. T curve is practically "flat" below $0.1\theta_D$. At high T, important deviations from the Bloch theory appear. The linear behavior breaks down because of higher-order corrections to the solution of the Boltzmann equation. In Cu and Ag, a strong upward deviation becomes prominent above 600 K, whereas in Pt and Pd, the deviation is downward as T

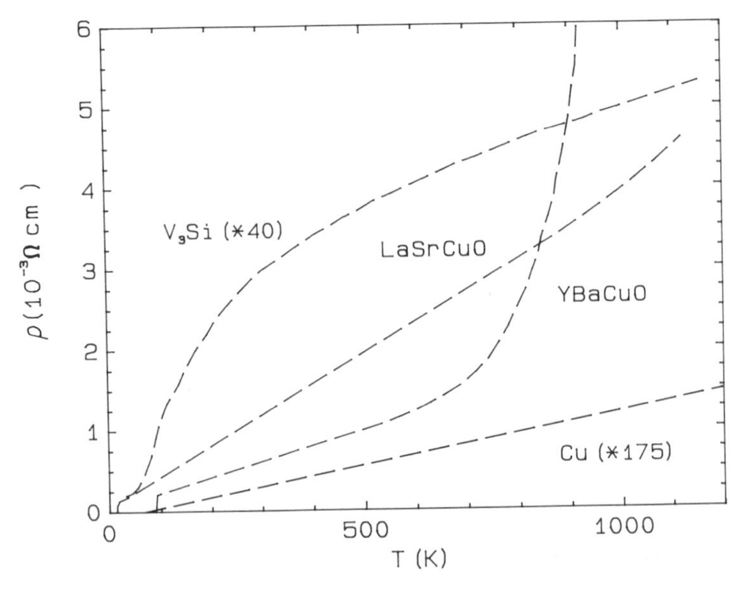

Fig. 5. Resistivity versus temperature for selected materials. From [45].

exceeds 700 K. In poor metals, in which phonon scattering is very strong at high T, ρ_L tends to saturate to a T-independent value when the mean free path ℓ approaches the de Broglie wavelength, i.e. when the "metallic" parameter $k_F\ell$ reaches ~1 (the Ioffe-Rege criterion). Thus, in most metals the linear T regime is rather limited. Near room temperature, the ρ_L vs. T profile may mimic a straight line over a wide range in T. But this line is sublinear (i.e. it has a negative intercept when extrapolated to $T=0$). Pt is perhaps the best-known commercially important example. When impurity scattering is present, the total resistivity is the sum $\rho_L + \rho_o$ where ρ_o is T independent (Mathiessen's rule [42]).

A remarkable feature of the high T_c oxides is the wide range in T over which ρ falls on a straight line passing through the origin, or is linear with a positive intercept [4,43]. The pervasiveness of this behavior in all the Cu-based high T_c oxides is also noteworthy. Even intentionally doped [43] samples retain the linear behavior up to fairly high levels of doping (10%). A recent study [44] of Ni-doped $La_{2-x}Sr_xCuO_4$ (x=0.16) shows that Mathiessen's rule is valid in this system. Thus, in the remainder of this section we will discuss only the T-dependent part of ρ. With Ni doping, a linear ρ_L vs. T has now been observed [44] down to 18 K (at the time of this writing). Since θ_D deduced from specific heat data [1-3] is ~450 K, this is in striking contradiction with Bloch-Gruneisen behavior. (The low-density correction to T_B does not affect the argument since k_Fa equals 1.1 for the level of Sr doping used.) At the high T end, Gurvitch and Fiory [45] (Fig. 7.5) report that the linear dependence extends to 1,100 K in ceramic $La_{2-x}Sr_xCuO_4$. Thus, the wide range in T over which ρ_L is linear clearly challenges conventional theories. (The inference [45] from ceramic data that $k_F\ell$ is alarming below the Ioffe Regel limit in the high-T limit should now be disregarded. Recent work on 123 single crystals [35] show that $k_F\ell$ is of the order of 30 near 100 K. At 1,000 K, $k_F\ell$ is of the order of 3.) Micnas et. al. [46] have argued that conventional Boltzmann theory with strictly 2D electrons provides a smaller lower bound for linear behavior than in 3D: ρ_L is linear in T for $T > T^* = 2hk_Fs$ where s is the sound velocity. However, using the value $k_F = 3 \times 10^7$ cm^{-1} and s = 5 x 10^5 cm/s, T^* computes to 230 K, which makes the theory irrelevant. (The condition $k_Fa \ll \pi$ required for their theory is also violated.) Allen et. al. [28] compute ρ_L vs. T from detailed band structure and an assumed phonon spectrum, and find a sublinear fit in the range 100 to 300 K. The anisotropy ρ_c/ρ_{ab} for 123 is also found to be much too small (15 vs. 200-300). Anderson and Zou [5] have argued that the linear ρ_L vs. T behavior is expected from a theory which treats the charge carriers as holons obeying boson statistics. Their assumptions have been questioned by Kallin and Berlinsky [47]. Experimentally, there are isolated reports of a linear ρ vs. T behavior extending down to 6K in a (possibly mixed phase) single-plane Bi 2021 sample. Interestingly, no convincing results exist which demonstrate that ρ saturates to a T-independent behavior in any reasonably "clean" high T_c system. Clearly, future work will shortly clarify the degree to which conventional Bloch-Boltzmann theory fails in accounting for this important transport property.

7.3. Specific Heat

The temperature dependence of the specific heat C(T) provides invaluable information concerning the nature and energy dependence of many excitations within condensed matter systems, e.g., phonons, electrons, spin waves, etc. [48,49]. It was the exponential temperature dependence of the electronic contribution to the specific heat for $T \ll T_c$ in conventional superconductors along with the lack of a latent heat of transformation at T_c which represented major facts influencing the development of the BCS theory of superconductivity [50]. Similarly, specific heat studies of the new oxides superconductors are providing valuable insight into both the normal and superconductivity behavior of these new and unusual systems. First, we will briefly review the anticipated behavior of the temperature dependence of the specific heat in both the normal and superconductivity state with particular attention given to a succinct summary of the expected temperature dependence of C(T) due to several individual contributions. This brief summary will prove useful in the discussions of some of the most recent data. This discussion will be followed by a review of the results from recent studies.

7.3.1. Normal State, C(T)

When we discuss measured C(T) behavior, we will exclusively be referring to C(T) measured at constant pressure, whereas theoretical discussions usually mean C(T) determined at constant volume. A distinction between C(T) at constant pressure and at constant volume is usually not important at low temperatures, but becomes more important at higher temperatures where the thermal expansion is appreciable. Because a clear separation between the lattice and electronic contributions to C(T) is extremely difficult at high temperatures ($T > \theta_D/12$) the discussion of individual contributions will primary focus on the low T limit of these contributions where $C_p \sim C_v$ and a separation of the individual contributions to the temperature dependence of C(T) is often possible.

There are several types of excitations which may contribute to the measured $C(T)$. In particular, $C(T) = C_e(T) + C_{ph}(T) + C_{sf}(T) + C_T + C_{sch}(T)$ where $C_e(T)$ is the electronic contribution, $C_{ph}(T)$ is the phonon term, $C_{sf}(T)$ is due to spin or magnetic fluctuations, C_T is due to ionic tunneling between nearly degenerate states, and $C_{sch}(T)$ represents a contribution due to a Schottky anomaly resulting from a multi-leveled system. Each contribution is separately discussed below.

$C_e(T)$, *The Electronic Contribution.* The electron contribution to $C(T)$ is given by γT for $T \ll T_F$, the Fermi temperature characterizing the electron gas. The electronic coefficient of $C(T)$, γ, for nearly free electrons is proportional to the density of electronic states at E_F, $N(E_F)$ and is given by $2\pi^2 N(E_F) k_B^2/3$. The linear T-dependence for $C_e(T)$ is independent of the dimensionality of the system. The above result does not explicitly reflect strong coupling corrections due to electron-phonon or electron-magnon interactions. The electron-phonon interaction leads to a correction of the bare density of states within an energy region of the order of the Debye energy around E_F. This correction results in an enhancement of the non-interacting electron band mass, i.e., a flattening of the energy dependence of the bands near E_F, and a renormalization of the electron specific heat coefficient given by

$$\gamma = \frac{m^*}{m_o} \gamma_o = \frac{2\pi^2}{3} k_B^2 N_o(E_F) \frac{m^*}{m_o} \tag{7.2}$$

where m^*/m_o is the mass enhancement ratio and m_o is the bare electron mass,

$$\frac{m^*}{m_o} = 1 + N_o(E_F) V_{ph} = 1 + \lambda_{ph} \tag{7.3}$$

and V_{ph} is a matrix element characterizing the strength of the electron-phonon interaction, $N_o(E_F)$ is the bare electronic density of states and λ_{ph} is the electron-phonon coupling parameter which also enters the BCS expression for T_c [51]. The above expression for m^*/m_o can be generalized to account for other mass enhancement mechanisms such as spin fluctuations. Thus, more generally m^*/m_o should be written as

$$\frac{m^*}{m_o} = 1 + N_o(E_F) V_{ph} + N_o(E_F) V_{sf} = 1 + \lambda_{ph} + \lambda_{sf} \tag{7.4}$$

where λ_{sf} is a mass renormalization coming from electron-electron interactions mediated by dynamic spin fluctuations [49,52,53]. The effects of spin fluctuations will be further discussed below.

$C_{ph}(T)$, *The Phonon Contribution.* If the lattice excitation spectrum is characterized by an energy dependent density of states $g_{ph}(\nu)$ such that $g_{ph}(\nu)d\nu$ is the fraction of lattice modes with frequency ranging from $\nu \to \nu + d\nu$, then the specific heat per mole at constant volume is given by

$$C^V_{ph} = 3nN_o k_B \int_0^\infty \frac{(h\nu/kT)^2 \, e^{h\nu/k_BT}}{(e^{h\nu/k_BT} - 1)^2} \, g_{ph}(\nu) d\nu \qquad (7.5)$$

where n is the number of ions per molecular unit, N_o is Avogadro's number, and k_B is the Boltzmann constant [48]. Thus, if the phonon density of states is known, $C^V_{ph}(T)$, i.e., $C(T)$ at constant volume, could be calculated and corrections resulting from thermal expansion could be included to provide an estimate for $C^P_{ph}(T)$ at constant pressure, i.e. the normally measured $C(T)$. However, in most cases $g_{ph}(\nu)$ is not known in sufficient detail. In the absence of a detailed knowledge of the phonon density of states, an approximate estimate of $C^V_{ph}(T)$ can be obtained by using the Debye model which assumes acoustic modes with a dispersion curve linear in phonon momentum. The Debye density of states is

$$g_{ph}(\nu) = \frac{3\nu^2}{\nu_D^3} \qquad (7.6)$$

where ν_D is the Debye frequency. Eqs. (7.5) and (7.7) lead to a Debye specific heat given by

$$C^V_{ph} = \frac{9nN_o k_B}{\nu_D^3} \int_0^{\nu_D} \frac{(h\nu/kT)^2 \, e^{h\nu/k_BT}}{(e^{h\nu/k_BT} - 1)^2} \, \nu^2 \, d\nu \qquad (7.7)$$

which can be simplified for $k_BT \ll h\nu_D$ and for $k_BT \gg h\nu_D$. At low temperatures ($k_BT \ll h\nu_D$), the integral may be extended to infinity, leading to

$$C_{ph}(T) = \left[\frac{12\pi^4}{5}\right] nN_o k_B \left[\frac{T}{\theta_D}\right]^3$$

$$= 1943 \; n \left[\frac{T}{\theta_D}\right]^3 \left[\frac{\mathrm{J}}{\text{mole-K}}\right] , \qquad (7.8)$$

where θ_D is the Debye temperature. Thus, the Debye model predicts a

T^3 law for C(T) at low temperatures, i.e., $0 < T < \theta_D/12$, where the upper limit defines the T/θ_D where $C_{ph}(T)$ deviates from T^3 by about one per cent. A T^5 term could be added to this long wavelength approximation for $C_{ph}(T)$ to partially compensate for deviations from the simple T^3 dependence.

$C_{sf}(T)$, *Spin Fluctuation Contribution.* The scaling of T_c across the d-transition metal series and the large enhancement of the nearly temperature independent magnetic susceptibility relative to band theory estimates for both d and f electron systems had previously suggested that exchange interactions with heavily damped spin waves lead to important renormalizations of normal state parameters and T_c [49,52-54]. These damped spin waves, called spin fluctuations, have been shown to significantly modify normal state behavior and contribute a pair weakening interaction [55,56] in many conventional phonon mediated superconductors. The exchange of spin excitations has also been discussed as the origin of the pairing interaction in the recently discovered superconducting heavy fermion systems [57-59] and is considered as a possible mechanism leading to the high T_c observed in the Cu-O based systems [60-62].

Spin fluctuations, or paramagnons, are highly damped excitations of the spin degrees of freedom of the system. The effects of spin fluctuations become increasingly apparent as a magnetic instability, such as ferromagnetism or antiferromagnetism, is approached [54]. As indicated above, these damped excitations lead to enhancements of the magnetic susceptibility and a modification to C(T) at low temperatures. The paramagnon contribution to the specific heat has been evaluated by Doniach and Engelsberg [49]. The development of such theories was first focused at explaining the unusual properties and superfluidity of ^{3}He and several f-electron systems. These excitations led to an enhancement of γ, the electronic coefficient of specific heat, and introduces a $T^3 \ln T$ correction. Including this $T^3 \ln T$-contribution, the electronic specific heat becomes

$$C_e(T) = \left[\frac{m^*}{m_o}\right] \gamma_o T + \delta\, T^3 \ln(T) \tag{7.9}$$

where m^*/m_o is a mass enhancement due to spin fluctuations and δ depends on electronic parameters such as the exchange interaction, $N(E_F)$, T_F, etc. The $T^3 \ln T$ term produces an upswing in $C(T)/T$ as $T \to 0$, similar to that observed for many of the high T_c oxide systems [63]. In addition to the enhancement of C(T), spin fluctuations also strongly influence the magnetic susceptibility. As $T \to 0$, $\chi(T)$ approaches $\chi(O) = \chi_o S$, where χ_o is the temperature independent non-interacting Pauli or band contribution to $\chi(T)$ and S is a spin fluctuation enhancement factor (Stoner factor). In terms of S, the enhancement of γ becomes $\gamma = \gamma_o \ln S$ or $m^*/m_o = \ln S$.

Spin fluctuations may play an important role in these new oxide superconductor because many of them, e.g., $(La,M)_2 CuO_4$ with M = Ba and Sr and $YBa_2Cu_3O_{7-\delta}$, are near magnetic instabilities (see Chap. 8) and recognizing this fact, pairing models closely linked to the spin degrees of freedom have emerged [6,60-62, Chap. 9].

$C_T(T)$, *Two-Level Tunneling Contributions.* During the last two decades, considerable attention has focused on the low T transport and thermodynamic properties of disordered systems [64-66]. These systems display anomalous properties when compared to typical crystalline systems and certain features of these systems resemble behavior reported for the new oxides. In crystalline materials, the phonon contribution to $C(T)$ and the thermal conductivity $K(T)$ are both proportional to T^3 as $T \to 0$ and the electronic contribution to $K(T)$ approaches zero with a linear temperature dependence. However, in many disordered systems, the lattice contribution to $C(T)$ is linear in T and $K(T)$ is quadratic in T, i.e., $C(T) \sim T$ and $K(T) \sim T^2$, as $T \to 0$. Such "glassy" behavior has also been reported for a few crystalline systems, e.g. β-alumina [67], Y-stabilized ZrO_2 [68], and Li_3N [69], and in some superconducting alloy systems such as ZrNb alloys [70]. The low temperature thermal properties of both the amorphous and crystalline systems discussed above are very similar even though the details of the structure and the nature of the bonding are very different.

These anomalous thermal properties have been modeled assuming a two-level excitation system representing the tunneling of atoms or groups of atoms from one potential well to a degenerate or nearly degenerate neighboring well. If one includes a small distribution in the size of the potential barriers and energy splitting between the two levels, one obtains a density of tunneling states which is nearly constant over a limited range of energy. If the phonon density of states in Eq. (7.5) is replaced by $g_{ph}(\nu) = n_o = \text{const}$, $C(T) = \beta T$ is obtained with β proportional to n_o, the density of tunneling states [71]. Thus, two level tunneling models should lead a nonelectronic contribution to $C(T)$ as $T \to 0$ which is linear in T, thus mimicking an electronic contribution. However, associated with this contribution, there should be a T^2 contribution to $K(T)$, along with a renormalization of ultrasonic attenuation and sound propagation parameters.

$C_{sch}(T)$, *A Schottky Contribution.* It is well known that a system of discrete excitations such as a two-level system with a ground state of degeneracy g_o, an excited state of degeneracy g_1, and an energy splitting Δ will lead to an additional contribution, a Schottky contribution, to $C(T)$ given by

$$C_{sch}(T) = \frac{R(\Delta/T)^2 \frac{g_o}{g_1} e^{\Delta/T}}{1 + \frac{g_o}{g_1} e^{\Delta/T}} \tag{7.10}$$

where R is the universal gas constant [72]. At $T \gg \Delta$, this expression can be approximated by

$$C_{sch}(T) \approx \frac{R g_o g_1}{(g_o + g_1)^2} (\Delta/T)^2 \quad . \tag{7.11}$$

A T^{-2} dependence of C(T) has been observed at low T for several of the recently discovered high T_c oxide systems, and has been associated with a Schottky contribution to C(T) as given by Eq. (7.11).

7.3.2. Superconducting State, C(T)

The two fluid model developed by Gorter and Casmir along with the phenomenological electrodynamic contributions by F. and H. London played an important role in the development of our understanding of superconductivity [73]. Of course, these earlier ideas were incorporated or placed on a firm foundation with the advent of the Ginzburg-Landau (GL) (See Chapter 2) and the Bardeen-Cooper-Schrieffer (BCS) [50] theories.

On very general grounds and recalling that a superconductor is a perfect diamagnet, i.e., $\chi_s = -1/(4\pi)$, the Gibbs free energy difference between the normal and superconducting state in a magnetic field H can be written

$$G_N - G_S = \left(\frac{H_c^2}{8\pi} - \frac{H^2}{8\pi}\right) V_s \tag{7.12}$$

where H_c is thermodynamic critical field and V_s is the volume of the superconductor. Using this free energy difference, the entropy ($S = -(\partial G/\partial T)_H$) and specific heat ($C = T\,dS/dT$), differences along the superconducting—normal state phase boundary are

$$S_n - S_s = -\left[\frac{V_s}{4\pi}\right] H_c \left[\frac{dH_c}{dT}\right] \tag{7.13}$$

and

$$C_n - C_s = -\frac{V_s T}{4\pi}\left[\left(\frac{H_c(d^2H_c)}{dT^2}\right) + \left(\frac{dH_c}{dT}\right)^2\right] \quad . \tag{7.14}$$

In the absence of a magnetic field, there is no entropy change or latent heat at T_c but there is a jump in the specific heat given by

$$C_n - C_s = -\frac{V_S}{4\pi}\left[T_c\left(\frac{dH_c}{dT}\right)^2_{T=Tc}\right] \quad . \tag{7.15}$$

For $T < T_c$, the above relationships only apply for a Type-I superconductor; a more extensive discussion of the thermodynamics in a magnetic field is required for Type-II superconductors. Using results of the G-L theory, other relationships between measurable quantities can be established, e.g.,

$$C_n - C_s = -\frac{V_s T_c}{4\pi\kappa^2}\left(\frac{dH_{c2}^2}{dT}\right) \tag{7.16}$$

where κ is the GL parameter and is given by 0.96 $\lambda_L(0)/\xi_o$ in the clean limit ($\ell_{tr} \gg \xi_o$) and 0.75 $\lambda_L(0)/\ell_{tr}$ in the dirty limit (ℓ_{tr} is the electron mean free path, and it is assumed that $\ell_{tr} \ll \xi_o$).

Returning to Eq. (7.14) and knowing that the observed temperature dependence of the thermodynamic critical field of many conventional superconductors is well represented by a parabolic relationship, i.e. $H_c(T)/H_c(O) = 1 - (T/T_c)^2$, Eq. (7.14) leads to

$$C_n - C_s = \frac{H_c^2(0)}{2\pi T_c}\left[\frac{T}{T_c} - 3\left(\frac{T}{T_c}\right)^3\right] \quad . \tag{7.17}$$

The first term on the right can be associated with the normal state contribution $C_n(T)$ equal to $C_n(T) = (H_c^2(0)/2\pi T_c^2)T = \gamma T$. Thus there should be a relationship between $H_c(O)$, T_c and γ. The second term on the right of Eq. (7.17) represents $C(T)$ of the electrons in the superconducting state, i.e. $C_{es}(T) = [3H_c^2(0)/2\pi T_c](T/T_c)^3$. This simple thermodynamic argument, along with the observed parabolic dependence of $H_c(T)$, predicts a T^3 dependence for $C_{es}(T)$ as $T \to 0$. As we will see, both experiments and the BCS theory [50] indicate that $C_{es}(T)$ goes to zero more rapidly than T^3, and this difference is due to the slight difference between the assumed parabolic T-dependence of $H_c(T)$ and the actual dependence. However, near T_c this difference is small and the relationship between γ, T_c and $H_c(O)$ is approximately correct.

One of the significant accomplishments of the BCS theory is that it provided a correct description of the thermodynamic properties of superconductors. The BCS theory predicts a second order transition to the superconducting state with a discontinuity in the specific heat at $T = T_c$ given by

$$\frac{(C_s - C_N)}{\gamma T_c} = \frac{\Delta C}{\gamma T_c} = 1.43 \tag{7.18}$$

where γ is the electronic specific heat coefficient in the normal state. Values for $\Delta C/\gamma T_c$ close to 1.43 have been measured for numerous superconductors. However, this ratio depends on the strength of the electron-electron interactions responsible for the pairing. Note that this result does not depend on the details of the pairing interactions and would be obtained in a weak-coupling BCS theory independent of the origin of the pairing mechanism.

Typical values of $\Delta C/\gamma T_c$ along with λ_{ph}, the phonon mediated pairing parameter, are given in Table 7.1. It is clear from Table 7.1 that as the pairing interaction λ_{ph} increases the ratio $\Delta C/\gamma T_c$ also increases. Recently, Blezius and Corbotte [74] have shown that

Table 7.1. Elemental Superconductors[a,b,c]

Element	T_c(K)	λ_{ph}	$\Delta C/\gamma T_c$	θ_D(K)
BCS	–	–	1.43	–
Al	1.16	0.38	1.45	428
Zn	0.85	0.38	1.25-1.30	309
Ga	1.08	0.40	1.44	325
Cd	0.52	0.38	1.40	209
In	3.40	0.69	1.73	112
Sn	3.72	0.60	1.60	200
Tl	2.38	0.71	1.50	79
Hg	4.16	1.00	2.37	72
Pb	7.19	1.12	2.71	105
Ta	4.48	0.65	1.69	258
V	5.30	0.60	1.49	399
Nb	9.22	0.82	1.87	277

[a] R. Meservey and B. B. Schwartz, *Superconductivity*, edited by R. D. Parks (Dekker, New York, 1965), pg. 165.
[b] G. Gladstone, M. A. Jensen, and J. R. Schrieffer, *ibid*, pg. 780.
[c] W. L. McMillan, *Phys. Rev.* **167**, 331 (1986).

$\Delta C/\gamma T_c$ for an isotropic Eliashberg superconductor initially increases with λ_{ph}, reaches a maximum, i.e. $\Delta C/\gamma T_c$ max $\sim$ 3.7, then decreases below the weak-coupling BCS limit with a further increase in the electron-phonon coupling strength. Thus, based on this recent work, $\Delta C/\gamma T_c \sim 1.43$ is consistent with weak coupling BCS theory or very strong coupling extensions of the BCS theory. If an excitonic coupling is added phenomenologically to the electron-phonon coupling and modeled just by adding the effective electron-phonon and electron-exciton couplings, then $\Delta C/\gamma T_c$ for a system with a $T_c \sim$ 100 K decreases monotonically with increasing electron-phonon weight. For values used by Marsiglio and Carbotte [75], the weak coupling value of $\Delta C/\gamma T_c$ is obtained for interactions dominated by the virtual exchange of high energy excitons and $\Delta C/\gamma T_c$ approaches $\sim$0.6 for coupling dominated by electron-phonon interactions. For superconductors with $T \sim$ 100 K, the predicted decrease of $\Delta C/\gamma T_c$ with increasing percentage of electron-phonon coupling is a reflection of the reduction of $\Delta C/\gamma T_c$ due to strong coupling corrections, mentioned above.

For $T \ll T_c$, the BCS theory predicts that $C_{es}(T)$, the electronic contribution to $C(T)$ in the superconducting state, will be proportional to $\exp(-\Delta/kT)$ where Δ is the superconducting gap in the electronic density of states at E_F. Numerical calculations of $C_{es}(T)/\gamma T_c$ and other thermodynamic functions have been calculated within the weak coupling BCS limit and tabulated by Muhlschlegel [76]. Shown in Fig. 7.6 is $C(T)/T$ vs. T for Al [77]. A clear discontinuity in $C(T)$ at T_c is seen and $\Delta C/\gamma T_c \sim$ 1.40 where γT is obtained from a fit of the normal state $C(T)/T$ to $\gamma + \beta T^2$. In Fig. 7.6, the solid line is a fit of the electronic contribution to $C(T)$ to the BCS predicted behavior [$C_{es}(T) = a \exp(-\Delta/T)$] with $\Delta \approx$ 1.34 T_c, close to the value anticipated using the weak coupling BCS theory. Many other systems, e.g. Ca, Sn, In, Pb etc., have been studied and the results are consistent with the weak coupling BCS theory or appropriate extensions of this theory to account for strong coupling corrections.

7.3.3. Thermal Properties of the High T_C Oxides

This section will summarize some of the recent specific heat results on the new oxide superconductors and compare these results to the behavior discussed in the previous section (see Section 7.3.2). As indicated earlier, the specific heat has played a central role in the development of our understanding of conventional superconductivity and more exotic superfluidity as exhibited by ^{3}He [78] and heavy fermion systems [57-59,79,80]. As with previous sections, the results presented are only representative and the discussion is

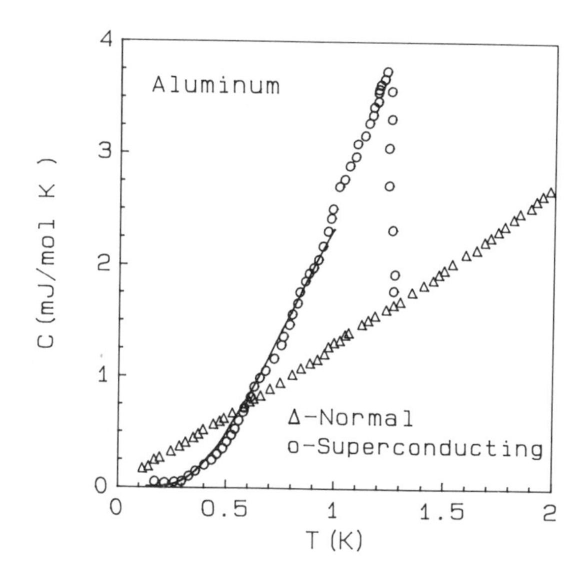

Fig. 6. Specific heat versus temperature for Al, o-normal state, Δ-superconducting state. The solid curve is a fit to C(T) $\propto$ exp($-\Delta/T$). Data from [77].

meant to establish certain trends without being consumed by details. For those readers interested in a more complete discussion of thermodynamic properties and, in particular, specific heat behavior in the new oxide superconductor there are several excellent reviews [63,81].

There are distinct differences between previous specific heat studies of conventional superconductors and those recently published for the oxide superconductors. First, the critical field necessary to suppress superconductivity is too large in the oxide superconductors to obtain reliable normal state data for $T < T_c$. This, coupled with the large phonon contribution to C(T) for $T \leq T_c$, makes it nearly impossible to establish the temperature dependence of $C_{es}(T)$, i.e., the electronic contribution to C(T) for $T \leq T_c$. Therefore, the discussion of C(T) for the new oxide systems is restricted to C(T) in the vicinity of T_c, where a C(T) anomaly is seen, and to C(T) as T→0, where a separation of electronic, phonon and other contributions to C(T) can usually be made and where clear predictions are available on the temperature dependence of the electronic contribution within the superconductivity state. The data near T_c, i.e., $\Delta C(T_c) = C_s(T_c-) - C_N(T_c+)$, is also complicated by two factors which contribute to a broadening of the transition. These factors include broad and, possibly, incomplete transitions due to the lack of phase purity and an intrinsic broadening of the transition due to thermodynamic fluctuations of the order parameter. The latter effect can provide valuable information concerning the nature of the order parameter and this topic will be briefly discussed in a later section. However, broadening due to phase purity introduces additional uncertainties in establishing $\Delta C(T_c)$.

7.3.3.1. $(La,M)_2CuO_4$, M = Ba and Sr

There have been several measurements of C(*T*) for $(La,M)_2CuO_4$ with M = Ba, Sr and Ca [82-85]. An anomaly at T_c has been observed in a few studies but this anomaly is broad and the size of the anomaly varies considerably. The lack of a sharper anomaly and the variation from sample to sample probably reflect problems associated with sample reproducibility and quality. Shown in Fig. 7.7 is C(*T*) vs *T* for $La_{1.8}Sr_{0.2}CuO_4$ [84]. In the inset is shown $\Delta C(T)$ vs *T* where $\Delta C(T) = C(T) - C_f(T)$ and $C_f(T)$ is the temperature dependence of C(*T*) obtained by fitting the data to a polynomial function but excluding the region near T_c, i.e., $T_c \pm 10$ K. Fitting $\Delta C(T)$ to a discontinuity which conserves the entropy change in the vicinity of T_c leads to $\Delta C(T_c) = 1.4$ J/mole-K and $T_c = 40$ K. This value of $\Delta C(T_c)$ is somewhat larger than the value of $\Delta C(T_c) \approx 0.37$ J/mole-K obtained for $La_{1.85}Sr_{0.15}CuO_4$ where a magnetic field was used to suppress the

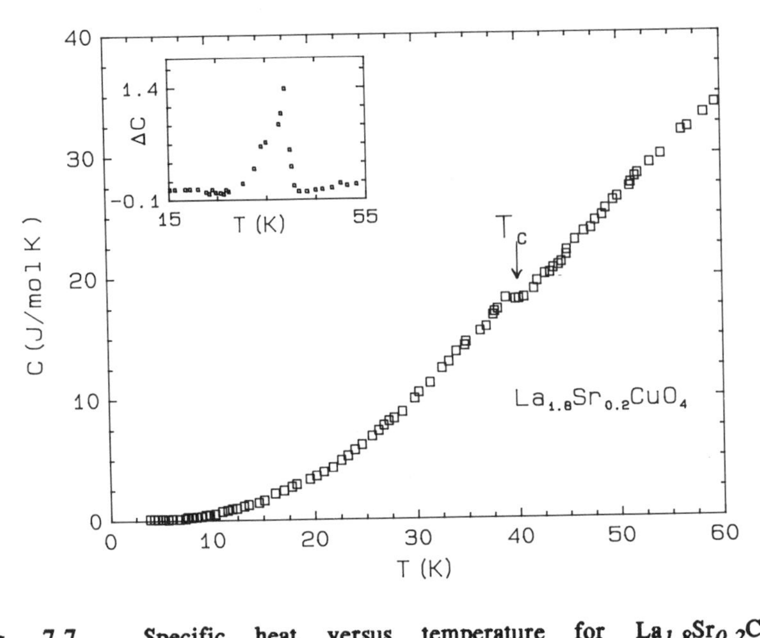

Fig. 7.7. Specific heat versus temperature for $La_{1.8}Sr_{0.2}CuO_4$. Inset: Specific heat anomaly in the vicinity of the superconducting transition (see text). Data from [84].

$C(T)$ anomaly and thus obtain an estimate of $C_N(T)$ near T_c [85]. The value of $\Delta C(T_c)$ = 1.4 J/mole-K is also larger than ΔC = 0.6 J/mole-K estimated from critical field data [86]. A reliable value of $\Delta C(T)$ has not been established for the 214 compounds, but these early results do establish that the superconductivity observed in these oxides is a bulk phenomena and a specific anomaly does occur at T_c.

Even though, there remains some controversy concerning the size of $\Delta C(T_c)$, it is worthwhile to use the above value of $\Delta C(T_c)$ = 1.4 J/mole-K to make some comparison to conventional superconductors and estimates of other 214 properties. An independent estimate of γ, the electronic specific heat coefficient, can be obtained from the Pauli contribution to the magnetic susceptibility, χ_p, i.e., χ_p/γ = the Wilson ratio $=\frac{1}{3}(\pi k_B/\mu_B)^2$ where μ_B = Bohr magneton. Setting $\chi_p = \frac{1}{3}(\pi k_B/\mu_B)^2\gamma$ does not appropriately address possible mass enhance effects such as spin fluctuations which would lead to $\chi_p = \frac{1}{3}(\pi k_B/\mu_B)^2\gamma_o(\ln S)/S$ where S is the Stoner factor. However, it has been demonstrated that χ_p/γ = Wilson ratio is a good approximation to the scaling of χ_p and γ for many highly correlated electron systems such as heavy fermion systems [79,80]. The measured value of $\chi(T)$ for $T \approx T_c$ is 4×10^{-4} emu/mole [83] and after subtracting estimates of the core, Landau and van Vleck contributions to $\chi(T)$, $\chi_p \approx 6.9\times10^{-4}$ emu/mole is obtained. This value of χ_p leads to an expected γ, γ_{sus} = 17 mJ/mole-K^2 where $\gamma_{sus} = 3(\mu_B/\pi k_B)\chi_p$. This γ value can be compared to γ_{BCS} obtained from $\gamma_{BCS} = \Delta C/1.43T_c$, i.e., using Eq. (7.17). For $La_{1.85}Sr_{0.15}CuO_4$ and using $\Delta C(T_c)$ = 1.4 J/mole-K [84], γ_{BCS} = 24 mJ/mole-K^2. Both γ_{sus} and γ_{BCS} are considerably larger than that obtained for typical metals, e.g. $\gamma \approx 0.7$ mJ/mole-K^2 for Cu, and larger than γ obtained from band theory calculation for the 214 compound [8,87]. Using a mean value of ~20 mJ/mole-K^2 and assuming the electronic properties of the CuO_2 planes are best described by a two-dimensional electron gas, i.e. $\gamma_{2D} = \pi m^* k_B^2/3\hbar^2\Delta$, where Δ is the spacing of the 2D system and m^* is the electron mass, a mass enhancement ratio m^*/m_o = 13 is obtained. The enhanced γ and m^*/m_o dramatically point to the importance of electron correlation effects in this system and if attributed to phonon induced mass enhancements [see Eq. 7.2], it would correspond to an unrealistically large electron-phonon coupling constant.

Substituting γ_x into $\Delta C = A\gamma T_c$ and using $\Delta C \approx 1.4$ J/mole-K yields A $\approx$ 2.00 as compared to the BCS value, 1.43. This slight enhancement of $\Delta C/\gamma T_c$ as compared to the weak-coupling BCS value is not significant considering the large uncertainties in some of the values used. This $\Delta C/\gamma T_c$ value indicates that the system could be characterized as a weak coupling superconductor thus requiring other mechanisms than solely electron-phonon mediated coupling to account for the high T_c's. A reliable determination of ΔC and χ_p and a better knowledge of electron correlations contributing to the mass

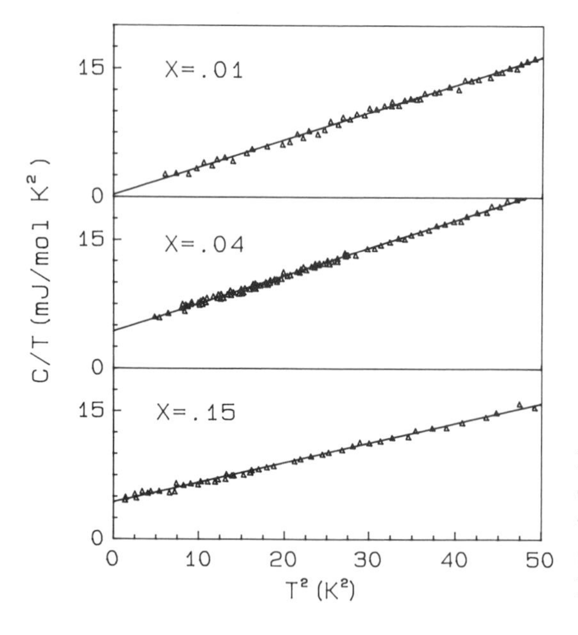

Fig. 7.8. Specific heat divided by temperature versus temperature squared for $La_{2-x}Ba_xCuO_4$. Data from [88].

enhancement is required to justify a more extensive analysis of these data.

Shown in Fig 7.8 is $C(T)/T$ vs T^2 for $La_{2-x}Ba_xCuO_4$ with x = 0.01, 0.04 and 0.15 [88]. $La_{1.85}Ba_{0.15}CuO_4$ is superconducting with T_c = 30 K whereas the other two samples remain normal. Recall from section 7.3.2, $C(T)/T \to 0$ as $T \to 0$ in the superconducting state. Shown in Fig 7.9 is the magnetic and superconducting phase boundary along with γ vs x for $La_{2-x}Ba_xCuO_4$. Very similar behavior is obtained for the $(La,Sr)_2CuO_4$ system [89]. As seen in Fig 7.9, $\gamma \approx 0$ for $x < 0.02$ but for $x \geq 0.02$, γ increases with increasing x, i.e., $\gamma \neq 0$, as the superconducting phase boundary is approached, and for x values within the superconducting phase. This unanticipated result is contrary to what would be expected based on a comparison to conventional superconductors and the BCS theory. However, the magnetic and superconducting phase boundary along with the γ vs x behavior displayed in Fig 7.9 is qualitatively consistent with what would be expected using a resonant valence bond model [17]. This point will be discussed further in Section 7.3.3.4.

7.3.3.2. $RBa_2Cu_3O_{7-\delta}$, R = Rare Earth and Yttrium

Most of the rare earths, except Ce, Pr, Pm and Tb, can be substituted for Y in $YBa_2Cu_3O_{7-\delta}$ without adversely affecting the superconducting properties [90]. Shown in Fig 7.10 is $C(T)$ vs T for

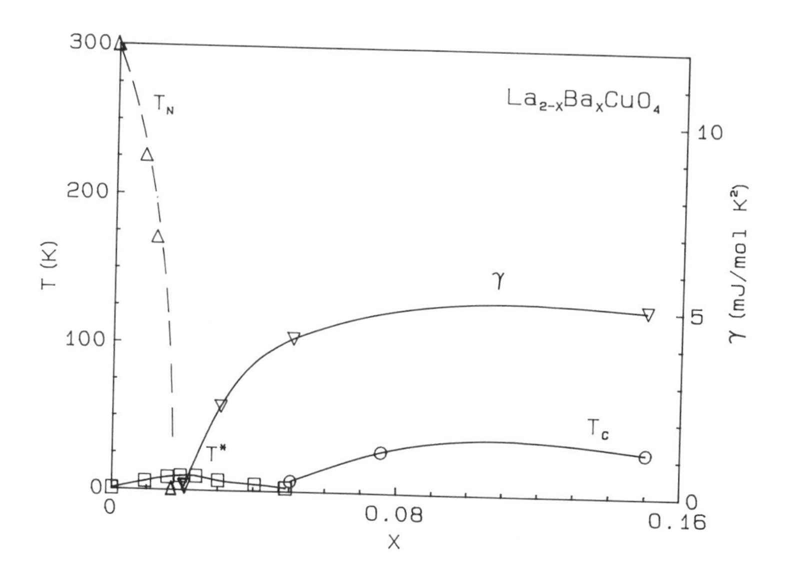

Fig. 7.9. Neel temperature T_N, superconducting transition temperature T_c, electronic specific heat coefficient γ, and "Spin-glass" temperature, T_g, versus Ba concentration in $La_{2-x}Ba_xCuO_4$. Δ – T_N, ∇ – γ, ○ – T_c, and □ – T_g. Data from [88].

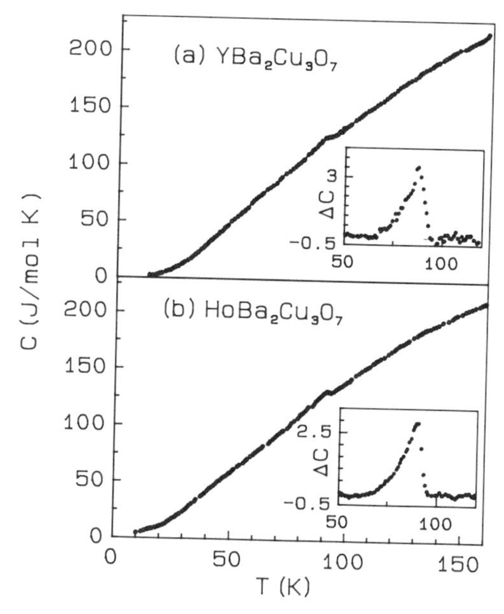

Fig. 7.10. Specific heat versus temperature for $YBa_2Cu_3O_7$ (a) and $HoBa_2Cu_3O_7$ (b). Insets: Specific heat anomaly in the vicinity of the superconducting transition (see text).

$YBa_2Cu_3O_{7-\delta}$ and $HoBa_2Cu_3O_{7-\delta}$ with $\delta \approx 0.0$ [91]. Shown in the inset of Fig 7.10 is $\Delta C(T)$ vs T where ΔC is obtained by fitting $C(T)$ vs T except in the vicinity of T_c and subtracting this fitted behavior from the measured $C(T)$ vs T. Note, a BCS-like discontinuity is seen in $C(T)$ vs T. Fitting ΔC for $YBa_2Cu_3O_7$ to a step discontinuity which conserves the entropy change near T_c leads to $\Delta C = 5.1$ J/mole-K or $\Delta C/T_c = 55$ mJ/mole-K^2. This value for $\Delta C/T_c$ is similar to values obtained in several other studies [63,81,92-99]. For $HoBa_2Cu_3O_7$, a sharper but slightly smaller specific heat jump is obtained. $\Delta C/T_c = 55$ mJ/mole-K leads to $\gamma_{BCS} = \Delta C/1.43T_c = 38$ mJ/mole-K^2. Magnetic susceptibility data for $YBa_2Cu_3O_7$ can be fit to $\chi(T) = \chi_o + C/(T + \theta)$ where χ_o is associated with the free carriers. $\chi_o \approx 2.8 \times 10^{-4}$ emu/mole has been obtained in many studies [91,100]. The Pauli contribution to χ_o can be estimated and leads to a $\gamma_x = 35$ mJ/mole-K^2. This value compares remarkably well with $\gamma_{BCS} = 38$ mJ/mole-K^2 obtained from $\Delta C(T_c)$ and suggest that the value of $\Delta C/\gamma T_c$ is close to 1.43, the weak coupling BCS value. Very similar values of $\Delta C/T_c$ have been obtained for many of the rare earth substituted 123 compounds [91,101], but for most of these systems an accurate determination of χ_o is precluded because of the large 4f-electron contribution to the magnetic susceptibility. As with the 214 compound, the 123 compound has a weak coupling BCS anomaly at T_c and a highly enhanced electronic contribution to the specific heat. Assuming the quasiparticles can be best represented as a 2D gas, $\gamma = 38$ mJ/mole-K^2 and $\Delta \approx 3.9$ Å leads to $m^*/m_o \approx 11$, which is similar to that estimated for the 214 compound. A very similar basal plane effective mass $m^*/m_0 \approx 10$ is obtained from the measurement of the anisotropy penetration depth [102]. A comparison of quantities derived from $\Delta C(T_c)$ and other properties are presented in Table 7.2. Junod et al. [96] measured the magnetization vs H near T_c for the 123 compound and integrated these data to obtain an estimate of the thermodynamic critical field, $H_c(0)$. These measurements were in agreement (within 10%) with the expected thermodynamic scaling between $\Delta C/T_c$ and dH_c/dT given by Eq. (7.15). Thus, it appears that 123, like 214, has a specific heat anomaly consistent with a weak coupling BCS theory and $\Delta C/T_c$ along with other quantities are reasonable consistent with GL predictions. In addition, the effective band mass for both systems is highly enhanced, $m^*/m_o \approx$ 10-13.

Shown in Fig 7.11 is $\Delta C/T_c$ vs T_c for $YBa_2(Cu_{1-x}Zn_x)_3O_7$ [103]. Zn doped into $YBa_2Cu_3O_7$ enters the Cu sites located in the CuO_2 planes and causes a rapid depression of T_c, i.e. $T_c \to 0$ for x ~ 0.10. [104,105]. Careful $C(T)$ measurements have been made on this system and even though $\Delta C(T_c)$ is reduced as T_c decreases it can be resolved for $T_c \geq 50$ K, i.e., $T_c(x)$ equal to approximately half its original value $T_c(0)$. For each Zn concentration studied, a $\Delta C(T_c)$ was calcu-

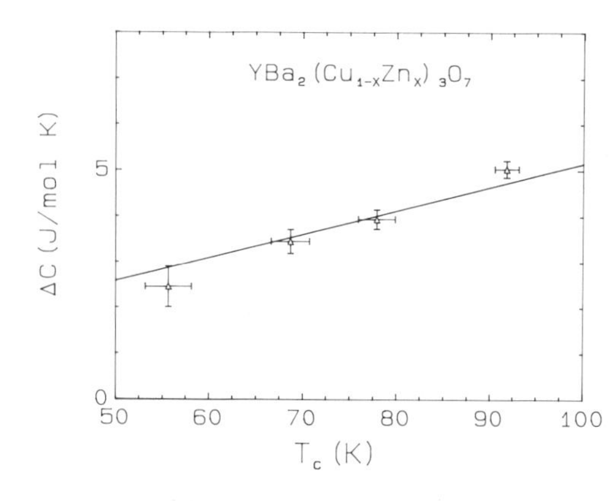

Fig. 7.11. Specific heat jump ΔC at T_c versus the superconducting transition temperature for $YBa_2(Cu_{1-x}Zn_x)O_7$. The solid curve is a fit to $\Delta C = 1.43\gamma_{BCS}T_c$, with $\gamma_{BCS} = 36$mJ/mole-K^2.

lated as described above and the anomaly at T_c was replaced by a sharp discontinuity in ΔC which conserved the entropy change near T_c. This analysis leads to a $\Delta C/T_c$ which is nearly constant. If the $\Delta C/T_c$ vs T_c shown in Fig 7.11 is fit to $\Delta C/T_c = 1.43\gamma_{BCS}$, one obtains a $\gamma_{BCS} = 36$ mJ/mole-K^2. The Pauli susceptibility was also measured and leads to a $\gamma_x \sim 35$ mJ/mole-K^2, nearly independent of Zn-concentration. These results strongly indicate that the 123 superconductors are weak coupled BCS superconductors with an highly enhanced electronic density of states, i.e. electron correlations are very strong in these two dimensional conducting systems as reflected by the large γ values obtained from ΔC and χ_p.

Shown in Fig 7.12 is $C(T)/T$ vs T^2 for $YBa_2Cu_3O_7$. An unexpected and still controversial result is the large electronic-like contribution to $C(T)$ as $T \to 0$. In addition to the electronic contribution, an upswing in $C(T)/T$ as $T \to 0$ is also reported in most studies. This upswing in $C(T)/T$ is qualitatively similar to that reported for spin

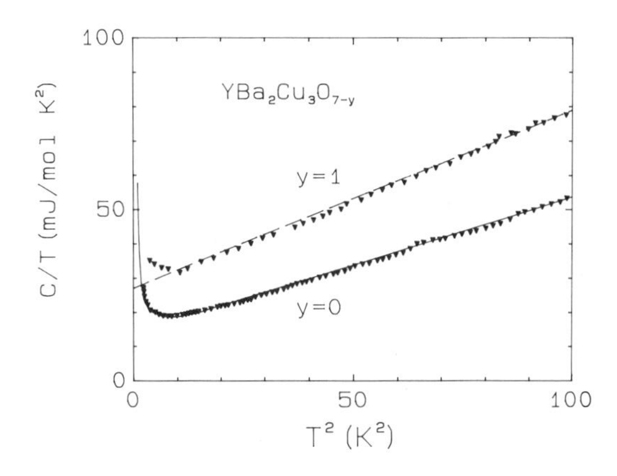

Fig. 7.12. Specific heat divided by temperatures versus temperature squared for $YBa_2Cu_3O_{7-y}$, with y = 0.0 and 1.0. The solid curve is given by $C(T) = A/T^{-2} + \gamma T + \beta T^3$, with A = 35 mJ-K/mole, γ = 12.9 mJ/mole-K^2, and β = 0.23 mJ/mole-K^4.

fluctuation systems, e.g. heavy fermion systems [79,80]. For these systems, most of the features seen in C(T) can be attributed to spin fluctuations of hybridized f-electrons. Attempts to fit C(T) for 123 to $\gamma T + \beta T^3 + \delta T^3 \ln T$, the expected dependence assuming ferromagnetic spin fluctuations, were not successful. The low temperature upswing which is characteristic of many studies is much too sharp to be attributed to spin fluctuations as described by a $\delta T^3 \ln T$ contribution. The origin of the upswing and the linear contribution has not been clearly established at this time. Measurements of C(T) on single crystals have also shown a linear contribution but did not display an upswing [106]. However, other single crystal studies have produced C(T) as T→0 similar to that reported for polycrystalline 123 [107] and shown in Fig. 7.12. The origin of the discrepancy between these two results is not clear, but the lack of an upswing in ref. [106] is significant. C(T) in Fig. 7.12 can be fit to $AT^{-2} + \gamma T + \beta T^3$ with A = 35 mJ-K/mole, γ = 12.9 mJ/mole-K^2, and β = 0.23 mJ/mole-K^4. The γ values determined from C(T)/T vs T^2 for the 123 compound have varied considerably, ranging from 4-20 mJ/mole-K^2 with most of the values between ~9-12 mJ/mole-K^2 [63,81]. This sample dependence of γ along with the apparent T^{-2} contribution to C(T) and measurements of C(T) in $H \neq 0$ have lead to speculation that a large fraction of γ and the upswing in C(T)/T can be attributed to minute second phases present in the sample, e.g. $BaCuO_{2+x}$ [63]. This certainly may account for a portion of γ and part of the upswing but there appears to be an intrinsic contribution to γ remaining after contributions due to second phases are appropriately accounted for.

As indicated earlier, a low temperature ionic tunneling contribution to C(T), which is proportional to T, is well known in disordered solids. Can this mechanism account for the unexpected C(T) behavior at T→0 reported for the 123 and 214 superconductors? For most disordered systems, this contribution usually leads to a smaller contribution to C(T) than reported here [108]. In addition, if ionic tunneling was the dominant mechanism contributing to this unexpected C(T) behavior, then there would be expected a strong oxygen dependence and a lack of magnetic field dependence to this behavior. It has been thought that ionic tunneling between occupied oxygen sites and vacancies in the vicinity of the Cu0 chains would be the most likely candidate for such a contribution. Vacuum annealing superconducting $YBa_2Cu_3O_{7-\delta}$ with δ ~ 0 removes oxygen from the chain sites (δ ~1.0), however the low temperature behavior of C(T) is not altered significantly. This lack of dependence of C(T) on oxygen stoichiometry and the size of C(T) as T→0 seems to rule out ionic tunneling as a major factor influencing the low temperature properties of the 123 compound.

Recent studies of $YBa_2(Cu_{1-x}Zn_x)_3O_7$ [105] and $(Y_{1-y}Pr_y)Ba_2Cu_3O_7$

[109] provide indirect evidence that the large γ's and upswing in $C(T)/T$ as $T \to 0$ may be due to intrinsic behavior. Both these systems lead to enhanced γ values which increase with increasing x and y and are much to large to be attributed to impurities phases or ionic tunneling contributions. For $YBa_2(Cu_{1-x}Zn_x)_3O_7$, $C(T)/T$ reaches $\sim$120 mJ/mole-K^2 at T = 1.5 K and these large γ values have been attributed to localization of carriers resulting from a Zn-induced disruption of the hybridization within the CuO_2 planes. Thus it seems reasonable that the nonzero γ values reported for the pure 123 compound reflect partially localized states at E_F brought about by disorder within the chains due to O vacancies or residual disorder within the planes. This would account for the sample dependence of γ that has been reported.

An alternative explanation of these results is that these systems are not BCS superconductors. There are several non-BCS models which have been proposed to account for the superconductivity and other properties in these oxide systems. These models also predict non-BCS behavior for $C(T)$ as $T \to 0$. In particular, the resonant valence bond (RVB) model proposed by Anderson and coworkers [5,17] has received considerable attention with regard to the anomalous $C(T)$ behavior reported for both the 214 and 123 superconducting compounds. Starting from the nearly half-filled Hubbard model, there are two types of excitations, holons of charge $+e$ and spin zero (bosons) and spinons of charge zero and spin ½ (fermions). The superconducting state results from a condensation of the holons and this state has a nonzero gap. However, the apparent gapless behavior evident in $C(T)$, infrared studies, etc. results from the spinons which are not part of the condensed state and contribute to $C(T)$ as $T \to 0$. These spinons lead to an electronic-like contribution to $C(T)$, i.e. a contribution proportional to T. Anderson and coworkers [17] have estimated the spin contribution of $C(T)$ and have obtained a value close to those reported for several of the high T_c oxides. However, the nature of the low-T excitations contributing to the apparent gapless behavior is not universally established [110,111]. Wang and Rice [110] calculated the specific heat of a RVB superconductor and obtained an estimate for the linear term with $\gamma \approx 20$ mJ/mole-Cu-K^2. However, this contribution is from gapless behavior in the holon excitation spectrum and spinon excitations have a gap. It is hoped that these issues will be resolved with further developments of the RVB model. The values calculated in both of the above studies are comparable to or larger than most of the reported values, and demonstrate the possible importance of this mechanism.

A clear resolution to the origin of the linear temperature contribution to $C(T)$ as $T \to 0$ is not available at this time. It's presence in the 214 and 123 systems and in other systems such as $BaPb_{1-x}Bi_xO_3$ [112] and $Ba_{1-x}K_xBiO_3$ [113] seem to suggest a more universal explanation of this unexpected result than provided by simply attributing it to impurity phases or ionic tunneling. This issue is further discussed in the next section.

7.3.3.3. Other Oxide Systems

Summarizing the results presented above, the 214 and 123 superconductors are characterized by 1) a specific heat anomaly at T_c which is consistent with weak coupling BCS estimates of $\Delta C(T_c)$, 2) the γ values derived from χ_p and $\Delta C(T_c)$ indicate that the electrons are highly correlated with effective masses $m^*/m_o \sim 10$-13 and 3) $C(T) = \gamma T + \beta T^3$ with $\gamma \neq 0$ at low temperatures. To what extent are these features also displayed by the other oxide superconductors, e.g., $Ba(Pb,Bi)O_3$, $(Ba,K)BiO_3$, Bi-Sr-Ca-Cu-Oxides and Tl-Ca-Ba-Cu Oxides?

The $BaPb_{1-x}Bi_xO_3$ was discovered in 1975 by A. W. Sleight and co-workers [114] and is one of the earliest high T_c oxides, $T_c \sim 13$ K for x ~ 0.20-0.25. More recently, higher T_c's have been obtained in the isomorphic system $Ba_{1-y}K_yBiO_3$, $T_c \sim 25$-30 K for $0.25 < y < 0.45$ [115,116]. These systems are the only non Cu-based oxides with $T_c > 12$ K and they crystallize in a cubic perovskite structure, i.e., do not possess the low dimensional CuO_2 planes and extreme anisotropy characteristic of the Cu-O based high T_c superconductors. However, they should be included as members of a possible new class of superconductors with high T_c and low coupling constants (see Sec. 7.4). $C(T)$ has been measured for $Ba(Pb,Bi)O_3$ [112] and $(Ba,K)BiO_3$ [113]. For both systems, $C(T)/T$ does not extrapolate to zero at low temperatures for those samples which are superconducting. For $Ba_{1-y}K_yBiO_3$, γ determined for $C(T)$ as $T \rightarrow 0$ is 0 for $y < 0.25$, where $T_c = 0$. However, within the superconducting region $\gamma \sim 1.2 \pm 0.2$ mJ/mole-K^2 was obtained, i.e., for $0.25 < y \leq 0.45$. These results resemble those presented earlier for 214 compounds (see Fig. 7.9) where $\gamma = 0$ in the nonsuperconducting, antiferromagnetic phase and nonzero γ values are obtained as the superconducting phase boundary is approached and within the superconducting region. $C(T)$ also has been measured in the vicinity of T_c for both $Ba(Pb,Bi)O_3$ [112] and $(Ba,K)BiO_3$ [117,118], and an anomaly is seen. For $Ba(Pb,Bi)O_3$, $\Delta C/T_c \sim 1.5 \pm 0.2$ mJ/mole-K^2 and leads to values consistent with weak coupling BCS predictions. Further work is required because the transition widths for $(Ba,K)Bi0_3$ have been too broad to permit an accurate determination of $\Delta C(T_c)$. Summarizing, these cubic superconductors have a linear temperature dependent contribution to $C(T)$ as $T \rightarrow 0$ in the superconducting state and $Ba(Pb,Bi)O_3$ has a $\Delta C(T_c)$ consistent with weak coupling BCS predictions. These results resemble those reported for both 214 and 123 oxide superconductors.

The most surprising and possibly one of the more important

results to emerge out of recent studies has been $C(T)$ measurements on the new Cu-O planar superconductors containing Bi-Sr-Ca-Cu-O and Tl-Ba-Ca-Cu-O. These new systems contain a variety of phases which primarily differ in the number of Cu-O planes per unit cell and, in general, T_c increases with the number of Cu-O planes per cell [119-121]. Consistent with the results reported above for the other oxide superconductors, $\gamma \neq 0$ for superconducting Tl-Ba-Ca-Cu-O samples [122] but $\gamma \sim 0$ for each of the Bi-Sr-Ca-Cu-O samples studied [122-127]. This latter result is inconsistent with results reported for all the other high T_c oxides discussed above and has important implications on the nature of the paired state. (See Sec. 7.3.3.4). For $T \sim T_c$, a $C(T)$ anomaly has been reported for both the Bi-Sr-Ca-Cu-0 and Tl-Ba-Ca-Cu-0 superconductors, but the transition widths in most samples studied have been too broad to permit an accurate determination of $\Delta C(T_c)$. A well resolved anomaly at T_c has been obtained for $(Bi,Pb,Sb)_2Sr_2Ca_2Cu_3O_{10}$ with ΔC at $T_c \sim 6.0$ J/mole K and $T_c = 109$ K [127]. The γ value obtained from $\Delta C = 1.43\ \gamma_{BCS}\ T_c$ and the temperature independent contribution to $\chi(T)$ is ~ 40 mJ/mole-K^2. This γ value leads to an effective mass enhancement comparable to that obtained for 124 and 123, i.e., $m^*/m_o \simeq 10$-13. Thus, other than the apparent absence of a γT contribution to $C(T)$ as $T \to 0$ for Bi-Sr-Ca-Cu-0, these systems resemble the 124 and 123.

7.3.3.4. An Overview of Thermal Properties

All the oxide superconductors with $T_c \geq 13$ K except Bi-Sr-Ca-Cu-oxides have nonzero electronic-like contributions to $C(T)$ as $T \to 0$. $\gamma \neq 0$ for $T \ll T_c$ has been attributed to a variety of possible contributions, e.g., incomplete transitions to the superconducting state, two level ionic tunneling, spin fluctuations, impurity phases etc. For many systems, an upswing in $C(T)/T$ as $T \to 0$ is observed in addition to the electronic-like term. The upswing seems to fit best the high temperature limit of a Schottky type contribution which leads to a $C_{sch}(T) \propto T^{-2}$ term for $T > \Delta$ where Δ is the splitting of the levels. In Zn-doped samples of $YBa_2Cu_3O_7$, this A/T^2 contribution becomes very large leading to values of $C(T)/T \sim 120$ mJ/mole-K^2 for $YBa_2(Cu_{0.86}Zn_{0.14})_3O_7$ [105]. Such large values of $C(T)/T$ are inconsistent with relating this anomalous behavior to impurity phases or ionic tunneling. Very large γ values are reported for impurity phases, i.e., $BaCuO_4$ and Y_2BaCuO_6 samples [128], but careful estimates of impurity phase contributions to $C(T)$ for 123 totally preclude attributing the nonzero γ values to such effects. In addition, similar behavior has been reported for Ba, Ca and Sr doped 214 superconductors which also brings into question the reliability of arguments solely attributing the measured γ values to impurity phases. One clear explanation of both the nonzero and the $C(T)/T$ upswings as $T \to 0$ is partially localized quasiparticle states. This localization is brought about by residual disorder which would disrupt the Cu-3d and O-2p hybridization in these strongly correlated electron systems. Results on $YBa_2(Cu,Zn)_3O_7$ and $(Y,Pr)Ba_2Cu_3O_7$ tend to support such arguments.

The most surprising result is the lack of an electronic contribution to $C(T)$ for $T \ll T_c$ in the Bi-Sr-Ca-Cu-O superconductors. $\gamma = 0$ for this system impacts heavily on many models proposed to explain the pairing within these systems. As discussed earlier, the RVB model predicts two types of excitations—a spinon, which has spin ½ but no charge, and a holon, which has positive charge e but no spin. It is the scattering of the holons by the spinons and the spinon density of states which leads to the linear temperature dependence of $\rho_{ab}(T)$. Superconductivity results from the pairing of holons, with or possibly without a quasiparticle gap. In addition, the spinons may also remain as low lying excitations and thus either the gapless holon excitation spectrum or the spinons could give rise to apparent gapless behavior as reflected in $C(T)$. Estimates of these contributions to $C(T)$ are sufficiently large to account for the observed linear temperature dependent contributions to $C(T)$. However, this behavior should be universal which makes the absence of a linear temperature-dependent term for the Bi 2212 and 2223 puzzling. Overall, the RVB model does quantitatively account for many of the magnetic and superconducting properties reported for the 123 and 214 superconductors. However, additional information is needed on these new oxide superconductors and more detailed predictions using the RVB model are required to help clarify the relevance of this model to these systems.

The absences of nonzero γ values for the Bi-Sr-Ca-Cu-Oxide superconductors also has important implications with respect to non s-state (BCS) pairing models. As pointed out by Chakraborty et. al. [126], the $\gamma = 0$ observed for Bi-Sr-Ca-Cu-O rules out gapless pairing and gap functions with lines of nodes. Such gapless behavior would lead to a $C(T)$ contribution linear in T. In addition, Chakraborty et. al. indicate that triplet pairing with point nodes is also unlikely. If this analysis is correct, these results strongly support simple s-state (BCS) pairing and $\gamma \neq 0$ reported for many of the other oxide superconductors must be accounted for using other models, e.g. partial localization of quasiparticles due to disorder. s-state pairing is also consistent with measurements of the temperature dependence of the magnetic penetration depth in single crystals of 123 [102].

With respect to $C(T)$ results in the vicinity of T_c, the situation is not completely clear. Nearly all the reliable data have been confined to the 123 system and here $\Delta C(T_c)$ obtained by replacing the anomaly by a $C(T)$ discontinuity which preserves the entropy change in the vicinity of T_c seems to be consistent with predictions based on a weak coupling BCS theory. Further work which clearly establishes the T and magnetic field dependence of $C(T)$ near T_c is required. Because of the short coherence lengths characterizing these new oxide superconductors, critical fluctuations should broaden the transition and careful measurements of these fluctuation contributions near T_c should provide valuable information on the nature of the superconductivity order parameter.

Prior to the discovery of the new high T_c oxides, contributions to the conductivity [129-131] and magnetic susceptibility [132,133] due to fluctuations of the Cooper pair density above and below T_c had been observed. These thermodynamic fluctuations [134] produce their most obvious modification of the mean field response above T_c where enhancements of the conductivity and diamagnetism are apparent. Understanding the nature of these fluctuations and the temperature dependence of their lifetimes provides valuable information concerning the dimensionality of the superconductivity fluctuations and the number of parameters required to specify the order parameter. Because of the large coherence length, characteristic of most conventional superconductors, i.e. $\xi(0) \sim 100\text{-}1000$ Å, fluctuation effects are more pronounced in lower dimensional systems, e.g., films and whiskers, and observations of these fluctuations were limited to mean field or Gaussian fluctuations. Fluctuation effects are enhanced in lower dimensions because of the reduced free energy difference between the superconducting and normal states due to the reduction of the fluctuating volume $\sim \xi^d(T)L^{3-d}$ where d is the dimensionality and L is a characteristic length scale for the system. Thus, the temperature dependence of the fluctuation conductivity for a BCS superconductor, first derived by Aslamazov and Larkin (AL) [135], depends on the dimensionality of the sample, and is given by

$$\sigma_G(T) = \begin{cases} 1/32 \ \left[e^2/h\xi(0)\right] \ \varepsilon^{-1/2} & \text{for 3-dimensions} \\ 1/16 \ \left[e^2/hd\right] \ \varepsilon^{-1} & \text{for 2-dimensions} \\ \pi/16 \ \left[e^2\xi(0)/hS\right] \ \varepsilon^{-3/2} & \text{for 1-dimension} \end{cases} \tag{7.19}$$

where $\varepsilon = (T-T_c)/T_c$ and the dimensionality is established by the dimensions that are small compared to $\xi(0)$, the coherence length. The specific heat will have both a mean field contribution $C_{MF}(T)$ which has been discussed in previous sections and a Gaussian fluctuation contribution $C_G(T)$, i.e., $C(T) = C_{MF}(T) + C_G(T)$. Standard models give

$$C_G(T) = C\pm \ |\varepsilon|^{-2+d/2} \tag{7.20}$$

with $C_+(C_-)$ equal to the amplitude for $\varepsilon > 0$ ($\varepsilon < 0$). The amplitude ratio $C_+/C_- = n/2^{d/2}$ where n is the number of components of the order parameter, i.e., $C_+/C_- = 2^{1-d/2}$ for a BCS superconductor with a complex order parameter, n = 2 [136, 137]. For the new oxide superconductors, modifications of mean field response due to fluctuations should be more apparent than in the conventional superconducting systems because of the short coherence lengths, e.g. $\xi_c(0) \sim 5$ Å and $\xi_{ab}(0) \sim 34$ Å for 123. The short coherence length decreases the fluctuating volume, making fluctuations more probable. The coherence length in the new oxide superconductors remains sufficiently large to make the observation of critical (logarithmic divergence) corrections unlikely. For n = 2, the Ginzburg criterion for the critical region predicts $|\varepsilon| \leq 10^{-3}$.

Manifestations of Gaussian fluctuations in the new oxide superconductors have been reported in measurements of the conductivity [138], magnetic susceptibility [139] and the specific heat [137,140]. Using parameters obtained for the anisotropic critical fields H_{c2} and transport properties, one would expect fluctuation behavior dominated by two-dimensional (2D) correlations well above T_c with a crossover to three dimensional (3D) behavior as T_c is approached due to the increase in $\xi_c(T)$, the coherence length perpendicular to the CuO_2 planes [141]. The paraconductivity, for both polycrystalline, thin films, and single crystals, shows 2D behavior well above T_c, but these results do not yield consistent behavior with respect to the crossover to 3D response as $T \to T_c$. Oh, et. al. [138] obtain 2D behavior for $\varepsilon > 0.06$ with a crossover to 3-D for $\varepsilon < 0.06$. However, Hagen et. al.'s [138] studies of the paraconductivity in single crystals displayed the 2D behavior at large ε but do not show a crossover to 3D. The amplitude factor for the paraconductivity appears to agree well with the AL prediction for 2D fluctuations indicating that other corrections to the AL theory may not be significant [142]. Measurements of the superconducting diamagnetic fluctuations in polycrystalline 123 yield 2D behavior for $T > 100$ K and 3D behavior for $T_c < T < 100$ K. These results give a similar crossover temperature to that obtained from paraconductivity measurements by Oh et. al. [138]. Lee et.al. [139] also estimated n, the number of order parameter components, and obtained a result consistent with s-wave pairing. Gaussian fluctuations have also been observed in $C(T)$ measured on single crystals [137,140]. These results could be accounted for by assuming 3D fluctuations with a non-BCS s-state pairing, i.e., number of order parameter components were 7 ± 2. However, more recent analysis of these results may be consistent with a smaller number of components of the order parameter.

The brief summary above should indicate that studies of fluctuation effects near T_c can provide rich insight into the dimensionality of the pairing and the nature of the order parameter describing the paired state. However, more work is required before definitive conclusions can be made.

7.4. The Role of Phonons

It has been well established that conventional metallic superconductivity, i.e., superconductivity in Pb, Sn, In, Al, Nb_3Sn, etc., can be adequately described by the BCS theory with phonon mediated pairing (see Chapter 1, 5 and 9). However, it is not clear that all the superconducting systems known prior to the recent discovery of high temperature superconductivity in the oxides were a result of phonon mediated pairing [143]. This controversy has primarily centered around the highly correlated, heavy electron systems [57-59,79,80], and the possibility of superconductivity in low-dimensional organic systems [144] and layered semiconductor- metal systems [145]. The possible division of superconductors into classes of systems each resulting from a different pairing mechanism is best illustrated in Fig. 7.13, where T_c is plotted against the density of electronic states as measured by γ, the electronic specific heat coefficient. Referring back to the BCS expression for T_c (see Chapter 2), T_c should be exponentially related to γ leading to broad bands of T_c vs. γ reflecting the varied pairing strengths within each class of superconductors. It would be expected that

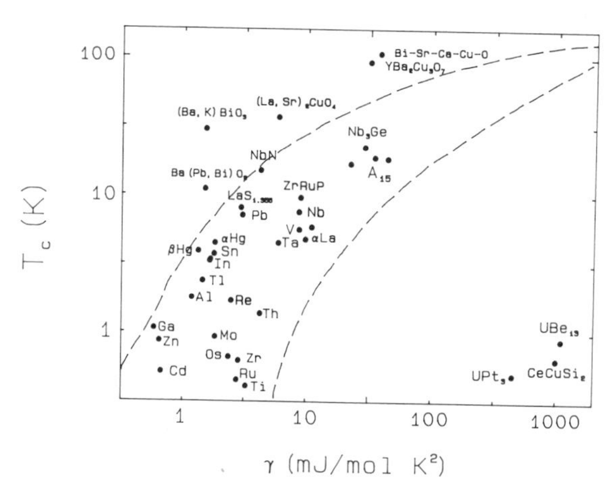

Fig. 7.13. Superconducting transition temperature versus electronic specific heat coefficient.

different classes of superconductors would group in separate or nearly separate regions of Fig. 7.13. This is dramatically reflected by the heavy electron systems which appear in the lower right corner removed from the bulk of the phonon mediated superconductors which have been highlighted by the somewhat arbitrarily defined dashed lines. The oxide superconductors are located at the upper edge or just above the band of T_c vs. γ for the phonon mediated superconductors. Are these oxide systems just highly coupled phonon-mediated superconductors or are these candidates of a new class of high T_c materials which will populate the upper left corner of Fig. 7.13? Approaches to new theories of superconductivity divide into two general classes. Those that propose an alternate pairing mechanism and then build that mechanism into a BCS framework [60,61,145,146] and those which rely on an alternate pairing mechanism and construct the ground state through a non-BCS approach [4,5].

This section primarily focuses on exploring the role played by phonons in the oxide superconductors. Most realistic strong coupling extensions of the BCS theory seem to exclude phonons as the sole mechanism leading to pairing at 100 K or higher. However, these models do not exclude a partial but important role being played by the phonons.

It was the sensitivity of T_c to isotopic mass of Hg [147] which gave the first strong evidence that the pairing in conventional superconductors was due to a phonon related mechanism. Much later, electron tunneling studies led to an even more explicit indication of the involvement of the phonons in the pairing [148]. In the search for evidence of phonon participation in the pairing for the recently discovered superconducting oxides, the isotope effect as provided the most direct evidence. The results of recent isotopic studies are discussed in the next section. Following that section, a discussion of the results of neutron scattering studies of the lattices dynamics in the new oxides and related compounds is presented. Finally, a brief summary of other measurements reflecting strong phonon coupling or anharmonicity is discussed.

7.4.1. The Isotope Effect

As indicated earlier, the strongest evidence prior to the formulation of the BCS theory that phonons were responsible for the superconducting interaction was the dependence of T_c on mean isotopic mass in Hg. [147]. For Hg, it was found that, $T_c \propto M^{-\beta}$ with $\beta \approx 0.5$ and M = the average isotopic mass. The dependence of T_c on isotopic mass is called the isotope effect. The BCS theory provides an immediate explanation of the isotope effect. The BCS expression for T_c is

$$T_c = \frac{\theta_D}{1.75} \exp\left[- \frac{1}{\lambda_{ph} - \mu^*} \right] \quad (7.21)$$

where θ_D is the Debye temperature that characterizes the phonon spectra and enters the BCS theory as a cut-off for the attractive phonon-mediated interaction, λ_{ph} is the effective electron-electron coupling mediated by phonons and μ^* is a parameter reflecting Coulomb correlations. Using an harmonic approximation, θ_D (or the Debye frequency ν_D) would be α $M^{-0.5}$ where M is the mean ionic mass. This mass dependence of θ_D substituted into Eq. (7.21) leads to $T_c \alpha M^{-\beta}$ with $\beta = \frac{1}{2}$. Listed in Table 7.2 is β for several well known superconductors. The isotopic exponents, β, for most of sp-electron (non-transition metal) systems are reasonably close to the BCS value but several systems, in particular the d-transition metal superconductors, have β-values that differ significantly from $\beta = \frac{1}{2}$. Modifications to the BCS equation for T_c relaxing the arbitrary cut-off at θ_D and including electron correlation effects can account for many of these deviations [149,150]. An important point to note is that an isotope effect strongly indicates that phonons are involved in the pairing and that the lack of an isotope effect cannot be interpreted as conclusive evidence of the absence of phonon mediated pairing.

Isotope studies have been conducted on nearly all of the high T_c superconducting oxides. Most of these studies have focused on the substitutions of ^{18}O for ^{16}O. Band structure calculations have demonstrated that the density of states at E_F are primarily due to the hybridization of Cu(3d) - O(2p) states within the $Cu0_2$ planes and many calculations of the electron-phonon interaction have suggested that oxygen-related modes were the most likely candidates to provide sufficient coupling to lead to an enhanced T_c within a conventional, phonon mediated BCS theory [152]. Also, oxygen provides the largest possible fractional change in mass of the constituents in these systems. Summarized in Table 7.3 are the results of numerous studies of an oxygen isotope effect in those oxides commonly associated with this new class of superconductors.

Raman and IR spectroscopy studies have shown that ^{18}O enters the lattice in an uniform way and shifts the phonon frequencies as expected using a harmonic model [158,160-165]. Shown in Fig 7.14 is β vs. T_c for the oxide systems. The β values in Table 7.3 and in Fig. 7.14 were determined by relating T_c to only the fractional change of the O-mass, i.e., $T_c \alpha M_i^{-\beta}$ where i identifies the individual constitutes. Such a comparison is probably more reasonable than computing β based on the fractional change of the complete unit cell mass. All of these systems have an oxygen isotope effect which is strong evidence that phonons are participating directly or indi-

Table 7.2. Isotope Effect in Elemental Superconductors

Superconductor	T_c(K)	β
Non-transition Elements[a]		
Zn	0.88	0.45 ± 0.01
Cd	0.56	0.50 ± 0.10
Hg	4.15	0.50 ± 0.03
Tl	2.36	0.50 ± 0.10
Sn	3.72	0.47 ± 0.02
Pb	7.23	0.48 ± 0.01
Transition Metals[a,b]		
Zr	0.49	0.00 ± 0.05
Mo	0.91	0.33 ± 0.05
Ru	0.49	0.00 ± 0.05
Os	0.67	0.20 ± 0.05
Re	1.70	0.23 ± 0.02
U(α)	1.10	−2.2 ± 0.20

a) R. Meservery and B. B. Schwartz in Superconductivity ed. by R. D. Parks, Marcel Dekker, Inc., New York (1969) pg. 126.
b) G. Gladstone, M. A. Jensen and J. R. Schrieffer, *ibid.*, pg. 769.

Table 7.3. Oxygen Isotope Effect for Oxide Superconductors

Superconductor	T_c(K)	β^*	Ref
Ba(Pb,Bi)0_3	11	0.60 ± 0.10	[153]
		0.22 ± 0.03	[154]
(Ba,K)Bi0_3	29	0.41 ± 0.03	[155]
		0.21 ± 0.03	[154]
(La,Ca)Cu0_4	20.6	0.70 ± 0.10	[153]
(La,Sr)Cu0_4	34-37	0.2 ± 0.1	[153]
		0.09 ± 0.37	[156]
		0.16 ± 0.02	[157,158]
		0.14 ± 0.008	[159]
$YBa_2Cu_30_7$	92	0.10 ± 0.04	[153]
		0.00 ± 0.02	[157,158]
		0.05	[160]
		0.0	[161]
		0.3 ± 0.1	[162]
		0.017 ± 0.00	[163]
Bi-Sr-Ca-Cu0	75	0.037 ± 0.004	[164]
	110	0.026 ± 0.002	[164]

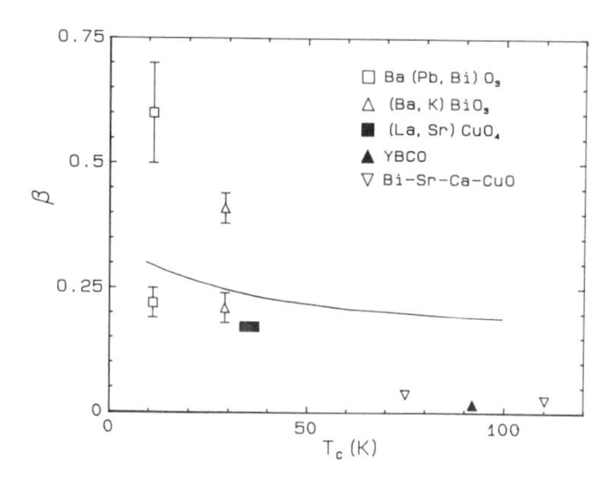

Fig. 7.14. Isotope effect exponent versus superconducting transition temperature for oxide superconductors. Solid curve - see text.

rectly in the pairing interaction.

As seen in Fig. 7.14, β appears to depend on T_c. The lower T_c systems have a slightly higher β value, but most values are significantly lower than the BCS value, i.e. $\beta = \frac{1}{2}$. If the pairing interaction is the sum of two contributions, e.g. phonons and excitons, and to a first approximation their spectral densities can be treated additively within an Eliashberg formalism [74,75], a strong T_c dependence to β reflecting the relative weight of excitons and phonons with $\beta \rightarrow 0$ for large exciton coupling is obtained. This kind of modeling gives a smooth continuous decrease in β vs. T_c and does qualitatively reflect the O-isotope effect results show in Fig. 7.14. A simplified model assuming phonon and excitonic mediated pairing and treating the excitation spectra for both excitations as Einstein-like modes can be easily developed, and leads to an isotope exponent $\beta = \lambda_{ph}/2(\lambda_{ph} + \lambda_{ex})$, where λ_{ph} (λ_{ex}) is the effective electron-electron coupling parameter due to phonon (excitonic) mediated pairing. Using this model, assuming $\nu_{ph} \approx 500$ K, $\nu_{ex} \approx 5000$ K, and setting the relative weight of phonon vs. excitonic pairing such that $\beta = 0.3$ for $T_c = 10$ K (an approximate average value reported for the $Ba(Pb,Bi)O_3$ system), the solid line in Fig. 7.14 is obtained. The decrease in β with increasing T_c obtained using this simplified model is not as dramatic as that observed experimentally. However, this analysis and that presented in [74] and [75] indicate that a decrease in β with increasing T_c qualitatively similar to that observed is expected if a phonon plus non-phonon mechanism contribute to the coupling. The lower β values for the Cu-based oxides as compared to this model may also reflect changes in β due to electron correlations, which is not included here and has been shown to significantly depress β for many d-band systems [151]. To draw more quantitative conclusions from these data would require considerable additional knowledge of the normal state properties for these systems then exists at this time.

Other isotope studies involving constituents other than oxygen have primarily been conducted on the 123 system. It is clear from the insensitivity of T_c to the mass of the ion at the Y site [90] that lattice vibrations associated with the Y site are not contributing to the pairing in any significant way. In addition, Cu [166,167] and Ba [166] isotope studies show no measurable dependence of T_c on isotopic mass. Some caution in interpreting these negative results for Ba and Cu is appropriate because of reduced relative change in mass associated with these studies as compared to the O-studies, i.e., $^{63}Cu \rightarrow {}^{65}Cu$ corresponds approximately a 3.5 times smaller mass change as compared to $^{18}O \rightarrow {}^{16}O$.

It is worth restating that the observation of an O-isotope effect does strongly indicate that electron-phonon coupling does contribute to the pairing within all the oxides studied, or at least the phonons are indirectly participating through a renormalization of some electronic or spin excitation which is primarily responsible for the pairing.

7.4.2. The Phonon Density of States in High T_C Oxides

If strong electron-phonon interactions are driving the large superconducting transition temperatures observed in the oxide systems then this interaction should not only manifest itself as a renormalization of electronic properties but also in phonon properties. Microscopic theories based on the Eliashberg equations have been developed to extend the BCS theory beyond the weak coupling limit [168-170]. These theories have tried to relate superconducting properties to a small number of normal state parameters. In particular, McMillan's expression for T_c [168] given by

$$T_c = \left[\frac{\langle \nu \rangle}{1.45}\right] \exp\left[-\frac{1.04\,(1 + \lambda)}{(\lambda - \mu^*(1 + 0.62\lambda))}\right] \tag{7.22}$$

where $\langle \nu \rangle$ is a characteristic phonon energy and μ^* is an effective Coulomb repulsion parameter, has been extremely useful in gaining a better feel for the evolution of T_c with changes in normal state parameters. The electron-phonon coupling parameter λ is given by

$$\lambda = 2 \int_0^{\infty} \frac{\alpha^2 g(\nu)}{\nu} d\nu \tag{7.23}$$

where α is an electron-phonon interaction and $g(\nu)$ is the energy dependent phonon density of states (PDOS). The success of this analysis along with the importance of α^2 and $g(\nu)$ is most dramatically seen in the comparison of $\alpha^2 g(\nu)$ obtained by de-convoluting electron tunneling (ET) data with $g(\nu)$ obtained from inelastic neutron scattering (INS). Shown in Fig. 7.15 is $g(\nu)$ obtained from INS and $\alpha^2 g(\nu)$ obtained from ET for *Ta* (T_c = 4.42K) [171]. Except for some variation in peak heights the correspondence is remarkable. Thus, a determination of $g(\nu)$ for these new oxides with estimates of α^2 would help clarify the role played by phonons in the pairing mechanism. Strong electron-phonon interactions should also be apparent in phonon lifetimes and thus reflected in the temperature dependence of the INS phonon linewidths. Measurements of $g(\nu)$ for the A-15 superconductors show temperature dependent softening with the softening being more pronounced for those systems with the larger T_c [172,173]. This softening is also observed in ultrasonic measurements [174] and is related to the structural transition which

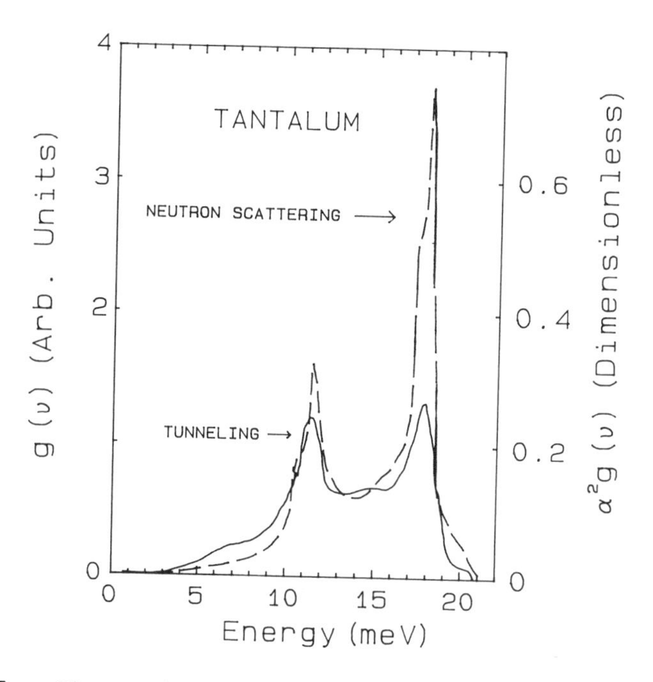

Fig. 7.15. Phonon density of states $g(\nu)$ versus energy ν. α^2 is the electron-phonon coupling parameter. The solid curve is $\alpha^2 g(\nu)$ obtained from S-N-S electron tunneling, and the broken curve is $g(\nu)$ obtained from inelastic neutron scattering. From [171].

occurs in the vicinity of T_c, e.g. for Nb_3Sn, $T_c \approx 17.8$ K and $T_s \approx 35$ K where T_s is the martensitic transition temperature [175]. Thus, the presence of a structural phase transitions in the vicinity of T_c, as in the tetragonal to orthorhombic transition in the doped 214 compound, may play a key role in providing an enhancement of phonon mediated pairing as occurs in several of the A-15 intermetallic systems.

Returning to Eq. (7.23), the phonon mediated electron-electron coupling can also be enhanced through changes of the electron-phonon interactions, i.e., $\alpha^2(\nu)$, which should be reflected in anomalous phonon lifetimes and/or changes in the phonon spectrum via $g(\nu)$. These changes should be apparent in the temperature dependence of the phonon dispersion curves and PDOS. Since the integral in Eq. (7.23) is over all phonon modes, any enhancement in $\alpha^2(\nu)$ must occur over sufficient phase space to make a significant contribution to λ. In addition to soft mode enhancement of T_c, high energy modes such as the speculated oxygen breathing modes [8,9,152,176] in the high T_c oxides, can also lead to an enhancement of T_c through a combination of strong electron-phonon coupling and increases in $\langle \nu \rangle$, the characteristic phonon energy in Eq. (7.22). Thus, if the large $T_c \approx 100$ K are to be explained by primarily a phonon mediated interaction, anomalous phonon properties would be anticipated.

The phonon dispersion curves and PDOS have been measured for several high T_c oxide systems and for many closely related non-superconducting systems, e.g., La_2NiO_4 [177]. The phonon properties are probably somewhat better understood in the 214 compound as compared to the 123 compound. This is partially because larger single crystals [20,178,179] have been available and other isostructural systems [177] have been studied. In contrast, most of the INS studies on the 123 compound have been on polycrystalline ceramic material. Raman scattering and other infrared studies also provide valuable information on phonon excitations and their temperature dependence. A brief summary of such results will be presented near the end of this section.

There have been several theoretical studies that have calculated electron-phonon interactions and/or discussed the role of phonons in the pairing mechanism for the oxide superconductors [8,9, 152,176,180]. For the 214 compound, the 2D electron energy bands in the basal plane are nearly half full leading to a near perfect nesting of the Fermi surface. As discussed earlier, this nesting leads to a Fermi surface instability involving Cu-O stretching modes (breathing modes) in the basal plane. These phonon modes have high frequencies because of the light oxygen mass. Theoretical analysis due to Weber [152] suggests that these modes are strongly coupled to the electrons leading to a strong electron-phonon coupling constant, i.e., $\lambda \approx 2$, and $T_c \approx 30$-40 K. The strong coupling to these breathing modes should lead to a renormalization of these modes near the Brillouin zone (BZ) boundary. An extension of this theoretical analysis to the 123 compound [180] does not lead to sufficient coupling to yield superconducting transition temperatures near 90 K. This, in part, is due to the lack of perfect nesting in the Fermi surface for the 123 system and thus a smaller electron-phonon coupling and greater structural stability. Summarizing, electron-phonon coupling may account for the coupling in the 214 compounds but there should be strong renormalization evident in the Cu-O stretching modes. However, such breathing mode driven coupling appears to be unable to account for the ~90 K transition temperatures in the 123 compound.

The 214 compound has a body-centered tetragonal structure at high temperatures with a second order transition to an orthorhombic structure at 530 K (see Chap. 4). This structural transition is caused by a tilting of the oxygen octahedra around the Cu-ion and is depressed with Sr-doping (see inset in Fig. 7.1). For $x \geq 0.2$ in $La_{2-x}Sr_xCuO_4$, the tetragonal phase is stable at all temperatures. The reduction of the structural transition temperature T_s with Sr-doping in the 214 drives T_s toward T_c resembling Nb_3Sn where T_s and T_c are reasonably close. INS studies on large single crystals [20,178,179] have shown that the tetragonal-to-orthorhombic transition is due to a soft transverse optical (TO) mode and is similar to that observed in many other perovskite systems [181]. Moreover, no evidence of the predicted softening of the oxygen breathing modes was found. With the exception of the soft TO mode, the phonon modes appear to be well behaved with no anomalous line broadening indicative of strong electron-phonon coupling. These findings are consistent with other theoretical modeling of the phonon properties [182] which predict stable breathing modes and soft mode behavior similar to that observed by INS [179]. Temperature dependent INS studies of polycrystalline 214 and Sr doped 214 by Renker et al [183] also show little evidence for strong renormalization of the generalized PDOS for these systems. Note, the generalized PDOS obtained for polycrystalline samples are weighted representations of the true PDOS with each mode multiplied by the ratio τ_i/M_i where τ_i is a bound scattering cross-section and M_i is the ionic mass. This ratio is nearly the same for La, Sr and, Cu but for O, this ratio is ~2.5 times larger. Thus, generalized PDOS should be used with caution in estimating other dynamical or thermodynamic properties. There is no significant shift in the PDOS vs. T for the 214 and only modest changes with Sr-doping. The oxide breathing modes are apparent at high energies (~75 meV) and do shift to lower energy with Sr doping indicative of increased electron-phonon coupling. However, the shifts are small and probably cannot explain the dramatic increase in T_c. With the exception of the soft TO mode [179], the phonon modes are well behaved with no anomalous line broadening indicative of strong electron-phonon coupling for any of the other phonon modes. It would appear that there is insufficient phonon mediated coupling in the 214 system to account for the high T_c observed.

These results should be contrasted with similar studies on $Ba(Pb,Bi)O_3$ ($T_c \approx 11$ K) . The possible importance of strong electron-phonon coupling due to oxygen breathing modes was first proposed to explain the high T_c reported for $Ba(Pb,Bi)O_3$ [184]. For this system, strong electron-phonon coupling is apparent in INS [185] and ET experiments [116,185,186]. In particular, low phonon energies at selected symmetry points coincide with phonon structures apparent in ET spectroscopy. Thus, it appears that phonon mediated coupling may be the primarily mechanism driving the pairing in this system. This would be consistent with the stronger isotope effect observed in this system as compared to the 214 and 123 compounds.

A comparison of the phonon properties for $(La,Sr)_2CuO_4$ with those obtained for La_2NiO_4, an isostructural, nonsuperconducting analog of 214, leads to some similarities and important differences [179]. La_2NiO_4 has a tetragonal-to-orthorhombic transition at $\sim$240 K [187] similar to that reported for 214. However, the softening of the TO mode responsible for this structural transition in 214 is not as dramatic in La_2NiO_4. The generalized PDOS are similar for the two systems and numerous properties point toward very anisotropic, nearly 2D, transport and Fermi surface behavior. However, whereas the 214 did not display any evidence for renormalization of phonon properties due to soft or strong coupling to the oxygen breathing modes, the Ni-O stretching mode [177] are strongly renormalized and are in agreement with theory [182]. Thus, what is evolving is that even through 214 and La_2NiO_4 are isostructural and resemble each other in many ways, there are subtle differences between these systems. La_2NiO_4 may not be the best analog system for comparison.

INS studies of the 123 system have been primarily on polycrystalline samples [188-194]. Thus detailed information on the phonon dispersion curves is not available. As mentioned earlier, Weber and Mattheiss [180] have indicated that it is unlikely that a $T_c \approx 90$ K can be obtained with only an electron-phonon mediated interaction. However, a positive isotope effect along with other measurements to be discussed in the next section clearly indicate that there are strong electron-phonon interactions and anharmonicities present. Shown in Fig. 7.16 is the generalized PDOS vs. energy for the 123 at 120 K and 12 K [189]. Based on model calculations [195], the features seen in the PDOS vs energy can be attributed to particular modes. Below 15 meV, the modes involving the Cu-chain ions and Ba-ions make the largest contribution with Y modes contributing in the range 15-20 meV. The strong features at $\sim$20 and 30 meV along with those above 45 meV appear to be attributed to oxygen vibra-

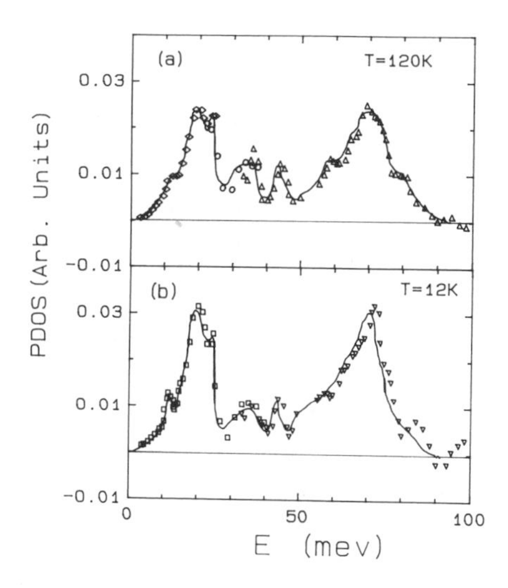

Fig. 7.16. Generalized phonon density of states versus energy ν for $YBa_2Cu_3O_7$, obtained from inelastic neutron scattering. Data from [189].

tions. The lower energy features (~20-30 meV) are due to transverse modes of oxygens coordinating the Cu-chain site and those above 45 meV are due to Cu-O bond bending modes, along with the oxygen breathing modes. It is these higher energy modes which could lead to enhanced T_c if the electron-phonon coupling is sufficient because of their large characteristic energy $\langle \nu \rangle$ entering Eq. (7.22). A comparison between the PDOS in Fig. 7.16a and 7.16b indicates that there are no significant changes in the PDOS with temperature. This result has been reproduced in several other studies and seems to rule out any strong temperature-dependent renormalization of the PDOS as seen in many A-15 superconductors. This lack of any significant T-dependence of the PDOS would argue against an enhancement in T_c due to an incipient structural transformation.

The PDOS vs. energy have also been measured for $YBa_2Cu_3O_{7-}$ with $0 \leq \delta \leq 1$ [188-192] and for $YBa_2(Cu_{0.9}Zn_{0.1})_3O_7$ [194]. Even though there are slight differences between the PDOS reported by various groups, the changes in PDOS with oxygen content are qualitatively similar. Contrary to the lack of a temperature dependence in the PDOS for 123, there are modest changes with oxygen stoichiometry. The largest difference occurs in the energy range ~20-35 meV. Changes in this energy range would be expected since model calculations attribute these to 0-modes for the bridging oxygen between the chains and planes and the oxygen chain site itself. Changes at higher energy (> 40 meV) primarily reflect changes in basal plane Cu-0 stretching modes. Associated with the decrease in oxygen content in the above studies is a structural change from an orthorhombic structure for $\delta \simeq 0$ to a tetragonal structure for $\delta = 1.0$. A more relevant comparison which may reflect the role played by phonons is a comparison of the PDOS for $YBa_2(Cu_{0.9}Zn_{0.1})_3O_7$, which has a depressed T_c ($T_c \simeq 0$) but remains orthorhombic with little dependence of the oxygen stoichiometry on Zn-doping [105]. The PDOS for the Zn-doped sample with $T_c \simeq 0.0$ as compared to the PDOS for the orthorhombic 123 are slightly enhanced at lower energies E < 20 meV and depleted in the 40-60 meV range [194]. However, these changes do not appear significant. Thus, INS does not show any clear signatures which would support significant phonon involvement in the pairing for the 214 and 123 oxide superconductors. For $Ba(Pb,Bi)O_3$ the situation is different, both INS and ET studies seem to indicate a major role played by phonons in the pairing interaction.

In addition to the neutron studies discussed above, complementary phonon data can be obtained from Raman and infrared studies. There have been numerous optical studies of phonon, electronic and magnon excitations in many of the superconducting oxides and several excellent reviews exist [196]. Optical spectroscopies have some distinct advantages and disadvantages when compared to neutron scattering spectroscopy. For example, larger samples are typically required in neutron studies, while optical spectroscopies are limited by selection rules to only a few phonon modes. In addition, limitations on penetration of the optical probe creates some concern over homogeneity of the sampled volume. This is particularly true in relationship to the uniformity near the surface of the oxygen content for systems such as the 123, whose oxygen content is sensitive to thermal histories and whose properties display a strong oxygen stoichiometry dependence. Also, many impurity phases present in ceramic samples have Raman cross-sections one to two orders of magnitude larger than the primary metallic phase [196-199] making Raman spectroscopy an extremely sensitive probe of impurity phases. The sensitivity to oxygen stoichiometry [200,201] and impurity phases made many of the early results on poorly characterized ceramic samples questionable. Such concerns became less of a problem in later studies on better characterized ceramic samples and single crystals. The single crystal Raman studies were critical to a reliable association of spectral features with particular phonon modes [196]. After the appropriate assignments to the spectral features was established, the higher energy resolution typical of optical spectroscopies vs. neutron studies has provided valuable information on the temperature and stoichiometry dependence of select phonon modes and their lifetimes.

Raman and IR spectroscopies have been reported on nearly all the known oxide superconductors, however, the available data is most complete for the 123 system. The extreme sensitivity of Raman spectra to many impurity phases commonly associated with the oxide superconductors and the sensitivity to oxygen stoichiometry and isotopic mass has made Raman spectroscopy a valuable characterization tool. As indicated earlier, the site-specific nature of Raman spectroscopy has played an important role in conjunction with measurements of the oxygen isotope effect [159,161,163-165]. Such measurements have already established that oxygen exchange experiments involve all the oxygen sites, not just those identified with the oxygen stoichiometry, and shifts in the mode frequencies have been consistent with simple $(m)^{-1/2}$ scaling of the modes.

The Raman spectra are similar for the Cu-based oxides studied and have provided an experimental check of lattice dynamic calculations. Raman spectroscopy has identified soft phonon modes associated with the orthorhombic-tetragonal transition in the 214 oxides and this softening is consistent with a tilting of the oxygen octahedra [202] and neutron results. With the exception of a Raman-active mode at 340 cm^{-1} (1 meV = 8.066 cm^{-1}) in $YBa_2Cu_3O_{7-\delta}$, the temperature and oxygen stoichiometry dependence of the mode frequency do not display any unusual features [196,204,205]. The oxygen concentration dependence of the mode at 340 cm^{-1} shows a break in slope at $\delta \simeq 0.5$, i.e. in the vicinity of $T_c \to 0$ [199]. In the fully oxygenated superconductor, softening of the 340 cm^{-1} phonon mode below T_c and H_{c2} has also been reported. This particular phonon mode is attributed to an out-of-phase, vertical displacement of the oxygen ions within the Cu-O planes. Modes associated with the Cu-O stretch vibrations seen at higher frequencies ($\sim$500 cm^{-1}) display no anomalous temperature dependence and harden with decreasing temperature as anticipated. In addition, Raman features associated with vibrations of the Ba ions do not show any anomalous dependence with decreasing temperature. The frequency and phonon linewidth of the mode in the vicinity of 340 cm^{-1} remains nearly constant between room temperature and T_c with approximately a 1% softening and increase in linewidth below T_c. This softening is quenched with a slight reduction in the oxygen stoichiometry [202] and is not seen in the isostructural, nonsuperconducting $PrBa_2Cu_3O_7$ system [206]. The onset temperature of the softening is also quenched with a magnetic field exceeding $H_{c2}(T)$ [207]. The origin of this softening is not clear but could be attributed to a self energy correction to the phonon frequencies due to superconducting pairing. Such a correction had been predicted for conventional superconductors and leads to a rather small effect, e.g. a shift of 0.07 cm^{-1} has been reported for a niobium acoustic phonon at $\sim$30 cm^{-1} [208]. Using a weak-coupling BCS approach, phonons with frequencies below the gap, i.e. $\nu < \Delta$ with $2\Delta \simeq 3.5\ kT_c$, should soften whereas phonons with $\nu > \Delta$ should harden [209]. The Raman active mode at 340 cm^{-1} is above the weak coupling BCS estimate of the gap and based on such an analysis should harden not soften. An extension of this analysis to include strong coupling corrections yields predictions semi-quantitatively consistent with the observed shifts [209]. Such analysis requires a coupling parameter $\lambda \simeq 2.9$ which would put these materials in the extreme strong coupling regime. In addition to the phonon softening of this particular Raman-active mode, softening has also been reported for several infrared active phonons with energy comparable to that reported for the Raman active mode [213]. The temperature dependence of the frequency for these modes is similar to that reported for the 340 cm^{-1} Raman mode and the softening is absent for the tetragonal, nonsuperconducting phase.

The anomalous temperature dependence of the mode frequency and linewidth of the Cu-O bending modes in the CuO_2 planes, without any anomalous feature evident in the other modes, points to the importance of the CuO_2 planes to the superconductivity. Analysis of this softening, assuming it can be attributed to renormalization of phonon frequencies due to the pairing, requires that the system be in the extreme strong coupling limit. This conclusion contradicts weak coupling arguments coming from thermodynamic studies. More recent studies, where it was assumed that softening is observed only when $\nu < 2\Delta$, placed limits on the superconducting gap which are consistent with weak coupling models [210].

Raman and IR spectroscopy have provided valuable information on many aspects of these new materials. They have helped to establish the degree of oxygen exchange in isotope studies, set limits on impurity phases, and provided information on phonon softening at a level which is probably not detectable using other phonon probes such as INS. More work is required before definitive statements can be made. Controlled studies of the phonon softening versus impurity doping, for example Zn and Pr substitution, may help clarify some of these points. Extension of these studies to the Bi- and Tl-based oxide superconductors has been limited.

7.4.3. Other Evidence of Strong Electron-Phonon Coupling and Anharmonicity

Other quantities have been measured which display unusual behavior, reflective of strong electron-phonon coupling or anharmonicity. As with many other areas of investigation, most of this work has concentrated on the 123 system. The temperature dependence of the thermal conductivity K(T) and the Mossbauer recoil-free fraction have features which are similar to behavior previously reported for several strong-coupled conventional superconductors. The results from such studies are briefly summarized below.

The Mössbauer effect has found wide applications in the study of structural and magnetic properties of solids. The Mössbauer absorption spectrum is characterized by the fraction of gamma rays that are absorbed through a recoil-free process, i.e., the intensity of the absorption line A, the position of the resonant absorption maximum δ, and the absorption linewidth Γ. The temperature dependence of both A, the recoil-free fraction, and δ, the isomer shift, are related to the PDOS and are given by

$$-\ln\left[A(T)\right] = k < x^2 >$$

and

$$\delta(T) = k' < v^2 > + \delta_o \quad , \tag{7.24}$$

where $<x^2>$ is the mean squared displacement or Debye-Waller factor associated with the absorbing ion, and $<v^2>$ is the mean squared velocity of the Mössbauer active ion. Mössbauer studies have provided valuable information on many of the conventional superconductors, in particular the A-15 compounds, e.g. V_3Ga [211]. Mössbauer studies have primarily concentrated on ^{57}Fe [212] and ^{119}Sn [91, 213,214] spectroscopy in the 123 compound. The temperature dependence of the recoil-free fraction of the ^{119}Sn Mössbauer absorption in the orthorhombic, superconducting phase of 123 shows a softening which appears for $T < 150$ K [91,213]. For the nonsuperconducting tetragonal phase ($YBa_2Cu_3O_6$) there was no evidence of such softening and the temperature dependence of the recoil-free fraction agreed with that obtained using the measured phonon spectrum obtained with neutrons. Similar phonon anomalies were observed for ^{119}Sn doped $(La,Sr)_2CuO_4$ [215] and for ^{119}Sn doped $Ba(Pb,Bi)O_3$ [216]. The softening of the mean displacement of the Sn ions which substitute on the Cu-site in 123 cannot be taken as direct evidence in favor of the electron-phonon mechanism giving rise to the superconductivity, but it does demonstrate that there are strong anharmonicities in the superconducting 123 phase which are not seen in the tetragonal, nonsuperconducting phase. Thus, these results provide additional evidence for an important role played by the Cu-O bonds in the high T_c superconductors, and indicate that there are significant renormalizations of the force constants and phonon properties. The indirect association of phonons in the pairing interactions due to a renormalization of the dielectric function could account for the phonon anomalies reported in the Mössbauer studies and the isotope effect.

In addition to the anomalies seen by Mössbauer spectroscopy, evidence for strong electron-phonon scattering has been seen in the temperature dependence of the thermal conductivity K(T). The thermal conductivity can be separated into an electronic and lattice contribution [217]. Assuming that Mathiesen's rule (see Chap. 1) is valid, K(T) can be written as

$$K = K_e + K_{ph} = [(K_e^{Ph})^{-1} + (K_e^{d})^{-1}]^{-1} + [(K^e_{ph})^{-1}]^{-1} \qquad (7.25)$$

where the subscript "e" and "ph" refer to electrons and phonons, respectively, and the superscripts refer to electron (e), phonon (ph) or defect (d) limited conductivity. An estimate of K_e is provided by the Wiedemann-Franz (WF) law, and is given by

$$K_e = L_o T/\rho \qquad (7.26)$$

where $L_o = 2.45 \times 10^{-8}$ WΩK^{-2} and ρ is the electrical resistivity.

K(T) has been measured for several of the high T_c oxides [218, 223]. For the 123 system, the thermal conductivity with decreasing T is nearly temperature independent above T_c, but increases sharply at T_c followed by a decrease for $T \ll T_c$. Using the Wiedemann-Franz law [Eq. (7.26)] along with electrical resistivity data, an estimate of an upper limit for $K_e(T)$ is obtained. This assumes the scattering is elastic. Inelastic scattering will result in a smaller Lorentz number, L_o and lower estimate of $K_e(T)$. Based on this analysis, the electronic contribution to K(T) is estimated to be ~ 10^{-2} of the total thermal conductivity. Thus, it appears that for $T \sim T_c$ most of the thermal current is carried by phonons. This is a reflection of the low carrier density characteristic of these systems.

If the K(T) for a BCS superconductor is dominated by electron conduction, i.e. K(T) $\simeq$ $K_e(T)$, then K(T) should dramatically decrease below T_c because those electrons which condense into Cooper pairs cannot carry entropy and therefore do not transport heat [224]. Thus, $K_e(T)$ should approach zero exponentially, i.e., exp $(-\Delta/kT)$. For a superconductor, where K(T) is dominated by phonon thermal transport but is electron limited, K(T) will increase below T_c as Cooper pairs form and no longer scatter phonons. Such behavior has been reported for several strong coupling conventional superconductors, e.g. Pb [225]. A comparison of this behavior to that observed for 123 along with estimates of the electronic contributions indicates that the enhancement in K(T) below T_c is due to phonon conduction limited by electron-phonon scattering. The phonon lifetime due to electron scattering is of the order of 10^{-15} s, implying strong electron-phonon coupling. Using these measurements along with Debye temperatures obtained from specific heat, enhanced T_c's close to those observed can be accounted for with strong coupling theory [223].

The above results indicate that the high T_c systems are characterized by strong electron-phonon coupling, but other measurements, e.g., thermodynamic studies and high temperature electrical transport measurements, appear to be inconsistent with strong coupling models. Many more microscopic studies will be required to clearly define the role played by phonons in both the normal and superconducting properties.

Acknowledgements

The authors have benefited from discussions on several points with P. W. Anderson and P. Schlottmann. J.E.C. thanks Y. Gao and S.-T. Ting for assistance in preparing the figures, and acknowledges support under the Defense Advanced Research Project Agency/Office of Naval Research, Contract No. 00014-88-K-0587. N-P. O. acknowledges support by the Office of Naval Research, Contract No. N-00014-88-K-0283.

References

[1] *Novel Superconductivity*, edited by S. A. Wolf and V. Z. Kresin (Plenum, New York, 1987).

[2] *Superconductivity in Highly Correlated Fermion Systems*, Proceedings of the Yamada Conference XVIII, edited by M. Tachiki, Y. Muto, and S. Maekawa (North Holland, 1987).

[3] *High Temperature Superconductors and Mechanisms and Materials of Superconductivity*, edited by J. Muller and J. L. Olsen (North Holland, 1988).

[4] P. W. Anderson, in *Frontiers and Borderlines in Many Particle Physics*, Varenna lectures, edited by J. R. Schrieffer and R. A. Broglia (North Holland, 1988).

[5] P. W. Anderson and Z. Zou, *Phys. Rev. Lett.* **60**, 132 (1988); *ibid.*, **60**, 2557 (1988).

[6] J. B. Goodenough, G. Demazeau, M. Pouchard, and P. Hagenmuller, *J. Sol. State Chem.* **8**, 325 (1973); N. Nguyen, J. Choisnet, M. Hervieu, and B. Raveau, *J. Sol. State Chem.* **39**, 120 (1981); N. Nguyen, F. Studer, and B. Raveau, *J. Phys. Chem. Solids* **44**, 389 (1984).

[7] J. G. Bednorz and K. A. Muller, *Z. Phys.* **B64**, 189 (1986).

[8] L. F. Mattheiss, *Phys. Rev. Lett.* **58**, 1028 (1987).

[9] J. Yu, A. J. Freeman, and J. H. Xu, *Phys. Rev. Lett.* **58**, 1035 (1987).

[10] J. D. Jorgensen, H. B. Schuttler, D. G. Hinks, D. W. Capone,

II, K. Zhang, M. B. Brodsky, and D. J. Scalapino, *Phys. Rev. Lett.* **58**, 1024 (1987).

[11] J. A. Wilson, F. J. DiSalvo, and S. Magajan, *Adv. Phys.* **24**, 117 (1975).

[12] For reviews, see P. Monceau, in *Electronic Properties of Inorganic Quasi-One-Dimensional Materials*, edited by P. Monceau (Reidel, 1985) Vol. **2**, p. 139; Claire Schlenker in *Low-Dimensional Conductors and Superconductors*, edited by D. Jerome and L. G. Caron, Vol. **155**, NATO ASI Series B, (Plenum, 1987).

[13] C. Herring, in *Magnetism*, edited by G. Rado and H. Suhl (Academic Press, 1966), Vol. IV.

[14] J. B. Torrance, H. J. Pedersen and K. Bechgaard, *Phys. Rev. Lett.* **49**, 881 (1982); K. Mortensen, Y. Tomkiewicz, and K. Bechgaard, *Phys. Rev.* **B25**, 3319 (1982).

[15] S. Uchida, H. Takagi, H. Ishii, H. Eisaki, T. Yabe, S. Tajima, and S. Tanaka, *Jpn. J. Appl. Phys.* **26**, L440 (1987).

[16] N. P. Ong, Z. Z. Wang, J. Clayhold, J. M. Tarascon, L. H. Greene, and W. R. McKinnon, *Phys. Rev.* **B35**, 8807 (1987).

[17] P. W. Anderson, *Science* **235**, 1196 (1987); P. W. Anderson, G. Baskaran, Z. Zou and T. Hsu, *Phys. Rev. Lett.* **58**, 2790 (1987); Z. Zou and P. W. Anderson, *Phys. Rev.* **B37**, 627 (1988).

[18] For a review, see D. Adler, *Solid State Physics* **21**, (1968).

[19] P. W. Anderson, *Solid State Physics* **14**, 99 (1963); *Phys. Rev.* **115**, 2 (1959).

[20] R. J. Birgeneau, C. Y. Chen, D. R. Gabbe, H. P. Jenssen, M. A. Kastner, C. J. Peters, P. J. Picone, T. Thio, T. R. Thurston, H. L. Tuller, J. D. Axe, P. Boni, and G. Shirane, *Phys. Rev. Lett.* **59**, 1329 (1987).

[21] M. W. Shafer, T. Penney, and B. L. Olson, *Phys. Rev.* **B36**, 4047 (1987).

[22] J. B. Torrance Y. Tokura, A. K. Nazzal, A. Bezinge, T. C. Huang, and S. S. P. Parkin, *Phys. Rev. Lett.* **61**, 1127 (1988).

[23] S. W. Cheong, et al., *Phys. Rev.* **B36**, 3193 (1987).

[24] Z. Z. Wang, J. Clayhold, N. P. Ong, J. M. Tarascon, L. H. Greene, W. R. McKinnon, and G. W. Hull, *Phys. Rev.* **B36**, 7222 (1987).

[25] P. Chaudari, et al., *Phys. Rev.* **B36**, 8903 (1987).

[26] S. W. Tozer, A. W. Kleinsasser, T. Penney, D. Kaiser, F. Holtzberg, *Phys. Rev. Lett.* **59**, 1768 (1987); T. Penney (private communication).

[27] A. Davidson, P. Santhanam, A. Palevski, and M. J. Brady, *Phys. Rev.* **B38**, 2828 (1988).

[28] P. B. Allen, W. E. Pickett, and H. Krakauer, *Phys. Rev.* **B36**, 3926 (1987); *ibid.*, *Phys. Rev.* **B37**, 7482 (1987).

[29] J. Clayhold, N. P. Ong, Z. Z. Wang, J. M. Tarascon, and P.

Barboux *Phys. Rev.* B39, 777 (1989).
[30] J. Clayhold, S. J. Hagen, Z. Z. Wang, N. P. Ong, J. M. Tarascon, and P. Barboux, *Phys. Rev.* B39, 7324 (1989).
[31] J. Clayhold, N. P. Ong, P. H. Hor, and C. W. Chu, *Phys. Rev.* B38, 7016 (1988).
[32] T. R. Dingle, T. K. Worthington, W. J. Gallagher, and R. L. Sandstrom, *Phys. Rev. Lett.* **58**, 2687 (1987); T. K. Worthington, W. J. Gallagher, and T. R. Dinger, *Phys. Rev. Lett.* **59**, 1160 (1987).
[33] D. L. Kaiser, F. Holtzberg, M. F. Chisholm, and T. K. Worthington, *J. Cryst. Growth* **85**, 593 (1987); and other articles in the same issue.
[34] Y. Iye, T. Tamegai, H. Takeya, and H. Takei, *Jpn. J. Appl. Phys.* **26**, L1057 (1987).
[35] S. J. Hagen, T. W. Jing, Z. Z. Wang, J. Horvath, and N. P. Ong, *Phys. Rev.* B37, 7928 (1988).
[36] L. I. Buranov, L. Ya. Vinnikov, G. A. Emeltchenko, P. A. Kononovich, V. N. Laukhin, Yr. A. Ossipyan, and I. F. Shchegolev, *JETP* **47**, 49 (1987).
[37] Y. Iye, T. Tamegai, H. Takeya and H. Takei, *Jpn. J. Appl. Phys.* **27**, L658 (1988).
[38] S. Martin, A. T. Fiory, R. M. Fleming, L. F. Schneemeyer, and J. V. Waszczak, *Phys. Rev. Lett.* **60**, 2194 (1988).
[39] J. H. Wang, Y. F. Yan, G. H. Cheng, X. Chu, D. N. Zheng, W. M. Yang, S. L. Jia, Z. Y. Ran, Z. H. Mai, Q. S. Yang and Z. X. Zhao (preprint).
[40] See, for example, K. Murata, T. Ukachi, H. Anzai, G. Saito, K. Kajimura and T. Ishigure, *J. Phys. Soc. Jpn.* **51**, 1817 (1982).
[41] K. Phan, et al., *Sol. State Commun.* **44**, 1351 (1982); S. Ono, *J. Phys. Soc. Jpn.* **40**, 498 (1976)
[42] J. M. Ziman, *Electrons and Phonons* (Clarendon Press, Oxford 1963), p. 357; A. H. Wilson, *The Theory of Metals*, 2nd edition (Cambridge, 1965), p. 262.
[43] J. M. Tarascon, L. H. Greene, W. R. McKinnon, G. W. Hull, and T. Geballe, *Science* **235**, 1373 (1987).
[44] J. Clayhold, N. P. Ong, Z. Z. Wang, J. Birmingham, J. M. Tarascon and P. Barboux (unpublished).
[45] M. Gurvitch and A. T. Fiory, *Phys. Rev. Lett.* **59**, 1337 (1987).
[46] R. Micnas, J. Ranninger, and S. Robaszkiewicz, *Phys. Rev.* **B36**, 4051 (1987).
[47] C. Kallin and A. J. Berlinsky, *Phys. Rev. Lett.* **60**, 2556 (1988).
[48] See, for example, *Specific Heats at Low Temperatures*, by E. S. R. Gopal, (Plenum, New York, 1966).
[49] S. Doniach and S. Engelsberg, *Phys. Rev. Lett.* **17**, 750 (1966).
[50] J. Bardeen, L. Cooper and J. Schrieffer, *Phys. Rev.* **108**, 1175

(1957).
[51] D. J. Scalapino, in *Superconductivity*, Vol. 1, edited by R. D. Parks, (Dekker, New York, 1969), Chapter 10, p. 503.
[52] N. F. Berk and J. R. Schrieffer, *Phys. Rev. Lett.* 17, 433 (1966).
[53] G. Gladstone, M. A. Jensen and J. R. Schrieffer in *Superconductivity*, Vol. 2, edited by R. D. Parks, (Marcel Dekker, Inc., New York, 1969), Chapter 13, p. 665 and references within.
[54] See for example D. C. Mattis, *Theory of Magnetism I. Statics and Dynamics*, (Springer-Verlag, Berlin, 1981); T. Moriya, *Spin Fluctuations in Itinerant Electron Magnetism*, (Springler-Verlag, Berlin, 1985).
[55] A. B. Kaiser, *J. Phys.* C3, 409 (1970).
[56] M. B. Maple and D. Wohlleben, *AIP Conf. Proc.*, No. 18, edited by C. D. Graham, Jr. and J. J. Rhyne, p. 447 (1974); M. B. Maple, L. E. DeLong and B. C. Sales in *Handbook on The Physics and Chemistry of the Rare Earths*, edited by K. A. Gschneidner, Jr. and L. Eyring, (North-Holland, Amsterdam, 1978), Vol. 1, p. 797.
[57] P. W. Anderson, *Phys. Rev.* B30, 1549 (1984).
[58] Articles in *Theory in Heavy Fermions and Valence Fluctuations, 1985*, edited by T. Kasuya and T. Sasa (Springer-Verlag, Berlin, 1985).
[59] T. M. Rice, *Jpn. J. Appl. Phys.* **26** Suppl 26-3, 1865 (1987).
[60] V. Emery, *Phys. Rev. Lett.* **58**, 2794 (1987).
[61] J. R. Schrieffer, X-G. Wen and S. C. Zhang, *Phys. Rev. Lett.* **60**, 944 (1988).
[62] A. J. Millis, S. Sachdev and C. M. Varma, *Phys. Rev* B37, 4975 (1988).
[63] R. A. Fisher, J. E. Gordon and N. E. Phillips, *J. Supercond.*, 1, 231 (1988).
[64] W. A. Phillips, *J. Non-Cryst. Solids* 31, 267 (1978); and articles in *Amorphous Solids: Low Temperature Properties*, edited by W. A. Phillips (Springer-Verlag, New York, 1981).
[65] C. E. Yu and A. J. Leggett, *Comm. on Cond. Matt. Phys.* 14, 231 (1988).
[66] R. C. Zeller and R. O. Pohl, *Phys. Rev.* B4, 2029 (1971).
[67] P. J. Anthony and A. C. Anderson, *Phys. Rev.* B16, 3827 (1977).
[68] W. N. Lawless, *Phys. Rev.* B22, 3122 (1980); D. A. Ackerman, D. Moy, R. C. Potter, A. C. Anderson and W. N. Lawless, *Phys. Rev.* B23, 3886 (1981).
[69] J. Ferre, J. Pommier, J. P. Renard and K. Knorr, *J. Phys.* C13, 697 (1980).
[70] W. Arnold, E. Gmelin, K. Guchelsberger, G. Weiss, H. van Lohneysen, *J. Phys. Colloq.* 41, 751 (1980).

[71] P. W. Anderson, B. I. Halperin and C. Varma, *Phil. Mag.* **25**, 1 (1972); W. A. Phillips, *J. Low Temp. Phys.* **7**, 351 (1972).
[72] W. Schottky, *Phys. Z* **23**, 448 (1972).
[73] See for example, B. S. Chandrasekhar, in *Superconductivity*, Vol. 1, edited by R. D. Parks, (Marcell Dekker, Inc., New York, 1969), Chapter 1; G. Rickayzen, *Theory of Superconductivity* (John Wiley and Sons, Inc., New York, 1965).
[74] J. Blezius and J. P. Carbotte, *Phys. Rev.* **B36**, 3622 (1987); *ibid, J. Low Temp. Phys.* **73**, 255 (1988).
[75] F. Marsiglio and J. P. Carbotte, *Phys. Rev.* **B36**, 3937 (1987); F. Marsiglio, R. Akis and J. P. Carbotte, *Sol. State Commun.* **64**, 905 (1987).
[76] B. Muhlshlegel, *Z. Phys.* **155**, 313 (1959).
[77] N. E. Phillips, *Phys. Rev.* **114**, 676 (1954).
[78] See for example, *Liquid and Solid Helium*, edited by C. G. Kuper, S. G. Lipson and M. Revzen (John Wiley and Sons, New York, 1975).
[79] G. R. Stewart, *Rev. Mod. Phys.* **56**, 755 (1984), and references within.
[80] P. E. Lee, T. M. Rice, J. W. Serene, L. J. Sham and J. W. Wilkins, *Comm. on Cond. Matter Phys.* **12**, 99 (1986), and references within.
[81] H. E. Fischer, S. K. Watson and D. G. Cahill, *Comm. on Cond. Matter Phys.* **14**, 65 (1988).
[82] A. Junod, A. Bezinge, D. Cattani, J. Cors, M. Decroux, O. Fischer, D. Genoud, L. Hoffman, J.-L. Jonda, J. Muller and E. Walker, *Jpn. J. Appl. Phys.* **26-3**, 1119 (1987).
[83] B. Batlogg, A. P. Ramirez, R. J. Cava, R. B. vanDover and E. A. Rietman, *Phys. Rev.* **B35**, 5340 (1987).
[84] G. Nieva, E. N. Martinez, F. dela Cruz, D. A. Esparza and C. A. D'Ovidio, *Phys. Rev.* **B36**, 8780 (1987).
[85] N. E. Phillips, R. A. Fisher, S. E. Lacy, C. Marcenant, J. A. Olsen, W. K. Ham and A. M. Stacy, *Jpn. J. Appl. Phys.* **26**, Suppl. **26-3**, 1115 (1987).
[86] D. K. Finnemore, R. N. Shelton, J. R. Clem, R. W. McCallum, H. C. Ku, R. E. McCarley, S. C. Chen, P. Klavins and V. Kogan, *Phys. Rev.* **B35**, 5319 (1987).
[87] P. B. Allen, W. E. Pickett, and H. Krakauer, *Phys. Rev.* **B37**, 7482 (1987); H. Krakauer, W. E. Pickett, and R. E. Cohen, *J. Superconductivity* **1**, 111 (1988).
[88] K. Kumagai, Y. Nakamichi, I. Watanabe, Y. Nakamura, H. Nakajima, N. Wada and P. Lederer, *Phys. Rev. Lett.* **60**, 724 (1988).
[89] M. Kato, Y. Maeno and Fujita, *Physica* **C152**, 116 (1988).
[90] Z. Fisk, J. D. Thompson, E. Zirngiebl, J. L. Smith and S. W. Cheong, *Sol. State Commun.* **62**, 743 (1980); P. Hor, R. L.

Meng, Y. Q. Wang, L. Gao, Z. J. Hurong, J. Beachtold, K. Forster and C. W. Chu, *Phys. Rev. Lett.* **58**, 1891 (1987).
[91] C.-S. Jee, T. Yuen, J. E. Crow, C. L. Lin and P. Schlottmann, in *Proc. of the Drexel Int. Conf. On High Temperature Superconductivity*, edited by S. M. Bose and S. D. Tyagi (World Scientific, Singapore, 1988) p. 133.
[92] M. V. Nevitt, G. W. Crabtree and T. E. Klippert, *Phys. Rev.* **B36**, 2398 (1987).
[93] S. E. Inderhees, M. B. Salamon, T. A. Friedmann and D. M. Ginsberg, *Phys. Rev.* **B36**, 2401 (1987).
[94] J. C. Van Miltenburg, J. Q. A. Koster, Y. K. Huang, A. A. Menovsky and H. Barten, *Physica* B+**C146**, 319 (1987).
[95] O. Beckman, L. Lundgren, P. Nordblad, L. Sandlund, P. Sveedlindh, T. Lundstrom and S. Rundquist, *Phys. Lett.* **A125**, 425 (1987).
[96] A. Junod, A. Bizinge and J. Muller, *Physica* **C152**, 50 (1988).
[97] M. Ishikawa, T. Takabatake and Y. Nakazawa, *Physica* **C148**, 332 (1987).
[98] M. Lang, T. Lechner, S. Riegel, F. Steglich, G. Weber, T. J. Kim, B. Luthi, B. Wolf, H. Rietschel, and B. Willhelm, *Z. fur Physik* **B69**, 459 (1987).
[99] K. Kadowaki, M. van Sprang, Y. K. Huang. J. Q. A. Koster, H. P. van der Meulen, Z. Tarnawski, J. C. P. Klasse, A. A. Menovsky, J. J. M. Franse, J. C. van Miltenburg, A. Schuijff, T. T. M. Palstra, R. DeRuiter, P. W. Lednor and H. Barten, *Physica* **B148**, 442 (1988).
[100] J. R. Thompson (Private Communication).
[101] M. E. Reeves, D. S. Citrin, B. G. Pazol, T. A. Friedman and D. M. Ginsberg, *Phys. Rev.* **B36**, 6915 (1987); T. Atake, Y. Takagi, T. Nakamura and Y. Saito, *Phys. Rev.* **B37**, 552 (1988).
[102] D. R. Harshman, G. Aeppli, E. J. Ansaldo, B. Batlogg, J. H. Brewer, J. F. Carolan, R. J. Cava, M. Celio, A. C. D. Chaklader, W. N. Hardy, S. R. Kreitzman, G. M. Luke, D. R. Noakes and M. Senba, *Phys. Rev.* **B36**, 2386 (1987); D. R. Harshman, *Phys. Rev.* **B36**, 2386 (1987); D. R. Harshman, L. F. Schneemeyer, J. V. Waszczak, G. Aeppli, R. J. Cava, B. Batlogg, L. W. Rupp, E. J. Ansaldo, R. F. Kiefl, G. M. Kuke, T. M. Riseman and D. L. William, *Phys. Rev.* **B39**, 851 (1989).
[103] C. S. Jee, Ph.D. Thesis (Temple University, 1988).
[104] G. Yia, M. Z. Cieplak, A. Garrin, F. H. Streitz, A. Bakshai and C. L. Chien, *Phys. Rev. Lett.* **60**, 1446 (1988); G. Xiao, M. Z. Ciepak, D. Musser, A. Gavrin, F. H. Streitz, C. L. Chien, J. J. Rhyne and J. A. Gotass, *Nature* **332**, 238 (1989).
[105] C.-S. Jee, D. Nichols, A. Kebede, S. Rahman, J. E. Crow, A. M. Ponte Conclaves, T. Mihalisin, G. H. Myer, I. Perez, R. E. Salomon, P. Schlottmann, S. H. Bloom, M. V. Kuric, Y. S. Yao

and R. P. Guertin, *J. Superconductivity* **1**, 63 (1988).

[106] S. von Molnar, A. Torressen, D. Kaiser, F. Holtzberg and T. Penney, *Phys. Rev.* **B37**, 3762 (1988).

[107] J. C. Lasjaunias, H. Noel, J. C. Levet, M. Potel and P. Gougeon, *Phys. Lett.* **A129**, 185 (1988).

[108] M. J. McKenna, A. Hikata, J. T. C. Elbaum, R. Kershaw and A. Wold, *Phys. Rev. Lett.* **62**, 1556 (1989).

[109] C.-S. Jee A. Kebede, D. Nichols, J. E. Crow, T. Mihalisin, G. H. Myer, I. Perez, R. E. Salomon and P. Schlottmann, *Sol. State Commun.* **69**, 379 (1989); A. Kebede, C.-S. Jee, J. Schwegler, J. E. Crow, T. Mihalisin, G. H. Myer, R. E. Salomon, P. Schlottmann, M. V. Kuric, S. H. Bloom and R. P. Guretin, *Phys. Rev.* B (to be published).

[110] S. A. Kivelson, D. S. Rokhsar, and J. P. Sethna, *Phys. Rev.* **B35**, 8865 (1987); S. Kivelson and D. S. Rokhsar, *Physica* **C153-155**, 531 (1988).

[111] Y. R. Wang and M. J. Rice, *Phys. Rev.* **B38**, 7163 (1988).

[112] T. Itoh, K. Kitazawa, and S. Tanaka, *J. Phys. Soc. Jpn.* **53**, 2668 (1984); B. Batlogg, *Physica* **B126**, 275 (1984).

[113] Y. Gao, J. E. Crow, G. Myer, R. E. Salomon and J. Schwegler (preprint).

[114] A. W. Sleight, J. L. Gillson and P. E. Bierstadt, *Sol. State Commun.* **17**, 27 (1975).

[115] R. J. Cava, B. Batlogg, J. J. Krajewski, R. Farrow, L. W. Rupp, Jr., A. E. White, K. Short, W. F. Peck and T. Kometani, *Nature* **332**, 814 (1988).

[116] D. G. Hinks, B. Dabrowski, J. D. Jorgensen, A. W. Mitchell, D. R. Richards, S. Pei and D. Shi, *Nature* **333**, 836 (1988).

[117] J. C. Ho, C. Y. Wu, D. G. Hinks, B. Dabrowski, Y. Zheng, A. W. Mitchell and C. Y. Huang, *Bull. APS* **34**, 693 (1989).

[118] J. E. Graebner, L. F. Schneemeyer and J. K. Thomas, *Bull. APS* **35**, 694 (1989).

[119] C. Michel, M. Hervieu, M. M. Bore, A. Grandin, F. Dislandes, J. Provost and B. Raveau, *Z. Phys.* **B68**, 421 (1987); H. Maeda, Y. Tanaka, M. Fukutomi and T. Asamo, *J. Appl. Phys. Lett. Jpn.* **27**, L209 (1988).

[120] Z. Z. Sheng and A. M. Hermann, *Nature* **332**, 55 (1988); *ibid* **332**, 138 (1988).

[121] A. W. Sleight, M. A. Subramanian and C. C. Torardi, *MRS Bulletin* **24**, 45 (1989), and references within.

[122] R. A. Fisher, S. Kim, S. E. Lacy, N. E. Phillips, D. E. Morris, A. G. Markelz, J. Y. T. Wei and D. S. Ginley, *Phys. Rev.* **B38**, 11942 (1988).

[123] M. Sera, S. Kondon, K. Fukuda and M. Sato, *Sol. State Commun.* **66**, 1101 (1988).

[124] K. Kumagai and Y. Nakamura, *Physica* **C152**, 286 (1988).

[125] S. L. Yuan, J. W. Ki, W. Wang, Q. Z. Ran, G. G. Zheng, W. Y. Guan and J. Q. Zheng, *Mod. Phys. Lett.* B2, 885 (1988).
[126] A. Chakraborty, A. J. Epstein, D. L. Cox, E. M. McCarron and W. E. Farneth, *Phys. Rev.* B39, 12267 (1989).
[127] Y. Gao, P. Schlottmann, J. E. Crow and N. D. Spencer (preprint).
[128] D. Eckert, A. Junod, A. Bezinge, T. Graf and J. Muller, *J. Low Temp. Phys.* **73**, 241 (1988).
[129] R. E. Glover, III, *Phys. Lett.* A25, 542 (1967).
[130] J. E. Crow, R. S. Thompson, M. A. Klenin and A. K. Bhatnagar, *Phys. Rev. Lett.* **24**, 371 (1970); W. E. Masker and R. D. Parks, *Phys. Rev* B1, 2164 (1970).
[131] W. L. Johnson and C. C. Tsuei, *Phys. Rev.* B13, 4827 (1976).
[132] A. Schmid, *Phys. Rev.* **186**, 527 (1969).
[133] J. P. Gollub, M. R. Beasley, R. S. Newbower and M. Tinkham, *Phys. Rev. Lett.* **22**, 1288 (1969); J. P. Gollub, M. R. Beasley, R. Gallorotti and M. Tinkham, *Phys. Rev.* B7, 3039 (1973).
[134] For a review of fluctuation effects in conventional superconductors, see W. J. Skocpol and M. Tinkham, *Rep. Prog. Phys.* **38**, 1049 (1975).
[135] L. G. Aslamozov and A. I. Larkin, *Phys. Lett.* A26, 238 (1968); *Sov. Phys. Sol. State* **10**, 875 (1968).
[136] S.-K. Ma, *Modern Theory of Critical Phenomena*, (Benjamn, New York, 1976) p. 82-93. Note: Eq. (3.59) is incorrect, see reference 137.
[137] S. E. Inderhees, M. B. Salomon, N. Goldenfeld, J. P. Rice, B. G. Pazol, D. M. Ginsberg, J. Z. Liu and G. W. Crabtree, *Phys. Rev. Lett.* **60**, 1178 (1988).
[138] T. K. Worthington, W. J. Gallagher and T. R. Dinger, *Phys. Rev.* B36, 7861 (1988); S. J. Hagen, T. W. Jing, Z. Z. Wang, J. Horvath and N. P. Ong, *Phys. Rev.* B37, 7928 (1988); B. Oh, K. Char, A. D. Kent, M. Naito, M. R. Beasley, T. H. Geballe, R. H. Hammond, A. Kapitulnik, and J. M. Graybeal, *Phys. Rev.* B37, 7861 (1988).
[139] W. C. Lee, R. A. Klemm and D. C. Johnston (preprint).
[140] M. B. Salamon, S. E. Inderhees, J. P. Rice, B. G. Pazol, D. M. Ginsberg and N. Goldenfeld, *Phys. Rev.* B38, 885 (1988).
[141] W. E. Lawrence and S. Doniach, *Proc. Twelfth Int. Conf. on Low. Temp. Phys. (Kyoto,* 1979), edited by E. Kankl (Kelgaku Publishing, 1971), p. 361.
[142] K. Maki, *Prog. Theor. Phys.* **40**, 193 (1968); R. S. Thompson, *Phys. Rev.* B1, 327 (1970); R. S. Thompson, *Physica* **55**, 296 (1971).
[143] See for example, V. L. Ginzburg and D. A. Kirzhnits, *High Temperature Superconductivity* (Consultants Bureau, New York

1982).

[144] W. A. Little, *Phys. Rev.* **A134**, 1416 (1964); D. Davis, H. Gutfreund and W. A. Little, *Phys. Rev.* **B13**, 4766 (1976), and references within.

[145] D. Allender, J. Bray and J. Bardeen, *Phys. Rev.* **B7**, 1020 (1973); *ibid* **B8**, 4433 (1973), and references within.

[146] C. M. Varma, S. Schmitt-Rink and E. Abrahams, *Sol. State Commun.* **62**, 681 (1987).

[147] E. Maxwell, *Phys. Rev.* **78**, 477 (1950); C. A. Reynolds, B. Serin, W. H. Wright and L. B. Nesbitt, Phys. Rev. **78**, 487 (1950).

[148] See for example, L. Solymar, *Superconducting Tunnelling and Applications*, (Chapman and Hall, London, 1972), and references within.

[149] J. C. Swhart, *Phys. Rev.* **116**, 45 (1959).

[150] P. Morel and P. W. Anderson, *Phys. Rev.* **125**, 1263 (1962).

[151] G. Gladstone, M. A. Jensen and J. R. Schrieffer, in *Superconductivity*, edited by R. D. Parks, (Marcel Dekker, Inc., New York, 1969) p. 665, and references within.

[152] W. Weber, *Phys. Rev. Lett.* **58**, 1371 (1987).

[153] H.-C. Zur Loye, K. J. Leary, S. W. Keller W. K. Ham, T. A. Faltens, J. N. Michaels and A. M. Stacy, *Science* **238**, 1558 (1987).

[154] B. Batlogg, R. J. Cava, L. W. Rupp, Jr., A. M. Mujsce, J. J. Krajewski, J. P. Remeika, W. F. Peck, Jr., A. S. Cooper and G. P. Espinosa, *Phys. Rev. Lett.* **61**, 1670 (1988).

[155] D. G. Hinks, D. R. Richards, B. Dabrowski, D. T. Marx and A. W. Mitchell, *Nature* **335**, 419 (1988).

[156] T. A. Faltens, W. K. Ham, S. W. Keller, K. J. Leary, J. N. Michaels, A. M. Stacy, H.-C. Zur Loye, D. E. Morris, T. W. Barbee, III, L. C. Bourne, M. L. Cohen, S. Hoen and A. Zettl, *Phys. Rev. Lett.* **59**, 915 (1987).

[157] B. Batlogg, R. J. Cava, C. H. Chen, G. Kourouklis, W. Weber, A. Jayaraman, A. E. White, K. T. Short, E. A. Rietman, L. W. Rupp, D. Werder and S. M. Zahurak, in *Novel Superconductivity*, edited by S. A. Wolf and V. Z. Krezin (Plenum Press, New York, 1987) p. 653.

[158] B. Batlogg, G. Kourouklis, W. Weber, R. J. Cava, A. Jayaraman, A. E. White, K. T. Short, L. W. Rupp, and E. A. Rietman, *Phys. Rev. Lett.* **59**, 912 (1987).

[159] L. C. Bourne, S. Hoen, M. F. Crommie, W. N. Creager, A. Zettl, M. L. Cohen, L. Bernardez, J. Kinney and D. E. Morris, *Sol. State Commun.* **67**, 707 (1988).

[160] H. Katayama-Yoshida, T. Hirooka, A. J. Mascarenyhas, Y. Okabe, T. Takahashi, T. Sasaki, A. Ochiai, T. Suzuki, J. I. Pankove, T. Ciszek and S. K. Dob, *Jpn. J. Appl. Phys.* **26**, L2085 (1987).

[161] B. Batlogg, R. J. Cava, A. Jayaraman, R. B. van Dover, G. A. Kourouklis, S. Sunshine, D. W. Murphy, L. W. Rupp, H. S. Chen, A. White, K. T. Short A. M. Mujsce and E. A. Rietman, *Phys. Rev. Lett.* **58**, 2333 (1987).
[162] M. Cardona, R. Liu, C. Thomsen, W. Kress, E. Schonherr, M. Bauer, L. Genzeland and W. Konig, *Sol State Commun.* **67**, 789 (1988).
[163] E. L. Benitez, J. J. Lin, S. J. Poon, W. E. Farneth, M. K. Crawford and E. M. McCarron, *Phys. Rev.* B**38**, 5025 (1988).
[164] H. Katayama-Yoshida, T. Hirooka, A. Oyamada, Y. Okabe, T. Takahashi, T. Sasaki, A. Ochiai, T. Suzuki, A. J. Mascarenhas, J. I. Pankove, T. F. Cixaek, S. K. Deb, R. B. Goldfarb and Y. Li, *Physica* C**156**, 481 (1988).
[165] G. A. Kourouklis, A. Jayaraman, B. Batlogg, R. J. Cava, M. Stavola, D. M. Krol, E. A. Rietman and L. F. Schneemeyer, *Phys. Rev.* B**36**, 8320 (1987).
[166] L. C. Bourne, A. Zettl, T. W. Barbee, III and M. L. Cohen, *Phys. Rev.* B**36**, 3990 (1987).
[167] Q. Li, Y.-N. Wei, Q.-W. Yan, G.-H. Chen, P.-L. Zhang, Z.-G. Shen, Y.-M. Ni, Q.-S. Yang, C.-X. Liu, T.-S. Ning, J.-K. Zhao, Y.-Y. Shao, S.-H. Han and J.-Y. Li, *Sol. State Commun.* **65**, 869 (1988).
[168] W. L. McMillian, *Phys. Rev.* **167**, 331 (1968).
[169] P. B. Allen and R. C. Dynes, *Phys. Rev.* B**12**, 905 (1975).
[170] P. B. Allen and B. Mitrovic, *Sol. State Phys.* **37**, 1 (1982).
[171] J. M. Rowell and R. C. Dynes, in *Phonons*, edited by M. A. Nusimovici (Flammarion Sciences, Paris, 1972) p. 150, and references within.
[172] B. P. Schweiss, B. Renker, E. Schneider and W. Reichandt in *Superconductivity in d- and f-Band Metals*, edited by D. H. Douglass (Plenum Press, New York, 1976) p. 189.
[173] G. Shirane, J. D. Axe and R. J. Birgeneau, *Sol. St. Commun.* **9**, 397 (1971).
[174] L. R. Testardi and T. B. Bateman, *Phys. Rev.* **154**, 402 (1967).
[175] B. W. Batterman and C. S. Barrett, *Phys. Rev. Lett.* **13**, 390 (1964).
[176] C. L. Fu and A. J. Freeman, *Phys. Rev.* B**35**, 8861 (1987).
[177] L. Pintschovius, J. M. Bassat, P. Odier, F. Gervais, B. Hennion and W. Reichardt, *Europhys. Lett.* **5**, 247 (1988); *Physicsa* C**153-155**, 276 (1988).
[178] P. Boni, J. D. Axe, G. Shirane, R. J. Birgeneau, D. R. Gabbe, H. P. Jenssen, M. A. Kastner, C. J. Peters, P. J. Picone and T. R. Thurston, *Phys. Rev.* B**38**, 185 (1988).
[179] R. J. Birgeneau and G. Shirane, to be published in *Physical Properties of High Temperature Superconductors*, edited by D. M. Ginsberg (World Scientific Publishing, 1989),

and references within.

[180] W. Weber and L. F. Mattheiss, *Phys. Rev.* **B37**, 599 (1988).

[181] G. Shirane, *Rev. Mod. Phys.* **46**, 437 (1974).

[182] R. E. Cohen, W. E. Pickett, H. Krakaver and L. L. Boyer, *Physica* **B150**, 61 (1988); *Phys. Rev. Lett.* **60**, 817 (1988).

[183] B. Renker, F. Grompf, E. Gering, N. Nucker, D. Ewert, W. Reichardt and H. Rietschel, *Z. Phys.* B - *Cond. Matter* **67**, 15 (1987).

[184] L. F. Mattheiss and D. R. Hamann, *Phys. Rev.* **B28**, 4227 (1983).

[185] W. Reichardt, B. Batlogg and J. P. Remeida, *Physica* **B135**, 501 (1985).

[186] B. Batlogg, J. P. Remeika, R. C. Dynes, H. Barz, A. S. Cooper and J. P. Gano in *Superconductivity in d- and f-Band Metals*, edited by W. Buckel and W. Weber (KFZ, Karlsruhe, West Germany 1982) p. 401.

[187] G. Aeppli and D. J. Buttery, *Phys. Rev. Lett.* **61**, 203 (1988).

[188] H. Rietschel, J. Fink, E. Gering, F. Gompf, N. Nocker, L. Pintschovius, B. Renker, W. Reichardt, H. Schmidt and W. Weber, *Physica* **C153-155**, 1067 (1988).

[189] J. J. Rhyne, D. A. Neuman, J. A. Gotaas, F. Beech, L. Toth, S. Lawrence, S. Wolf, M. Osofsky and D. U. Gubser, *Phys. Rev.* **B36**, 2294 (1987).

[190] L. Mihaly, L. Rosta, G. Coddens, F. Mezei, g. Hutiray, G. Kriza and B. Keszei, *Phys. Rev.* **B36**, 7137 (1987).

[191] B. Renker, F. Gompf, E. Gering, G. Roth, W. Reichardt, D. Ewert and H. Rietschel, *Physica* **C153-155**, 272 (1988).

[192] B. Renker, F. Gompf, E. Gering, G. Roth, W. Reichardt, D. Ewert, H. Rietschel and H. Mutka, *Z. Phys.* B-*Cond. Matter* **71**, 437 (1988).

[193] B. Renker, F. Gompf, E. Gering, D. Ewert, H. Rietschel and A. Dianoux, *Z. Phys.* B-*Cond. Matter* **73**, 309 (1988).

[194] F. Gompf, B. Renker and E. Gering, *Physica* **C153-155**, 274 (1988).

[195] P. Buresch and W. Buhrer, *Z. Phys.* B-*Cond. Matter* **70**, 1 (1988).

[196] See for example, C. Thomsen and M. Cardona, in *Physical Properties of High-Temperature Superconductors*, edited by D. M. Ginsberg, World Science, Singapore (1989) and references within.

[197] H. Rosen, E. M. Engler, T. C. Strand, V. Y. Lee and D. Bethune, *Phys. Rev.* **B36**, 726 (1988).

[198] M. Udagawa, *Jpn. J. Appl. Phys.* **26**, L858 (1987).

[199] Z. V. Popovic, C. Thomsen, M. Cardona, R. Liu, G. Stanisic, R. Kremer and W. Konig, *Sol. St. Commun.* **66**, 43 (1988).

[200] C. Thomsen, R. Liu, M. Bauer, A. Wittlin, L. Genzel, M. Cordona, E. Schonherr, W. Bauhofer and W. Konig, *Sol. St.*

Commun. **65**, 55 (1988).

[201] R. M. MacFarlane, H. J. Rosen, E. M. Engler, R. D. Jacowitz and V. Y. Lee, *Phys. Rev.* **B38**, 284 (1988).

[202] M. Krantz, H. J. Rosen, R. M. MacFarlane and V. Y. Lee, *Phys. Rev.* **B38**, 4992 (1988).

[203] W. H. Weber, C. R. Peters and E. M. Logothetis (preprint).

[204] A. Wittlin, R. Liu, M. Cardona, L. Genzel, W. Konig, W. Bauhofer, Hj. Mattausch and A. Simon, *Sol. St. Commun.* **64**, 477 (1987).

[205] C. Thomsen, R. Liu, A. Wittlin, L. Genzel, M. Cardona and W. Konig, *Sol. St. Commun.* **65**, 219 (1988).

[206] C. Thomsen, R. Liu, M. Cardona, U. Amador and E. Moran, *Sol. St. Commun.* **67**, 271 (1988).

[207] T. Ruf, C. Thomsen, R. Liu and M. Cardona, *Phys. Rev.* **B38**, 11985 (1988).

[208] S. M. Shapiro, G. Shirane and J. D. Axe, *Phys. Rev.* B12, 4899 (1975).

[209] R. Zeyher and G. Zwicknagl, *Sol. St. Commun.* **66**, 617 (1988).

[210] D. Tanner (private communication).

[211] C.W. Kimball, L.W. Weber, and F.Y. Fradin, *Phys. Rev.* **B14**, 2769 (1976).

[212] C. Blue, K. Elgaid, I. Zitkovsky, P. Bookhand, D. McDaniel, W.C.H. Joiner, J. Oostens, and W. Huff, *Phys. Rev.* B37,5905 (1988).

[213] T. Yuen, C.L. Lin, J.E. Crow, G.H. Myer, R.E. Salomon, P. Schlottmann, N. Bykovetz, and W.H. Herman, *Phys. Rev.* **B37**, 3770 (1988).

[214] P. Boolchand, R.N. Enzweiler, I. Zitkovsky, J. Wells, W. Bresser, D. McDaniel, R.L. Meng, P.H. Hor, C.W. Chu, and C.Y. Huang, *Phys. Rev.* B37, 3766 (1988).

[215] J. Giapintzakis, J.M. Matykiewicz, C.W. Kimball, A.E. Dwight, B.D. Dunlap, M. Seaski, and F.Y. Fradin, *Phys. Lett.* **A121**, 307 (1987).

[216] C.W. Kimball, A.E. Dwight, S.K. Farrah, T.F. Karlow, D.J. McDowell and S.P. Taneja, in *Superconductivity in d- and f-Band Metals: 1982*, edited by W. Buckel and W. Weber (KFA, Karlsruhe, 1982), pg. 409.

[217] See for example, R. Berman, *Thermal Conduction in Solids*, (Oxford University Press, Oxford, 1979).

[218] C. Uher and J.L. Cohn, J. Phys. C **21**, L957 (1988).

[219] M. Nunez Requeiro, D. Castello, M.A. Izbizky, D. Esparza, and C.D. Ovidio, *Phys. Rev.* B36, 8813 (1987).

[220] Y. Bayot, F. Delannay, C. Dewitte, J-P. Erauw, X. Gonze, J-P. Issi, A. Jonas, M. Kinany=Alaoui, M. Lambricht, J-P. Michenaud, J-P. Minet, and L. Piraux, *Solid State Commun.* **63**, 983 (1987).

[221] D.T. Morelli, J. Heremans, and D.E. Swets, *Phys. Rev.* **B36**, 3917 (1987).
[222] F. Steglich, U. Ahleim, D. Ewert U. Gottwick, R. Held, H. Kneissel, M. Lang, U. Rauchschwalbe, B. Renker, H. Rietschel, G. Sparn, and H. Spille, *Physica Scripts.* **37**, 901 (1987).
[223] J. Hermans, D.T. Morelli, G.W. Smith, and S.C. Strite III, *Phys. Rev.* B37, 1604 (1988).
[224] D.M. Ginsberg and L.C. Hebel, in *Superconductivity*, edited by R.D. Parks (Dekker, New York, 1969) pg. 229.
[225] P. Lindenfeld, *Phys. Rev. Lett.* **6**, 613 (1961).

CHAPTER 8

Magnetic Properties

Jeffrey W. Lynn

8.1. Introduction

The superconducting oxides exhibit a rich variety of cooperative phenomena, including metal-insulator transitions, antiferromagnetism, and of course superconductivity. From the standpoint of superconductivity the central issue is to identify the nature of the superconducting state. One aspect concerns the basic character of the superconducting state itself, that is, whether it develops in the usual BCS way as a result of an instability of the Fermi liquid, or perhaps entails a different mechanism such as RVB. The second basic question involves the identification of the interaction which is responsible for the electron pairing. A review of the theoretical models for the superconducting state and the various pairing interactions is given in the next chapter.

The magnetic properties of these oxide materials have been of particular interest since the discovery in the $(La_{2-x}Sr_x)CuO_{4-\delta}$ and $\mathscr{R}Ba_2Cu_3O_{6+x}$ ($\mathscr{R}$ = rare earth) systems that the Cu ions carry an unpaired spin, with the consequent possibility that the magnetic fluctuations may be responsible for the superconductivity. At small x the Cu spins order antiferromagnetically in the semiconducting phase, with Neel temperatures T_N as high as 500 K. This is a very high ordering temperature for an insulating magnet. Moreover, above the *3-d* ordering temperature very strong spin correlations persist within the Cu-O planes, with a magnetic energy scale which is an order-of-magnitude larger than T_N would indicate. The existence of these large magnetic energies has fueled speculation that this is the energy scale needed for high superconducting transition temperatures T_c. This possibility has been further supported by the observations that the magnetic fluctuations survive into the superconducting phase which is found at higher x, and also by the discovery that the temperature dependence of these fluctuations is directly affected by the superconducting transition. However, these simplistic arguments have a number of pitfalls, and the question of the origin of the superconductivity has by no means been settled yet. Nevertheless, even if magnetism is not the origin of the superconductivity in these high T_c materials, it is quite clear that the Cu-O layers are intimately involved in both the magnetism and superconductivity, and the striking magnetic behavior these materials display is of fundamental interest in its own right.

A second aspect concerns the magnetism of the rare earths. Some of the very first data on the $\mathcal{R}Ba_2Cu_3O_{6+x}$ compounds showed that substitution of the trivalent rare earths[1] had little effect on the superconducting properties [1-10], and this has also been found to be the case for the $\mathcal{R}Ba_2Cu_4O_8$ system as well [11]. Hence the rare earth sublattice is electronically isolated from the superconducting (Cu-O) sublattices, leading to magnetic ordering of the rare earth ions at very low temperatures ($\lesssim$ 2 K) just like conventional "magnetic-superconductor" systems (see Chapter 1). The crystal field splittings of the rare earth ions have been found to be sensitive to the oxygen concentration, but the rare earths order nevertheless, and in the metallic regime we find the coexistence of long range antiferromagnetic order with superconductivity. The tetravalent rare earths (Ce, Pr, Tb), on the other hand, are more strongly coupled to the rest of the electronic system. The superconducting transition temperature is found to decrease strongly with increasing substitution of these elements, and the magnetic ordering temperatures are much higher than for the trivalent rare earths.

In the present chapter we review the magnetic ordering observed in this class of materials, both on the Cu sites in the insulating regime, and on the rare earth sites. We also review the nature of the very strong spin-spin interactions in the Cu layers, which leads to the remarkably high Neel temperatures, and to the even larger energetics for the spin dynamics.

8.2. Cu-O Magnetism

The first evidence for the magnetic ordering of the Cu ions in these systems was obtained for $La_2CuO_{4-\delta}$ by susceptibility [12], neutron scattering [13-20], and muon precession [21,22] experiments. The magnetic ordering temperature T_N turns out to be extremely sensitive to the oxygen concentration, with T_N varying from 0 to ~300 K as δ

[1] Nd, Sm, Eu, Gd, Dy, Ho, Er, Tm, Yb and Lu.

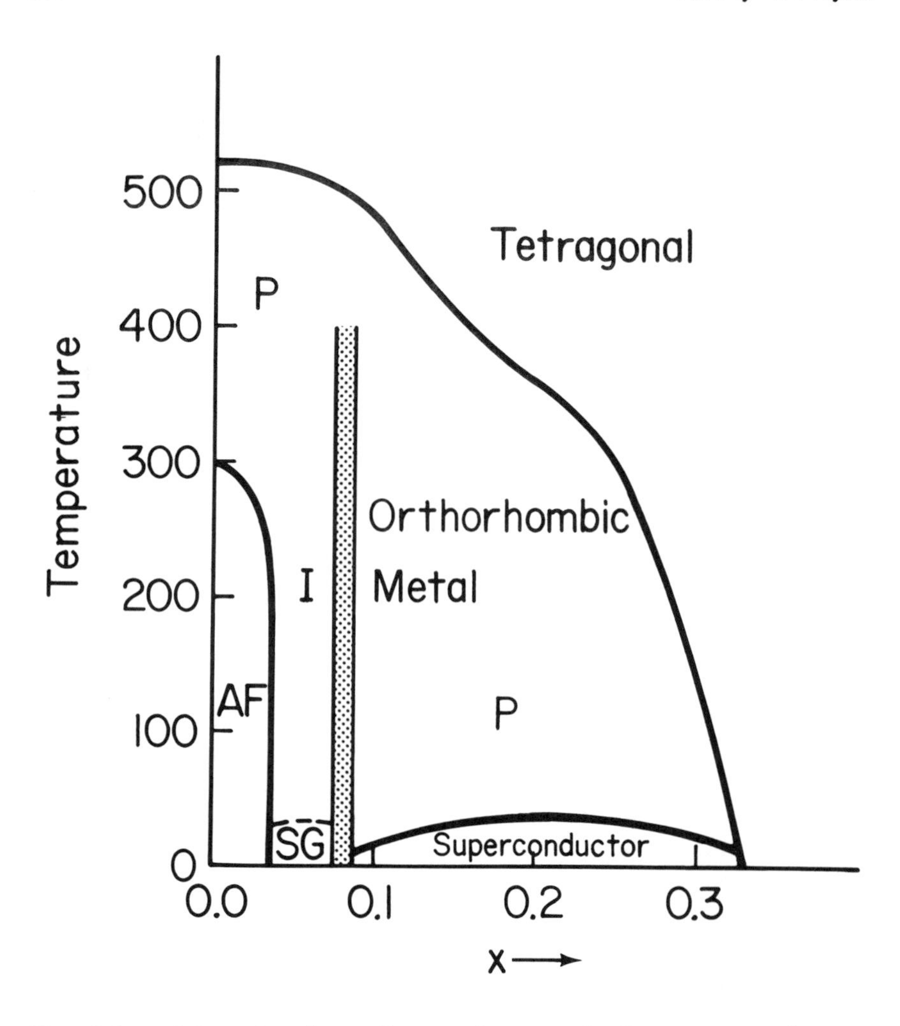

Fig. 8.1. Schematic phase diagram for $La_{2-x}Sr_xCuO_{4-\delta}$ as a function of Sr (or Ba) concentration x. Shown is the region of antiferromagnetic order (AF), spin glass behavior (SG), and paramagnetic (P) behavior. The insulating phase is designated by (I).

ranges from 0 to 0.03. Because of this extreme sensitivity to oxygen concentration it has been difficult experimentally to determine the phase diagram precisely. The properties of $La_{2-x}Sr_xCuO_{4-\delta}$ also vary dramatically as a function of Sr (or Ba) concentration x, but the variation is gradual enough that the phase diagram has been mapped out in detail. The properties of the $\mathcal{R}Ba_2Cu_3O_{6+x}$ system are also very sensitive to the oxygen concentration, and the general behavior as a function of x is quite similar for the two systems. We start our discussion with the magnetic ordering in the $La_{2-x}Sr_xCuO_{4-\delta}$ system.

8.2.1. La_2CuO_4 System

The magnetic and superconducting behavior of the $La_{2-x}Sr_xCuO_{4-\delta}$ system is shown schematically in Fig. 8.1 [12-25].[2] Pure La_2CuO_4 ideally should be ionically neutral, with La^{3+}, Cu^{2+}, and O^{2-}, although experimentally this has not been unambiguously established as just discussed. Theoretically at $\delta = 0$ the material is expected to be a half-filled Hubbard insulator, with a fully ordered magnetic state, while experimentally initial indications were that $\delta = 0$ corresponded to $T_N = 0$, with T_N increasing rapidly with δ and maximizing at $\delta = 0.03$ [12].[3] $\delta < 0$ (the superoxygenated state), on the other hand, produces a superconducting state [25]. For the present discussion we will assume that $x = 0$ corresponds to the antiferromagnetic insulator state, with the maximum T_N. With increasing Sr^{2+} (or Ba^{2+}) concentration there are uncompensated oxygen holes in the system, which initially produce an insulator-to-metal transition, and then superconductivity. In addition to the changing electronic properties, there is a second-order (i.e. continuous) orthorhombic-tetragonal (*OT*) structural phase transition: In the tetragonal phase $a = c$, and $b > a$, while as T is decreased below T_{OT} the difference between a and c becomes progressively (and smoothly) larger. This *OT* transition is strongly dependent on the Sr concentration as can be seen in the figure. Finally, we see that the magnetic properties also depend strongly on x, with the 3*d* Neel temperature decreasing rapidly with x, and only spin-glass-like behavior surviving for Sr concentrations above a few per cent. The value of the ordered moment also decreases with increasing x, but the magnitude of the moment (μ_{eff}) does not vary substantially with x, and large paramagnetic moments are observed in the superconducting regime.

The observed spin structure is shown in Fig. 8.2, and consists of ferromagnetic sheets of spins in the *b-c* plane, with adjacent sheets arranged antiparallel [13-20]. The magnetic moments are found to reside on the Cu ions, and to date no evidence of a magnetic moment on the planar O ions has been found [17-18]. The Cu-Cu exchange interaction within the *a-c* planes is mediated by the

[2]A theoretical phase diagram is shown in Fig. 9.1 [26].

[3]A small La deficiency can compensate for the oxygen deficiency, which is why it has been difficult to assign an exact stoichiometry to measured physical properties such as T_N.

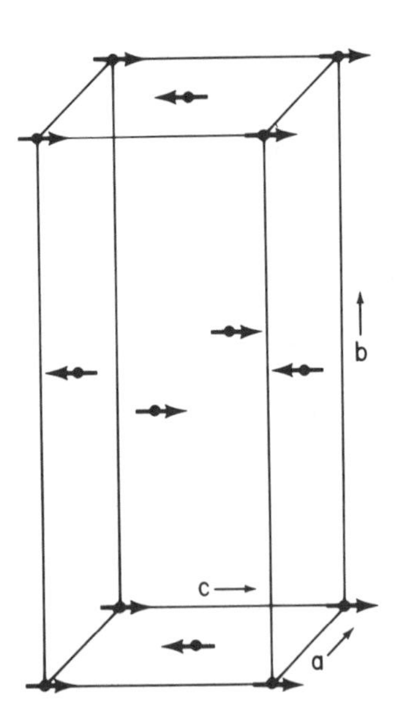

Fig. 8.2. Antiferromagnetic structure for the Cu spins in $La_2CuO_{4-\delta}$. b corresponds to the pseudotetragonal axis. Note that the structure consists of planes of spins ferromagnetically aligned (in the b-c plane), with adjacent planes aligned antiferromagnetically.

intervening O ions, and the directional Cu-O bonding yields an exchange interaction which is much larger than the exchange interaction between planes. This anisotropy of the exchange interactions is amplified by the fact that b is substantially larger than c. In addition, note that each spin within the plane layer has four spins above, and four spins below, which are almost the same distance away (since $a \cong c$). Hence the interactions between layers tend to cancel. In fact, for the related system K_2NiF_4, which is tetragonal, the cancellation is complete, and that system behaves as an ideal 2d antiferromagnet [27]. For the La_2CuO_4 material, due to the orthorhombic distortion, the cancellation is not complete, but the effective coupling between layers is much weaker than the in-plane interactions, which leads to strong correlations and 2-d like behavior [16].

To quantitatively interpret diffraction data a detailed comparison of a model calculation with the data must be made. The integrated intensity for a magnetic Bragg reflection is given by [28]

$$I_M = C \left[\frac{\gamma e^2}{2mc^2} \right]^2 \langle 1 - (\hat{\tau} \cdot \hat{M})^2 \rangle \; |F_M|^2 \qquad (8.1)$$

where the neutron-electron coupling constant in parenthesis is $-0.27 \cdot 10^{-12}$ cm, $\hat{\tau}$ and $\hat{M}$ are unit vectors in the direction of τ and the spin direction, respectively, and the orientation factor $\langle 1 -$

$\langle(\hat{\tau}\cdot\hat{\mathbf{M}})^2\rangle$ must be calculated for all possible domains.[4] This orientation factor originates from the fact that the neutron-electron interaction results from a dipole-dipole coupling. C is an instrumental constant which includes the resolution of the measurement. The magnetic structure factor F_M is given by

$$F_M = \sum_{j=1}^{N} \langle\mu_z\rangle_j \, f_j(\tau) \, e^{-W_j} \, e^{i\tau\cdot\mathbf{r}_j} \tag{8.2}$$

where $\langle\mu_z\rangle_j$ is the thermal average of the aligned magnetic moment of the magnetic ion at the jth site at position $\mathbf{r}_j$, W_j is the Debye Waller factor for the jth atom, and the sum extends over all magnetic atoms in the unit cell. $f(\tau)$ is the magnetic form factor, which is the Fourier transform of the atomic magnetization density. The scattering can be put on an absolute scale, whereby the saturated value of the magnetic moment can be obtained, by comparison with the nuclear Bragg intensities.

The observed intensity then is directly related to the square of the thermal average of the magnetic moment, $\langle\mu_z\rangle$, which is the order parameter for the phase transition. Fig. 8.3 shows the magnetic Bragg intensity for a crystal which has a 3-*d* ordering temperature at ~200 K [16]. The scattering increases continuously with decreasing temperature below the ordering temperature in a manner which is typical of a second-order phase transition. The ordered moment at low temperatures was determined to be 0.35 ± 0.05 μ_B for this sample.

Above the 3-*d* ordering temperature there are very strong spin correlations which persist within the Cu-O layers, as we will discuss in Sec. 8.2.3 below. Because of the large distance between planes, and the cancellation of interactions as discussed, the effective exchange coupling J_b between planes is much weaker than the exchange interactions J_{ac} ($J_b \ll J_{ac}$) within the planes. In this case the 3-*d* ordering is really driven by the 2-*d* correlations, which we can understand by the following argument. At any particular temperature $T > T_N$, there are areas within the planes which are correlated, with correlation length ξ_p, and hence a correlated region has a typical area ξ_p^2. Now the spins in the plane above (and below) are also highly correlated. If the correlations

[4]For simplicity we have assumed that the magnetic structure is collinear (i.e. all the spins point along a common axis). In fact the magnetic structures observed in these materials are not always collinear.

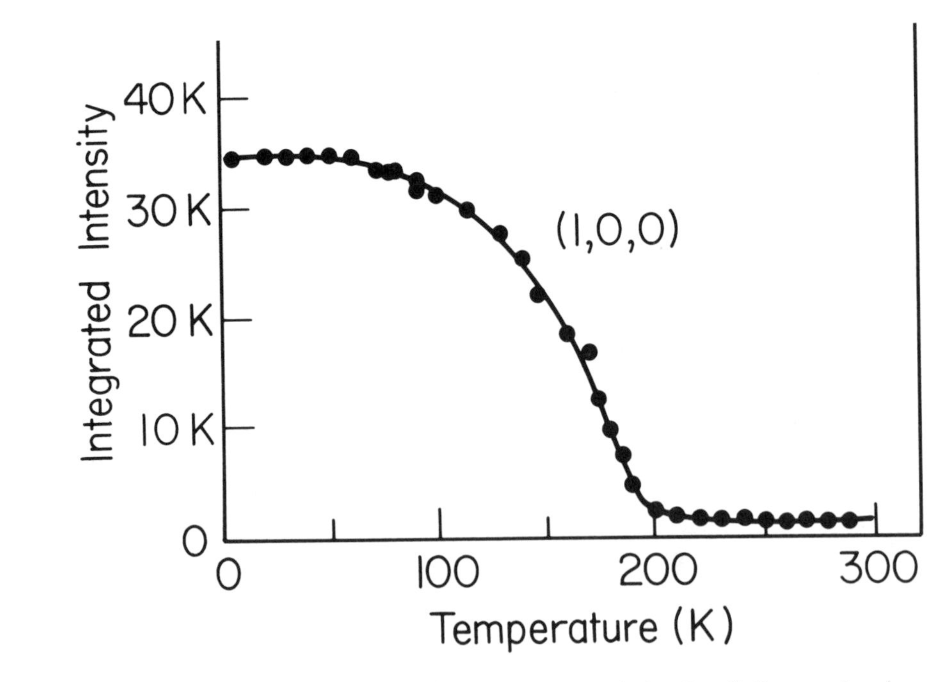

Fig. 8.3. Magnetic Bragg scattering observed in $La_2CuO_{4-\delta}$, showing a Neel temperature of $T_N = 200$ K (after Shirane, et al. [16]).

between planes are out-of-phase (and J_b is antiferromagnetic) then an energy $J_b \times$ *number of correlated ions* is gained, while if they are in-phase then it costs the same amount of energy. Thus there is a strong tendency for the system to become correlated between planes even though the individual exchange interaction is weak. A rough estimate for the Neel temperature is given by $kT_N \sim J_b \xi_p^2$ (see [27]). Of course, below T_N the energetic arguments remain the same, but the length scale ξ_p needs to be replaced by the average domain size in the plane.

We are now in a position to understand some of the magnetic aspects of Fig. 8.1. The substitution of Sr in the system produces holes, which disrupts the magnetic interactions within the planes [20]. This reduces the correlation range, and thus T_N drops rapidly with x. The magnetic frustration which is caused by the holes in fact quickly destroys the ordering, and leads to spin glass like behavior at larger x [20, 22, 29]. The magnetic ordering is suppressed long before the superconducting (or metallic) state appears. However, there are still large paramagnetic moments present in the range of x where superconductivity exists.

8.2.2. $RBa_2Cu_3O_{6+x}$ (R = Rare Earth)

The magnetic and superconducting behavior of the $\mathscr{R}Ba_2Cu_3O_{6+x}$ system as a function of oxygen concentration is quite similar to the behavior of the $La_{2-x}Sr_xCuO_{4-\delta}$ system as a function of Sr concentration. Again, evidence for magnetic ordering of the Cu in this system was obtained both by muon spin precession experiments

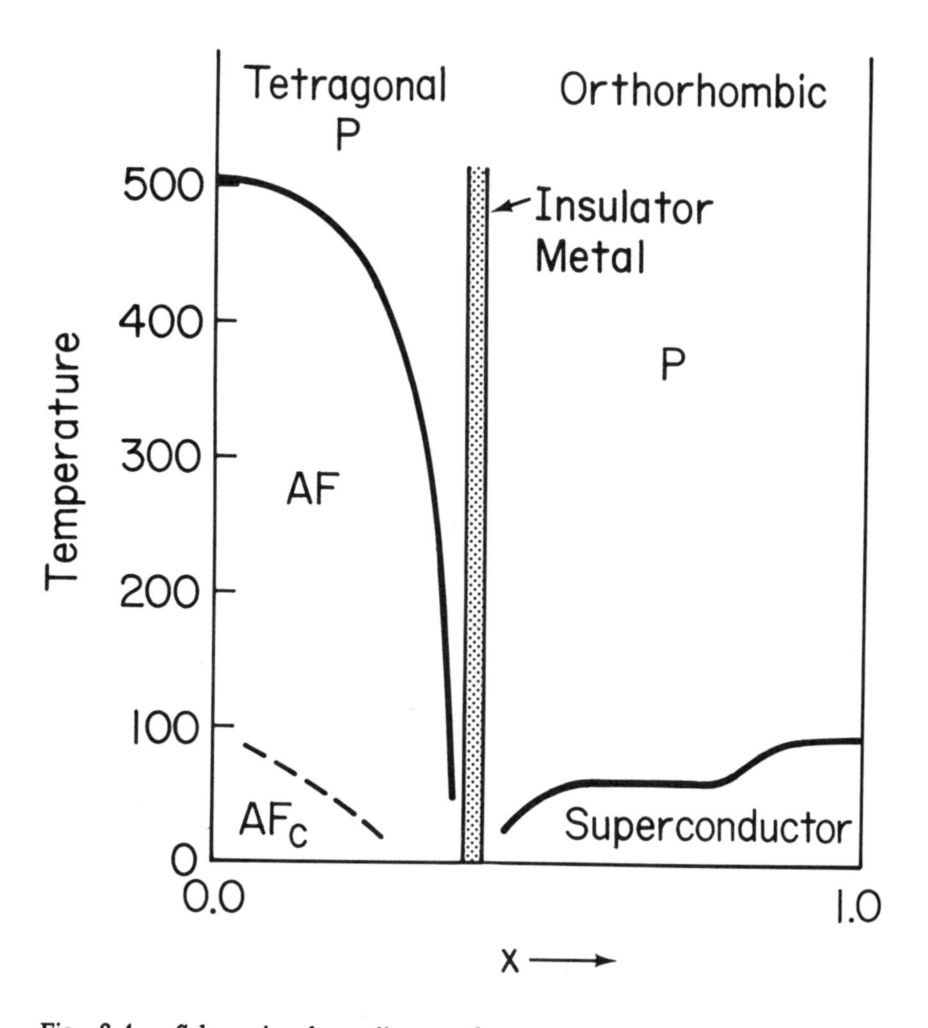

Fig. 8.4. Schematic phase diagram for $\mathscr{R}Ba_2Cu_3O_{6+x}$ as a function of oxygen concentration x on the "chain" sites. The paramagnetic (P) and antiferromagnetic (AF) phases are shown, as well as an antiferromagnetic phase found at lower temperatures (AF_c), where the spins on the Cu chains order.

[30,31] and neutron scattering experiments [32-43].[5] Before proceeding to the details of the magnetic ordering, we note a few of the basic properties of this class of materials. All the physical properties are very sensitive to the oxygen concentration as shown in Fig. 8.4 [35-45,25]. The $\mathcal{R}Ba_2Cu_3O_{6+x}$ compounds are orthorhombic for $x > x_{OT}$ (Orthorhombic↔Tetragonal),[6] with $a \approx$ b and c $\approx$ 3a, while they are tetragonal for $x < x_{OT}$ (see Chapter 4). This orthorhombic-to-tetragonal structural transition is accompanied by a metal-to-insulator (semiconducting) transition, and also divides the antiferromagnetic region from superconductivity. Thus in the small x regime we have a tetragonal antiferromagnetic insulator, while at large x we have an orthorhombic superconductor. The phase boundaries shown in Fig. 8.4 depend to some extent on the details of the sample preparation and treatment, and hence $x = x_{OT}$ should be considered only an approximate dividing line.

In each chemical unit cell there are three copper-oxygen layers of ions which are stacked along the c-axis. Two of these layers have oxygen ions between the Cu ions in both the a and b crystallographic directions (the Cu-plane layers), and the oxygens cannot be removed. The basic chemical binding is such that each Cu ion gives up two electrons, a 4s electron, and one electron from the 3d shell. This hole in the 3d shell of the Cu^{2+} ion has a spin associated with it, and is responsible for the magnetic moment. The third Cu layer only has O ions along one axis. This is the so-called "chain" layer, and the oxygen concentration can be readily varied in this layer from full occupancy ($x = 1$) to full depletion ($x = 0$). With no oxygen in this layer the Cu ions nominally are singly ionized, and hence carry no moment, while holes in the 3d shell would form as oxygen is added. However, it should be kept in mind that there is very strong hybridization of the copper and oxygen wave functions in all three layers. The consequent strong electron correlations are an essential part of the electronic description of both the magnetism and the superconductivity, and the correct electronic character is still a matter of debate. It is these same electron correlations which cause so much difficulty with the one-electron band structure models of these materials (see

[5]In the La_2CuO_4 system the susceptibility χ clearly reveals the magnetic phase transition. However, the peak in χ originates from the antisymmetric exchange interaction, which produces a canting of the spins (see, e.g. [19]). No such peak in χ occurs for the $R Ba_2Cu_3O_{6+x}$ system.

[6]The orthorhombic-to-tetragonal phase transition occurs at $x_{OT} \approx$ 0.4-0.5 for most of the rare earths, depending to some extent on the sample treatment.

Chapter 5).

Both the magnetic and superconducting properties are a very sensitive function of the oxygen concentration x in this "chain" layer as shown in Fig. 8.4. In the fully oxygenated case ($x = 1$) the system is a 95 K superconductor for all the trivalent rare earth elements $\mathscr{R}$. In the magnetic regime ($x \lesssim x_{OT}$) there are two separate transitions to long range antiferromagnetic order of the Cu spins which have been observed. The high temperature transition involves ordering in the Cu-plane layers, and has a Neel temperature T_{N1} which is ~500 K at $x = 0$, and monotonically decreases to zero for $x \sim x_{OT}$. At the lower transition the Cu spins in the "chain" layers also become ordered. This transition temperature T_{N2} is very sensitive to x, and apparently also to other parameters of the sample preparation techniques as well as discussed below.

8.2.2.1. High Temperature Transition ($x < x_{OT}$)

On cooling from the high temperature paramagnetic phase, the Cu spins in the Cu-plane layers develop three-dimensional (*3-d*) long range antiferromagnetic order [32-40], while any spins in the Cu chain layer remain disordered, presumably due to the weaker coupling caused by the reduced amount of oxygen in this layer. The spin arrangement of this phase is very simple. The nearest-neighbor spins within the tetragonal Cu-plane layers are aligned antiparallel, with the moment direction constrained to lie in the a-b plane. If we proceed along the c-axis, we find that the spins are also arranged antiferromagnetically in a + - + - + - sequence as shown in Fig. 8.5. Because there are two Cu-plane layers per chemical unit cell, the magnetic and chemical periodicities are the same along the c-axis, and hence in a (neutron) scattering experiment the magnetic Bragg index ℓ will be integral. In the tetragonal plane, on the other hand, the magnetic unit cell is twice the chemical cell in both directions, and we therefore find Bragg indices which are half-integral. Hence this magnetic phase is characterized by magnetic reflections such as ($h/2$, $k/2$, ℓ). The different domains were found to be equally populated, and hence the preferred spin direction within the tetragonal plane has not been determined yet.

The strongest reflection for this magnetic structure turns out to be the ($\frac{1}{2}\frac{1}{2}2$), and its temperature dependence is shown in Fig. 8.6. These data have been taken on a single crystal of $NdBa_2Cu_3O_{6.1}$ [35, 39, 40], which has a measured transition temperature T_{N1} = 430 ± 5 K. No hysteresis was observed on the warming and cooling cycles for any of these magnetic phase transitions. However, T_{N1} has also been found to increase rapidly with applied pressure [43].

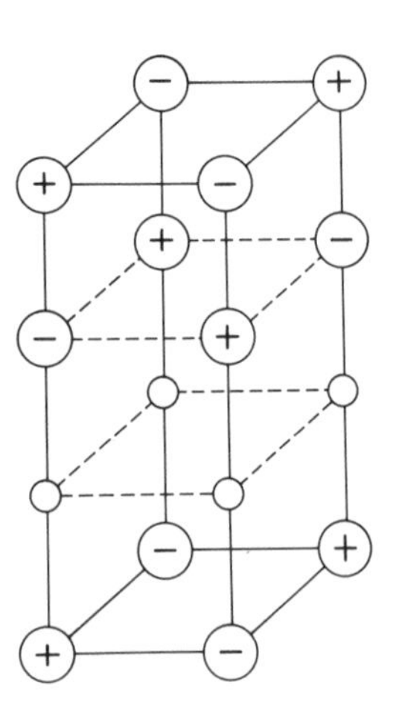

Fig. 8.5. Magnetic spin configuration for the Cu moments in the high temperature ordered phase ($T_{N2} < T < T_{N1}$). The large circles represent the copper ions in the Cu planes, and the smaller circles represent the Cu in the "chain" layer. In this phase there is no ordered moment in the chain layer.

The temperature dependence of this order parameter is quite unusual. The intensity increases on cooling from the paramagnetic state in the usual way, but below ~300 K the rate of increase appears to be somewhat smaller than expected. This "flattening" of the curve has been observed in a number of crystals. The maximum intensity occurs at about 100 K, with an ordered moment $M_p \cong (0.64 \pm 0.06)\ \mu_B$ on the Cu ions in the Cu planes (labeled M_p). There is no ordered moment on the Cu ions in the chain layers (designated M_c). The downturn in the intensity below 100 K is associated with the second phase transition, and will be discussed in the next section.

The value of the magnetic moment M_p at small x has consistently been observed to the ~0.6 μ_B, which is substantially smaller than the 1 μ_B moment expected for a localized Cu^{2+} ion. This lower value for the maximum moment may be due to the enhanced thermal fluctuations associated with a two-dimensionally ordered system. In the present case the ordering we have described is of course three-dimensional in nature, but the in-plane exchange is much larger than the exchange between planes, and the *2-d*-like fluctuations may be important. In fact larger ordered moments have been observed if the chain sites order as discussed in the next section. The moment may also be reduced because of charge transfer between the Copper and oxygen ions. The ordered moment has been found to be quite sensitive to the value of the oxygen concentration [36], decreasing with increasing x, with M_p(max) proportional to the ordering temperature T_N.

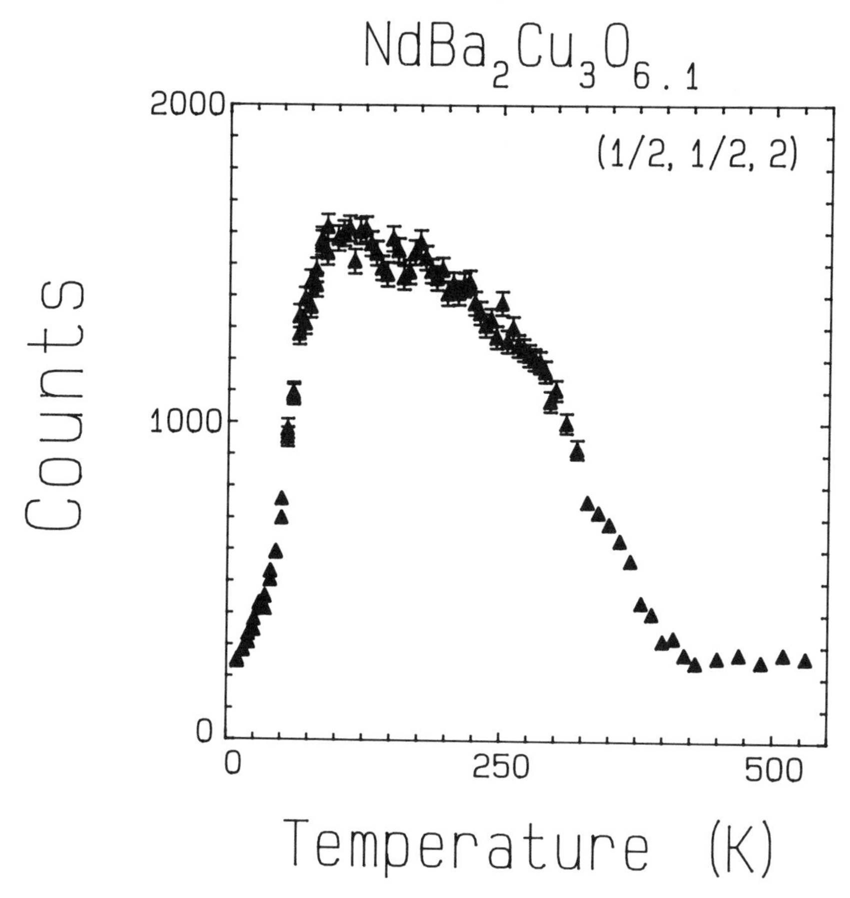

Fig. 8.6. Magnetic intensity of the (½½2) antiferromagnetic Bragg reflection for $NdBa_2Cu_3O_{6.1}$ as a function of temperature. T_{N1} for this crystal is 430 K. The sharp downturn in intensity at low temperatures is accompanied by new Bragg peaks at positions such as (½½³⁄₂) as discussed in the next section. See reference [35].

In addition to the magnitudes of the magnetic moments, the neutron measurements yield information about the spatial distribution of the magnetization density, which is contained in the magnetic form factor $f(\tau)$. This information can be used to determine which ions carry a magnetic moment, and detailed information about the electronic wavefunction can be obtained if the data are precise enough. One of the important questions to address is whether any of the oxygen ions carry a significant magnetic moment. The most extensive data to date is on the single crystal of $NdBa_2Cu_3O_{6.1}$, where over 40 independent magnetic reflections have been measured [40]. The data were found to be consistent with the

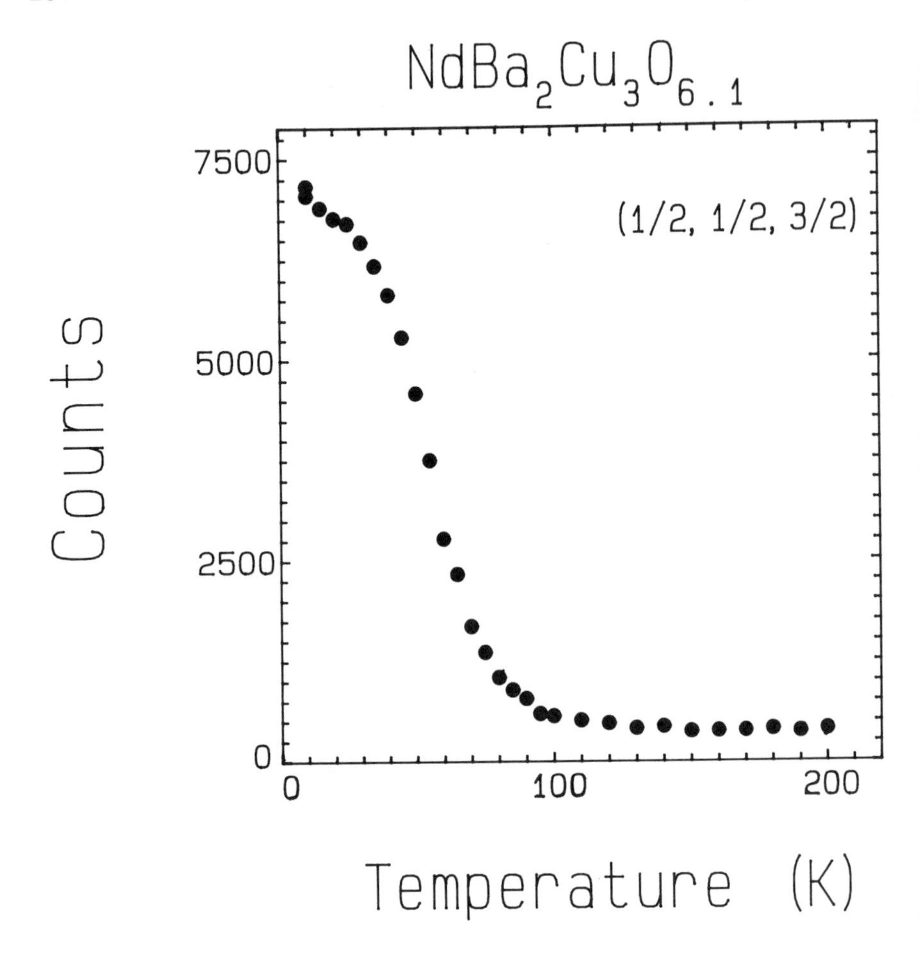

Fig. 8.7. Temperature dependence of the $(\frac{1}{2}\frac{1}{2}\frac{3}{2})$ peak for a single crystal of $NdBa_2Cu_3O_{6.1}$. Note that the increase in these ℓ half-integral peaks at $T_{N2} \approx 80$ K coincides with the sharp decrease in intensity of the whole-integral peaks as shown in Fig. 8.6. See reference [35].

assumption of moments only on the Cu-plane ions, with a 3d (Cu) magnetic form factor. No oxygen moment was observed. Polarized beam measurements on $YBa_2Cu_3O_{6.3}$ also have revealed no evidence of a magnetic moment on the oxygen ions [41].

8.2.2.2. Low Temperature Transition

The sharp downturn in the observed $(\frac{1}{2}\frac{1}{2}2)$ intensity shown in Fig. 8.6 indicates that a new type of ordered phase is developing at low temperatures [34,35]. This new phase is characterized by magnetic peaks in which all three indices are half integral. Fig. 8.7 shows the temperature dependence of the ($\frac{1}{2}\frac{1}{2}\frac{3}{2}$) peak, which exhibits typical behavior for a magnetic order parameter, and an onset temperature ~80 K [35, 39, 40]. Comparing the data in Figs. 8.6 and 8.7 shows that the downturn in the $(h/2,k/2,\ell)$ reflections coincides with the development of intensity at reflections such as $(h/2,k/2,\ell/2)$.

This new magnetic phase turns out to be associated with magnetic ordering on the chain sites. However, unlike the high temperature phase discussed in the previous section, where the same spin configuration has been observed in many different samples independent of oxygen concentration, the chain ordering appears to be very sensitive to the chemical and metallurgical state of the sample. Hence the phase boundary as a function of oxygen concentration has not been established, and indeed it is not even clear if a universal curve is appropriate. We will first describe this phase transition, and then discuss some of the inconsistencies at the end of this section.

At low temperature the whole-integral peaks have almost no intensity, and hence the ground state spin configuration is to a good approximation represented by the half-integral peaks alone. The spin structure which has been determined for this sample is shown in Fig. 8.8, and consists of a collinear arrangement of spins, with the spin direction in the tetragonal plane [35]. The

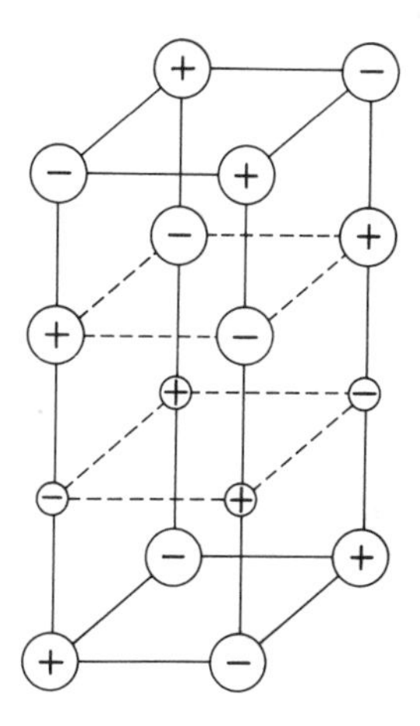

Fig. 8.8. Basic spin structure for the Cu moments at $T = 0$. There is a substantial moment (~$\frac{1}{2}$ μ_B) on the Cu ions in the oxygen deficient planes (small circles).

nearest-neighbor spins within the Cu planes (designated M_P, for planes) are again aligned antiparallel as they are in the high temperature magnetic phase. However, the oxygen-deficient Cu moment (designated M_C, for chains) is nonzero below T_{N2}. Within the chain layer the nearest-neighbor spins are aligned antiparallel just like the Cu-plane layers. As we proceed along the c-axis, the nearest neighbor spins also alternate their spin direction from layer to layer. However, since we now have three Cu layers per unit cell participating in the magnetic structure, the magnetic unit cell is doubled along the c-axis direction because the sense of the spins is reversed in going from one chemical unit cell to another as shown in the figure.

The values of the ordered moments obtained from these data [40] at low temperatures are $M_P = 0.81 \pm 0.08\ \mu_B$ for the Cu planes (assumed equal), and $M_C = 0.35 \pm 0.05\ \mu_B$. This compares with $M_P \sim 0.6\ \mu_B$ for the maximum moment found above T_{N2}, with the spin structure shown in Fig. 8.5. The essential difference between the two structures is of course the moment M_C which develops below T_{N2}. The large value of M_C which has been found on the Cu "chain" layers demonstrates that the Cu^{1+} state is not an appropriate description of the electronic configuration of the "chain" layers in these materials. However, the magnitude of the chain moment depends not only on oxygen concentration, but apparently on other details of the sample characteristics as well as discussed below. At intermediate temperatures ($0 < T < T_{N2}$) the spin configuration is a superposition of the "pure" configurations shown in Figs. 8.5 and 8.8, which results in a noncollinear structure. A noncollinear structure is essential if we make the physically reasonable assumption that the magnitudes of the moments on the two Cu-O planes are the same.

The upper (T_{N1}) and lower (T_{N2}) transition temperatures for three single crystal samples studied are shown in Table 8.1. Over this range of oxygen concentration both transition temperatures are proportional to x. However, the behavior of the chain spins turns out to be very complicated. The ratios of the intensities for the observed peaks, for example, are different for these three samples,

Table 8.1 Upper and lower transition temperatures for three values of x.

Sample	T_{N1}(K)	T_{N2}(K)
$NdBa_2Cu_3O_{6.1}$ [35,39]	430	80
$YBa_2Cu_3O_{6.2}$ [34]	400	40
$NdBa_2Cu_3O_{6.35}$ [35,39]	230	10

which means that the spin configuration and/or the relative values of the moments are not the same. They are also dependent on temperature [40]. Thus the specific spin configuration may vary as a function of x, temperature, and/or the chemical and metallurgical state of the sample. In fact, in polycrystalline samples for the same nominal oxygen concentration frequently half-integral peaks are not observed at any temperature [37, 38, 41], which suggests that perhaps strain may play an important role in the magnetic behavior of the chain sites. Moreover, in one single crystal of $NdBa_2Cu_3O_{6+x}$, *only* half-integral peaks have been observed [42], with a transition temperature of 385 K and with the chain spins pointing parallel (rather than antiparallel as indicated in Fig. 8.8) to the Cu spins in adjacent plane layers.

Thus the behavior of the spins on the chain sites appears to be quite sensitive to the state of the sample. One possibility is that there is a significant substitution of the rare earth onto the Ba sublattice in some samples, which affects the electronics and hence the chain magnetism. Rutherford backscattering measurements on the Nd crystals of Table 8.1 did not reveal any significant Nd concentration on the Ba site [40], but such a substitution may explain the data of [42]. A second possibility is that short range or long range order of the oxygen on the chain sites is influencing the magnetic behavior, as it does for the superconducting properties [45]. Several alloy studies have been carried out to study the influence of the chain sites on the magnetism. Co, for example, substitutes preferentially for the Cu chain ions, and Co is found to strongly enhance the half-integral type ordering [46, 47]. Magnetic ordering has also been observed via Mossbauer experiments in systems where the Cu ions are replaced by Fe ions [48, 49].[7]

It is of central importance to identify the mechanism which is controlling the magnetism of these chains, since the chain layer controls all the properties of the system—magnetism, metallic behavior, and superconductivity—and the magnetic behavior on the chain ions may be the best indicator of the electronic state of this layer. Clearly the magnetic behavior of the chain spins depends on more than just the oxygen concentration, and considerably more work will be needed to clarify these phenomena.

8.2.3. Dynamics and Fluctuations

In addition to investigations of the magnetic order of the Cu spins, considerable information has been obtained about the spin dynamics

[7] A second magnetic phase transition has also recently been found at lower temperature in La_2CoO_4 [50].

both by neutron [16, 20, 51-56] and Raman [57-61] scattering. For the La_2CuO_4 and $YBa_2Cu_3O_6$ ordered antiferromagnets, the measurements reveal that the exchange interactions within the Cu-O planes are very much larger than between the layers, which gives rise to *2-d* magnetic behavior, and an overall energy scale which is much larger than the ordering temperatures would suggest. More importantly, these strong spin correlations (and the associated magnetic moments) are also observed in the $La_{2-x}Sr_xCuO_4$, $YBa_2Cu_3O_7$, and $Bi_2Sr_2CaCu_3O_8$[8] superconducting phases. The correlation range is found to be much shorter in the superconducting state, but the amplitude of the magnetic moment is not diminished substantially.

We start our discussion by considering the insulating materials such as $YBa_2Cu_3O_6$, which display 3-*d* long range antiferromagnetic order below T_N. The interactions within the Cu-O layers, though, are much stronger than between layers, which bestows a 2-*d* character to the spin excitations and strong spin correlations within the planes. These 2-*d* correlations will give rise to a ridge of scattering along rods in reciprocal space [16,27], and in fact for a system which possesses true 2-*d* long range order, we will have Bragg rods (rather than Bragg points in the 3-*d* case) which are indicative of the order. Emanating from these rods are the dispersion relations for the elementary excitations as depicted in Fig. 8.9. The reduced wave vector **q** for the spin wave excitation lies parallel to the Cu-O planes, while for a purely 2-*d* system the excitation energies are independent of the component of **q** perpendicular to the planes. The slope of the dispersion relation, as well as the energy of the excitations at the zone boundary, is controlled by the exchange energy. Neutron inelastic scattering measurements can directly determine the dispersion relations, and in principal can yield the most detailed information. Raman scattering, on the other hand, measures a two-magnon scattering process, which is related to the density of magnetic states. However, Raman scattering has the distinct advantage that data can be obtained on very small samples. Either type of measurement will provide a determination of the magnetic exchange, and in fact both types of measurements have given important and consistent results.

The first information on the exchange energy and dimensionality of the magnetic system was provided by low energy neutron scattering measurements in La_2CuO_4 [16]. Because the excitations are very dispersive, constant energy scans were performed where the energy transfer is held fixed and **q** is varied through the spin wave

[8]Magnetic order and spin freezing has recently been observed in $Bi_2Sr_2YCu_2O_x$ and $Bi_2SrCaYCu_2O_x$, respectively [62], and magnetic order has also been found in $TlBa_2YCu_2O_7$.

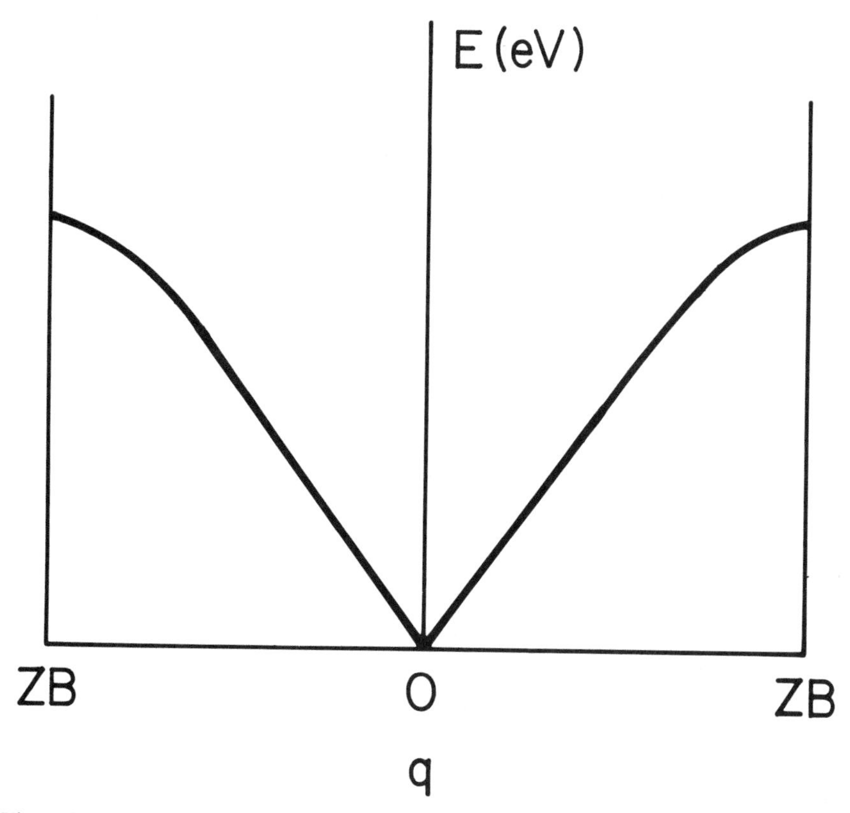

Fig. 8.9. Schematic of the spin wave dispersion relation for the *2-d* Cu-O planes. q represents the projection of the wave vector **q** in the Cu-plane direction. In the ideal *2-d* case the spin wave energy is independent of the component of q perpendicular to the Cu-plane direction.

dispersion surface. The slope of the dispersion relation is in fact so large that initially it was not possible to resolve the +q and -q excitations, but more recent experiments have yielded a slope of the dispersion relation of (0.85 ± 0.03) eV-Å [55a]. Light scattering experiments also found a large magnetic energy scale, with a peak in the magnetic scattering at an energy of ~0.4 eV [57, 59, 60]. Along the **q** direction perpendicular to the planes, on the other hand, no significant dispersion or correlations were found, confirming the 2-*d* character of the excitations [16, 20]. Finally, above the 3-*d* Neel temperature, no elastic scattering along the rods was observed, showing that the system is not really ordered two-dimensionally. The correlation range within the planes, however, was very large, exceeding 200 Å. This situation is analogous to a (2-*d*) fluid, and hence the system was referred to as a *Quantum Spin Fluid* [16].

Similar energetics and behavior have been found in the $YBa_2Cu_3O_6$ system in both Raman scattering [58, 60] and neutron scattering experiments [53]. The Raman data show a peak in the scattering at a somewhat smaller energy of ~0.35 eV, while the inelastic neutron scattering results yield a lower limit for the slope of the dispersion relation of 0.5 eV-Å, with the same kind of two-dimensional character to the excitations.

An essential question to address is whether these energetic magnetic fluctuations, or indeed any magnetism at all, survive the transformation into the superconducting region of the phase diagrams shown in Figs. 8.1 and 8.4. Raman scattering results have been carried out in the $La_{2-x}Sr_xCuO_4$, $YBa_2Cu_3O_{6+x}$, and $Bi_2Sr_2CaCu_3O_8$ superconducting systems, and they show that the magnetic fluctuations do indeed survive, and are still quite energetic. Neutron scattering results obtained on the $La_{2-x}Sr_xCuO_4$ system [20] have shown that the range of the magnetic correlations decreases rather dramatically in going from $x = 0$ ($\xi \sim 400$ Å) into the superconducting region, where $\xi \sim 10$ Å. However, the amplitude of the magnetic fluctuations is not substantially diminished in the superconducting state.

The magnetic fluctuations which are observed in the superconducting region of the phase diagram have been found by Raman scattering [60] and small angle neutron scattering measurements [56] to be suppressed by superconductivity. Fig. 8.10 shows the change in the low frequency spectra of a $YBa_2Cu_3O_{6+x}$ sample as the temperature is lowered through the superconducting phase transition. The spectra have been subtracted from the spectra obtained above T_c, so that a positive value indicates a decrease of intensity. There are two points to note. One is that there is a sudden decrease in intensity which occurs at T_c, which appears to be independent of frequency over the range from 30 to 500 cm^{-1}.[9] The origin of this effect is unknown. Second is that there is an additional reduction in intensity below about 130 cm^{-1}, which is the expected region of the BCS gap value.

A second observation of a suppression of the scattering in the superconducting state has been obtained via small angle neutron scattering measurement on $ErBa_2Cu_3O_7$ [56]. The reduction in the scattering is shown in Fig. 8.11, where the top part of the figure shows the change in the small angle neutron scattering intensity, and the bottom portion of the graph shows the measured magnetization as the superconductivity sets in. One expected suppression effect is the screening of the magnetic fluctuations by the superconducting electrons, at wavelengths $\gtrsim \lambda_L$, where λ_L is the London penetration depth. This suppression comes from the electromagnetic

[9] 1 meV = 8.066 cm^{-1}.

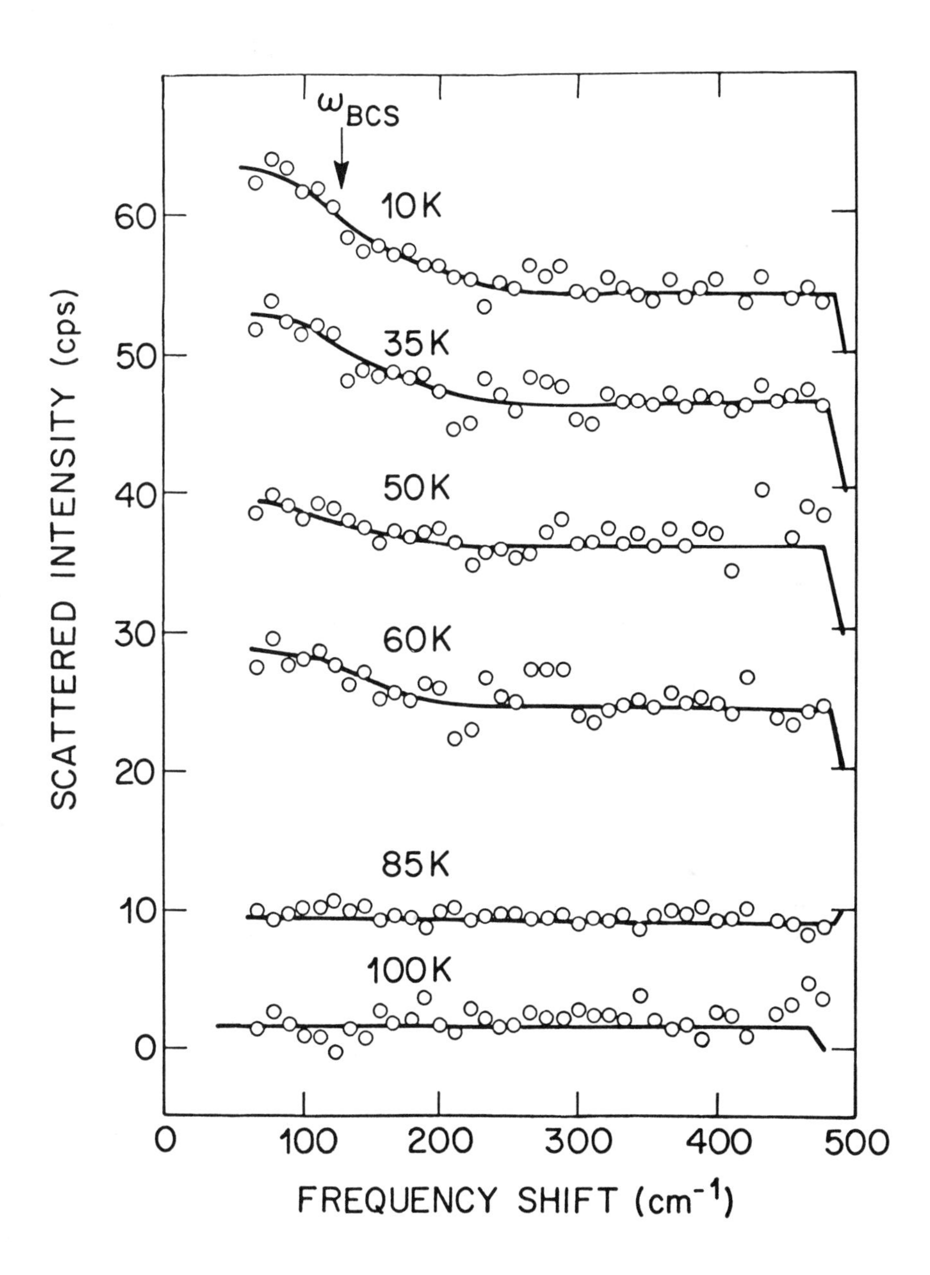

Fig. 8.10. Net Raman spectra in superconducting (T_c = 60 K) $YBa_2Cu_3O_{6+x}$. The data have been subtracted from an averaged spectra taken above T_c, so that positive values indicate a decrease in intensity. The arrow indicates $\omega = 3.5\ kT_c$, the gap expected on the basis of (isotropic) BCS weak-coupling theory (Lyons and Fleury [60]).

interactions: There is a magnetic field **B(q)** associated with the magnetic fluctuations, and the supercurrents will screen these fluctuations at small wave vectors. However, the length scale which is obtained from the data is only 22 Å, which is much too short to

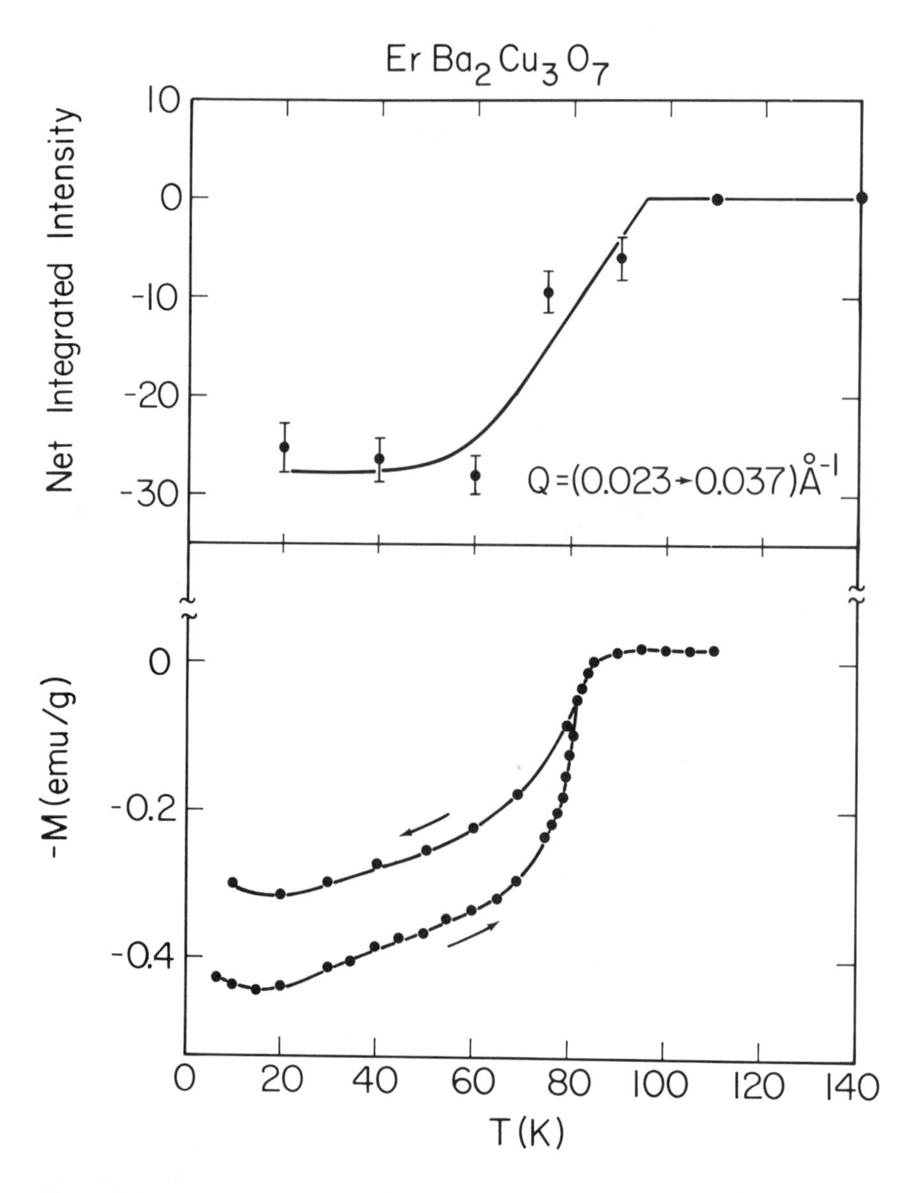

Fig. 8.11. Reduction of the neutron scattering at small wave vectors as a function of temperature (a). The scattering is observed to decrease below the superconducting phase transition in a manner similar to the bulk magnetization (b) [56].

be interpreted as a London penetration depth. A second effect is expected from the reduction of the spin susceptibility of the superconducting electrons themselves [63, 64], which is caused by the formation of Cooper pairs. The length scale associated with this effect is the coherence length ξ. In the traditional "magnetic superconductors" the superconducting electrons do not have a spin component, and hence the conduction electron spin susceptibility is too small to be observed. But in the oxide superconductors the Cu ions carry a spin in the superconducting phase, and this is now thought to be the cause of the reduction shown in Fig. 8.11. However, further experimentation is clearly desirable.

All these observations of the magnetic scattering in the superconducting regime strongly support theoretical models which are based on magnetic fluctuations being responsible for the superconducting pairing. One viewpoint of how a high superconducting transition temperature might be achieved is to draw an analogy with the electron-phonon case. To achieve a high T_c in the conventional materials, one would like to increase the electron-phonon interaction $\lambda_{e\text{-}p}$. However, as $\lambda_{e\text{-}p}$ becomes larger, the lattice softens, and eventually a structural instability occurs which reduces or eliminates the density of electron states at the Fermi surface and destroys the superconductivity. With a magnetic degree of freedom, on the other hand, we can have a soft excitation in the system which can interact with the electrons, but without the detrimental effects on the structure, and this might be a way to produce a higher T_c than would be possible in the electron-phonon case.

If we carry the analogy to completion, then in the limit of very strong magnetic interactions we have an insulating phase which is magnetically ordered. The strong electron interactions have again yielded an insulating state as in the large $\lambda_{e\text{-}p}$ case, but with a magnetic energy scale which is an order of magnitude larger than the energy scale for the phonons. In addition, the combination of the large energy scale and the ordered state yields a very low density of magnetic states in the low energy regime. As the magnetic interactions are decreased in strength and the correlations disrupted, on the other hand, we lose the ordered magnetic state and the system becomes metallic, but the magnetic fluctuation spectrum is still very energetic compared to the phonons. Moreover, the soft spin waves yield a high density of magnetic fluctuations in the low energy regime. If this scenario is correct, then it would seem that the disordered magnetic state is an essential ingredient to have superconductivity, which is of course the observed situation. However, it is not possible to conclude that magnetic fluctuations are the origin of the pairing. What is clear is that magnetism must play a prominent part in any theoretical description of the superconducting oxides.

8.3. Rare Earth Magnetism

Some of the very first data on the *1-2-3* compounds showed that substitution of the trivalent rare earths had little effect on the superconducting properties [1-10]. Hence the usual Abrikosov-Gorkov spin depairing mechanism [65], which operates via the rare-earth ion—superconducting electron exchange interaction, is very small. This electronic isolation of the rare earth sublattice also results in a very small exchange interaction between the rare earth ions themselves, yielding transition temperatures which are a few degrees K or less just like conventional "magnetic-superconductor" systems (see Chapter 1). This relationship between magnetism (via dipolar interactions) and superconductivity has attracted considerable interest for many years. In the ternary magnetic superconductors such as $ErRh_4B_4$ and $HoMo_6S_8$, the magnetic and superconducting electrons also reside on separate sublattices, and hence there is only a weak electromagnetic coupling between these two cooperative phenomena which nevertheless gives rise to some interesting competitive behavior at very low temperatures [66]. The present oxide superconductors should be model systems for studying these interactions in detail.

Since the rare earth exchange is small, dipolar interactions will play a very important role in the magnetic interactions. The magnetic properties will also be strongly influenced by crystalline electric field splittings of the rare earths.[10] Considerable work has already been carried out [68-72] to determine the crystal field splittings, which we can very briefly summarize as follows. The overall splitting can be quite large ($\sim$100 meV), and rather complicated due to the low symmetry. The splittings also depend on the oxygen concentration, both because of the presence of the extra oxygen and because of the change in symmetry from tetragonal to orthorhombic. This can affect the transition temperatures as well as the nature of the ordering.

In the case of the $ErBa_2Cu_3O_7$ system, the magnetic transition occurs at $T_N \approx 0.6$ K, where dipolar interactions are expected to dominate. Since there is only one magnetic ion per chemical unit cell, and $c \approx 3a$, the interactions are expected to be much stronger in the *a-b* plane, and the 2-*d* nature of the scattering which was observed [56, 73] was expected. The observed scattering is shown in Fig. 8.12, and corresponds to chains of moments which are aligned

[10]For the heavy rare earth ions, J is a good quantum number. Then an electric field gradient in the crystal will couple to the orbital part of the wavefunction of the 4f electrons, splitting the J-fold degeneracy of the free ion. The exception is Gd, for which $L = 0$ [67].

ferromagnetically along the *b* axis, while adjacent chains are aligned antiparallel to form an overall antiferromagnetic configuration. Over the temperature range initially explored, no magnetic correlations were in evidence along the c-axis. More recent studies on powders at lower temperatures [74, 75] show the expected three-dimensional ordering, while the the expected two-

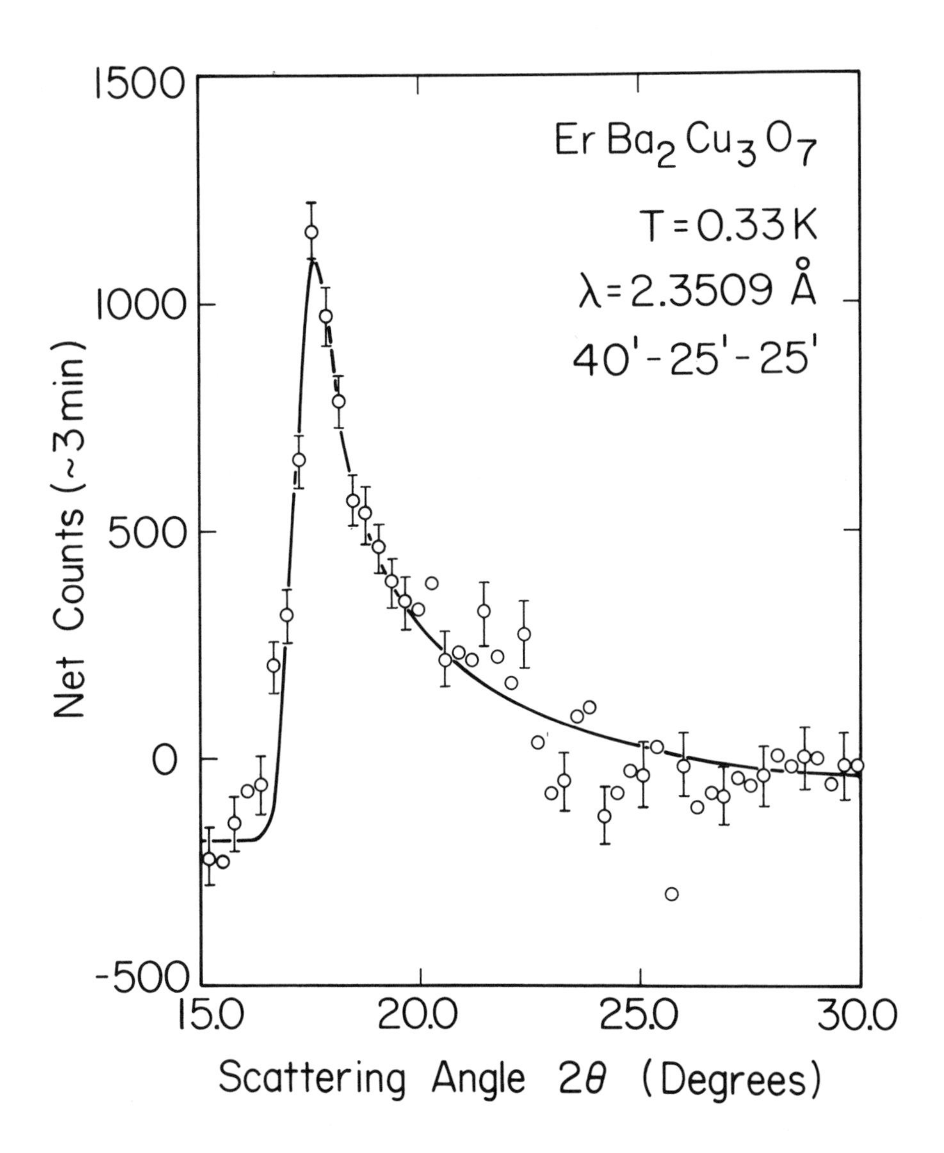

Fig. 8.12. Diffraction data taken on $ErBa_2Cu_3O_7$ at 0.33 K, after subtraction of background. The asymmetric lineshape is typical of a two-dimensionally ordered system. [56].

dimensional behavior has been observed directly in single crystal samples [75]. Indeed the order parameter is observed to obey the (exact) Onsager solution for the 2-d Ising model, and a sharp rod of scattering is found in the scattering [75], which is a direct demonstration of the 2-d behavior. Two-dimensional behavior has also been observed in specific heat studies [88,89,91,82], and it is likely that all the heavy rare earth systems exhibit magnetic behavior which is low-dimensional in nature. Two-dimensional ordering or three-dimensional correlations have also been observed in $HoBa_2Cu_3O_7$ below the ordering temperature of 0.15 K [76]. In the case of Ho, there is a very strong nuclear hyperfine interaction and a large nuclear moment, and thus this phase transition could be driven by the ordering of the nuclear spins.

For the $DyBa_2Cu_3O_7$ [77, 76], $GdBa_2Cu_3O_7$ [78-81] and $NdBa_2Cu_3O_{6+x}$ [82] (powder) materials, only three dimensional behavior has been observed so far. The magnetic ordering temperatures, determined by specific heat and neutron diffraction, are given in Table 8.2 for all the rare earth systems investigated to date. The moment direction for these materials is along the c-axis, in contrast to the Er system, and the ordering temperatures are substantially higher for the Dy and Gd systems. Indeed they cannot be accounted for simply by including dipolar interactions. There is only one rare earth atom per chemical unit cell, and the magnetic structure is as simple as it can be, with magnetic neighbors in all three directions aligned antiparallel as shown in Fig. 8.13. Thus the unit cell is doubled in all three directions, and hence all

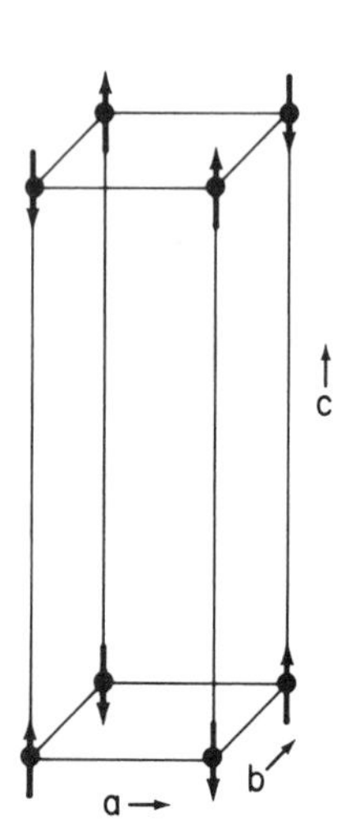

Fig. 8.13. Magnetic structure for the Nd, Gd, Dy and Pr rare earth ions in $\mathcal{R}Ba_2Cu_3O_{6+x}$.

Table 8.2. Magnetic properties for rare earth ions in $\mathscr{R}Ba_2Cu_3O_{6+x}$.

Compound	T_N (K)	$\mu(T=0)$ (μ_B)	Spin Direction	Q	μ_{eff} (μ_B)	μ_z (free ion)	μ_{eff} (free ion)
$PrBa_2Cu_3O_7$	17[ab]	0.74[b]	c-axis[b]	½½½[b]	2.7[c]	3.2	3.58
$NdBa_2Cu_3O_7$	0.52[de]	1.07[e]	c-axis[e]	½½½[e]	3.1[dg]	3.27	3.62
					3.6[f]		
$NdBa_2Cu_3O_6$	1.5[e]	0.76[e]	c-axis[e]	½½½[e]			
$SmBa_2Cu_3O_7$	0.6[dh]				1.2[d]	0.71	0.85
					0.73[g]		
$EuBa_2Cu_3O_7$					0[d]	0	
$GdBa_2Cu_3O_7$	2.6[i]	7.4[j]	c-axis[jkl]	½½½[j]	7.9[dm]	7	7.94
	2.24[dj m-t]						
$GdBa_2Cu_3O_{6.5}$	2.24[uv]	6.9[u]	c-axis[u]	½0½[u]			
$GdBa_2Cu_3O_6$	2.2[rv]		c-axis[v]	½½½[v]			
$DyBa_2Cu_3O_7$	0.95[howx]	6.9[wx]	c-axis[w]	½½½[wx]	10.2[o]	10	10.65
	0.8[f]				10.6[fm]		
					9.7[g]		
$HoBa_2Cu_3O_7$	0.15[† ox]	~2[x]		½00[x]	10.4[fgom]	10	10.61
$ErBa_2Cu_3O_7$	0.60[roy]	4.9[u]	a-b plane‡	½00[y]	9.4[fgo]	9	9.58
	0.87[z]						
$TmBa_2Cu_3O_7$					8.1[dg]	7	7.56
$YbBa_2Cu_3O_7$	0.35[1]	1.7[1]			5.35[f]	4	4.54
					3.52[g]		

† This system has either *2-d* order or *3-d* magnetic correlations.
‡ The magnetic moment direction has not been established. The crystal field results indicate the a-b plane as the easy axis, and this is also suggested by [56].

a) Jee, et al [104]
b) Li, et al [83]
c) Dalichaouch, et al [105]
d) Maple; Ferreira; Lee; Zhou [96-99]
e) Yang, et al [82]
f) Brown, et al [8]
g) Xiao, et al [3]
h) Ramirez, et al [92]
i) Lin, et al [87]
j) McK.Paul, et al [78]
k) Wortman, et al [86]
l) Smit, et al [88]
m) Thompson, et al [6,7,103]
n) Kadowaki, et al [85]
o) Dunlap, et al [96,99]
p) Van den Berg, et al [89]
q) Ho, et al [84]
r) Simizu, et al [91]
s) Reeves, et al [90]
t) Nakamura, et al [102]
u) Chattopadhyay, et al [74,79]
v) Mook, et al [80,81]
w) Goldman, et al [77]
x) Fischer, et al [76]
y) Lynn, et al [56,73,75]
z) Yang, et al [93]
1) Hodges, et al [94]

three magnetic Bragg indices are half-integral. This magnetic structure then can be represented by a wave vector **Q** of ($\frac{1}{2}\frac{1}{2}\frac{1}{2}$). In the superconducting state there is coexistence of long range antiferromagnetic order with superconductivity, in an identical fashion to the behavior found in many of the Chevrel phase and $\mathcal{R}$-rhodium-boride systems [66].

Some of the magnetic phase transitions have been studied in the oxygen deficient (semiconducting) state as well as the superconducting state. For the Gd system, the two transitions were found to be identical, and no change in the ordering has been observed at any x. However, results on a single crystal of $GdBa_2Cu_3O_{6.5}$ [79] show a chain-like ordering similar to that found in the Er system. A similar duality of spin structures is observed in the Er system, and is likely a result of the weak interactions along the c-axis.

The magnetic transition has been studied in some detail as a function of oxygen concentration for the $NdBa_2Cu_3O_{6+x}$ system, where the ordering temperature is found to be very sensitive to the oxygen concentration. Fig. 8.14 shows the temperature dependence of the magnetic ordering for $NdBa_2Cu_3O_6$ and $NdBa_2Cu_3O_7$ [82]. The fully oxygenated system orders at 0.5 K, with a sharp anomaly typical of a 2-d Ising system observed in the specific heat, while the oxygen deficient material has an ordering temperature which is three times higher. The value of the ordered moment, though, is quite small for both oxygen concentrations. The crystal field splittings do depend on oxygen content, but the origin of this large change in T_N has not been pinpointed yet.

The tetravalent rare earths (Ce, Pr, Tb), on the other hand, are more strongly coupled to the rest of the electronic system, affecting both the superconducting behavior and the magnetism. The Pr material forms [106] the same orthorhombic structure as $YBa_2Cu_3O_7$, but is thought to be strongly mixed valent, and close to the tetravalent ionic state. The superconductivity is found to be strongly suppressed as a function of Pr concentration [105, 107-109], with a behavior which is consistent with the classical Abrikosov-Gorkov depairing theory [110], and superconductivity is lost for Pr concentrations greater than 60% for Y [105, 107-109].

The magnetic ordering temperatures are also much higher than for the trivalent rare earths. $PrBa_2Cu_3O_7$, for example, which is orthorhombic and semiconducting, orders magnetically at $T_N \approx 17$ K [83]. The magnetic structure is identical to the trivalent rare earths as shown in Fig. 8.13. The spin direction is also along the c-axis direction, but with a magnetic moment of only 0.74 μ_B at low temperatures. This is much smaller than the free ion moment, and points to a mixed valent character, especially in view of the broad distribution of magnetic inelastic scattering observed in this

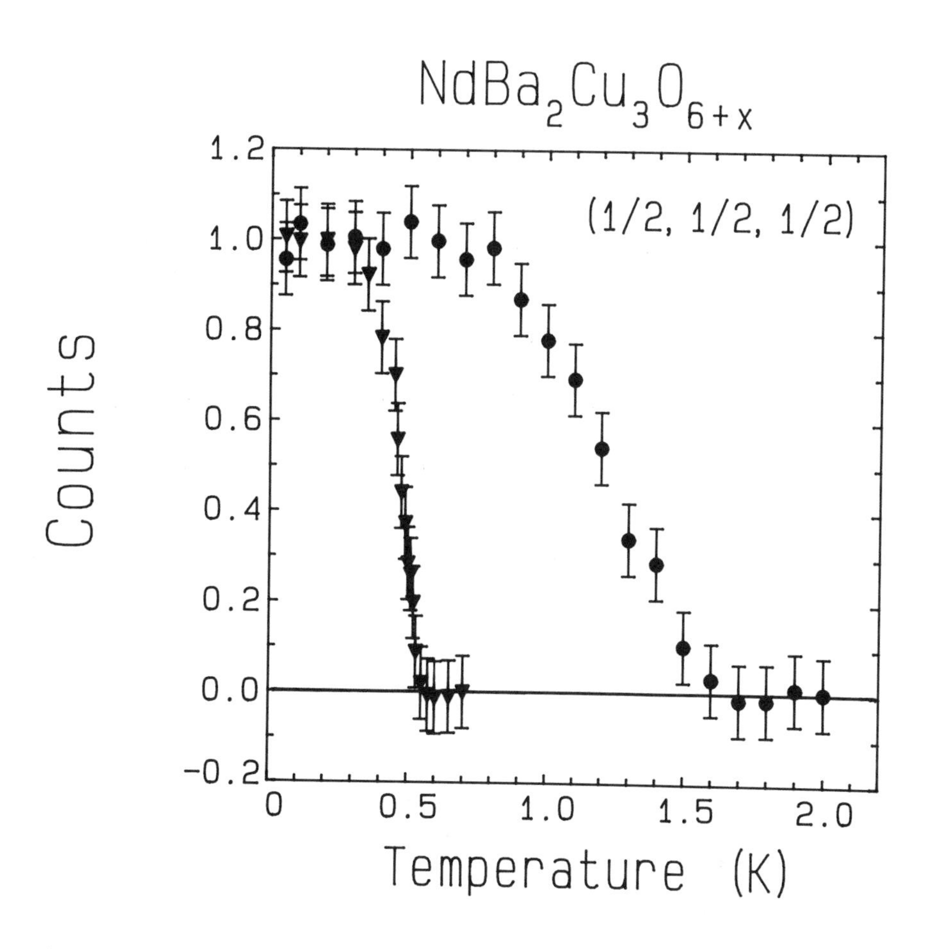

Fig. 8.14. Magnetic Bragg intensity associated with the Nd ordering in $NdBa_2Cu_3O_7$ and $NdBa_2Cu_3O_6$ [82]. The removal of oxygen causes a threefold increase in the antiferromagnetic transition temperature.

material. One of the most interesting properties, however, is the large electronic contribution to the specific heat by the 4f electrons. The electronic coefficient determined from the low temperature specific heat measurements for $T < T_N$ is $\gamma \sim 200$ mJ/mole-K^2, which is comparable to many heavy fermion systems. These observations indicate that the *f*-electrons in these tetravalent materials are strongly hybridized, and play an essential role in the electronic properties.

A tremendous amount of information has been obtained about the magnetism in these new oxide systems, both for the Cu ions and the rare earth ions. The time scale to obtain this information has been incredibly short compared to the usual time for the conduct of research. Our understanding has progressed rapidly, but it is clear that much more work needs to be done.

Acknowledgments

I would like thank my collaborators who have worked on these systems with me, in particular Wen-Hsien Li, who has been involved in all of these studies with me. I would also like to thank G. Shirane and K. B. Lyons for allowing me to use figures 3 and 10, respectively. The research at the University of Maryland is supported by the National Science Foundation under grant DMR 86-20269.

References

[1] P. H. Hor, R. L. Meng, Y. Q. Wang, L. Gao, Z. J. Huang, J. Bechtold, K. Forster, and C. W. Chu, *Phys. Rev. Lett.* **58**, 1891 (1987).

[2] Z. Fisk, J. D. Thompson, E. Zirngiebl, J. L. Smith and S-W. Cheong, *Sol. State Commun.* **62**, 743 (1987).

[3] G. Xiao, F. H. Streitz, A. Garvin and C. L. Chien, *Sol. St. Commun.* **63**, 817 (1987).

[4] J. P. Golben, S-I. Lee, Y. Song, T. Noh, X-D. Chen, J. R. Gaines and R. Tettenhorst, *Phys. Rev.* **B35** (8705 (1987).

[5] J. M. Tarascon, W. R. McKinnon, L. H. Greene, G. W. Hull and E. M. Vogel, *Phys. Rev.* **B36**, 226 (1987).

[6] J. R. Thompson, S. T. Sekula, D. K. Christen, B. C. Sales, L. A. Boatner and Y. C. Kim, *Phys. Rev.* **B36**, 718 (1987).

[7] J. R. Thompson, D. K. Christen, S. T. Sekula, B. C. Sales and L. A. Boatner, *Phys. Rev.* **B36**, 836 (1987).

[8] S. E. Brown, J. D. Thompson, J. O. Willis, R. M. Aikin, E. Zirgiebl, J. L. Smith, Z. Fisk, and R. B. Schwarz, *Phys. Rev.* **B36**, 2298 (1987).

[9] F. Zuo, B. R. Patton, D. L. Cox, S. I. Lee, Y. Song, J. P. Golben, X. D. Chen, S. Y. Lee, Y. Cao, Y. Lu, J. R. Gaines, J. C. Garland and A. J. Epstein, *Phys. Rev.* **B36**, 3603 (1987).

[10] Y. LePage, T. Siegrist, S. A. Sunshine, L. F. Schneemeyer, D. W. Murphy, S. M. Zahurak, J. V. Waszczak, W. R. McKinnon, J. M. Tarascon, G. W. Hull and L. H. Greene, *Phys. Rev.* **B36**, 3617 (1987).

[11] D. E. Morris, J. H. Nickel, J. Y. T. Wei, N. G. Asmar, J. S. Scott, U. M. Scheven, C. T. Hultgren, A. G. Markelz, J. E. Post, P. J. Heaney, D. R. Veblen, and R. M. Hazen (preprint).

[12] D. C. Johnston, J. P. Stokes, D. P. Goshorn and J. T.

Lewandowshi, *Phys. Rev.* **B36**, 4007 (1987).

[13] D. Vaknin, S. K. Sinha, D. E. Moncton, D. C. Johnston, J. M. Newsam, C. R. Safinya, and H. E. King, Jr., *Phys. Rev. Lett.* **58**, 2802 (1987).

[14] S. Mitsuda, G. Shirane, S. K. Sinha, D. C. Johnston, M. S. Alvarez, D. Vaknin, and D. E. Moncton, *Phys. Rev.* **B36**, 822 (1987).

[15] T. Freltoft, J. E. Fischer, G. Shirane, D. E. Moncton, S. K. Sinha, D. Vaknin, J. P. Remeika, A. S. Cooper, and D. Harshman, *Phys. Rev.* **B36**, 826 (1987).

[16] G. Shirane, Y. Endoh, R. J. Birgeneau, M. A. Kastner, Y. Hidaka, M. Oda, M. Suzuki, and T. Murakami, *Phys. Rev. Lett.* **59**, 1613 (1987).

[17] T. Freltoft, G. Shirane, S. Mitsuda, J. P. Remeika, and A. S. Cooper, *Phys. Rev.* **B37**, 137 (1988).

[18] C. Stassis, B. N. Harmon, T. Freltoft, G. Shirane, S. K. Sinha, K. Yamada, Y. Endoh, Y. Hidaka, and T. Murakami, *Phys. Rev.* B **38**, 9291 (1988).

[19] M. A. Kastner, R. J. Birgeneau, T. R. Thurston, P. J. Picone, H. P. Jenssen, D. R. Gabbe, M. Sato, K. Fukuda, S. Shamoto, Y. Endoh, K. Yamada, and G. Shirane, *Phys. Rev.* **B38**, 6636 (1988).

[20] Y. Endoh, K. Yamada, R. J. Birgeneau, D. R. Gabbe, H. P. Jenssen, M. A. Kastner, C. J. Peters, P. J. Picone, T. R. Thurston, J. M. Tranquada, G. Shirane, Y. Hidaka, M. Oda, Y. Enomoto, M. Suzuki, and T. Murakami, *Phys. Rev.* **B37**, 7443 (1988). See also G. Shirane, R. J. Birgeneau Y. Endoh, P. Gehring, M. A. Kastner, K. Kitazawa, H. Kojima, I. Tanaka, T. R. Thurston, and K. Yamada, *Phys. Rev. Lett.* **63**, 330 (1989).

[21] Y. J. Uemura, C. E. Stronach, D. C. Johnston, M. S. Alvarez, and D. P. Goshorn, *Phys. Rev. Lett.* **59**, 1045 (1987).

[22] D. R. Harshman, G. Aeppli, G. P. Espinosa, A. S. Cooper, J. P. Remeika, E. J. Ansaldo, T. M. Riseman, D. Ll. Williams, D. R. Noakes, B. Ellman, and T. F. Rosenbaum, *Phys. Rev.* **B38**, 852 (1988).

[23] P. Gutsmiedl, G. Wolff, and K. Andres, *Phys. Rev.* **B36**, 4043 (1987).

[24] Y. Kitaoka, K. Isida, T. Kobayashi, K. Amoya, and K. Asayma, *Physica* **C153-155**, 733 (1988).

[25] Additional references and details about the superconducting properties can be found in Chapters 6 and 7.

[26] A. Aharony, R. J. Birgeneau, A. Coniglio, M. A. Kastner, and H. E. Stanley, *Phys. Rev. Lett.* **60**, 1330 (1988).

[27] R. J. Birgeneau, H. J. Guggenheim, and G. Shirane, *Phys. Rev.* **B1**, 2211 (1970).

[28] G. E. Bacon, *Neutron Diffraction*, 3rd ed. (Oxford University Press, Oxford, 1975).

[29] For a review, see S. M. Shapiro, in Chap. 5 of *Spin Waves and Magnetic Excitations*, edited by A. S. Borovik-Romanov and S. K. Sinha in Vol. 22.2 of *Modern Problems in Condensed Matter Sciences* (North Holland, Amsterdam 1988). J. W. Lynn and J. J. Rhyne, *ibid*, Chap. 4.

[30] N. Nishida, H. Miyatake, D. Shimada, S. Okuma, M. Ishikawa, T. Takabatake, Y. Nakazawa, Y. Kuno, R. Keitel, J. H. Brewer, T. M. Riseman, D. Ll. Williams, Y. Watanabe, T. Yamazaki, K. Nishiyama, K. Nagamine, E. J. Ansaldo, and E. Torikai, *Jpn. J. Appl. Phys.* **26**, L1856 (1987).

[31] J. H. Brewer, E. J. Ansaldo, J. F. Carolan, A. C. D. Chaklader, W. N. Hardy, D. R. Harshman, M. E. Hayden, M. Ishikawa, N. Kaplan, R. Keitel, J. Kempton, R. F. Kiefl, W. J. Kossler, S. R. Kreitzman, A. Kulpa, Y. Kuno, G. M. Luke, H. Miyatake, K. Nagamine, Y. Nakazawa, N. Nishida, K. Nishiyama, S. Ohkuma, T. M. Riseman, G. Roehmer, P. Schleger, D. Shimada, C. E. Stronach, T. Takabatake, Y. J. Uemura, Y. Watanabe, D. Williams, T. Yamazaki, and B. Yang, *Phys. Rev. Lett.* **60**, 1073 (1988).

[32] J. M. Tranquada, D. E. Cox, W. Kunnmann, H. Moudden, G. Shirane, M. Suenaga, P. Zolliker, D. Vaknin, S. K. Sinha, M. S. Alverez, A. J. Jacobson, and D. C. Johnston, *Phys. Rev. Lett.* **60**, 156 (1988).

[33] W-H. Li, J. W. Lynn, H. A. Mook, B. C. Sales and Z. Fisk, *Phys. Rev.* **B37**, 9844 (1988).

[34] H. Kadowaki, M. Nishi, Y. Yamada, H. Takeya, H. Takei, S. Shapiro, and G. Shirane, *Phys. Rev.* **B37**, 7932 (1988).

[35] J. W. Lynn, W-H. Li, H. A. Mook, B. C. Sales, and Z. Fisk, *Phys. Rev. Lett.* **60**, 2781 (1988).

[36] J. M. Tranquada, A. H. Moudden. A. I. Goldman, P. Zolliker, D. E. Cox, G. Shirane, S. K. Sinha, D. Vaknin, D. C. Johnston, M. S. Alvarez, A. J. Jacobson, J. T. Lewandowski and J. M. Newsam, *Phys. Rev.* **B38**, 2477 (1988).

[37] D. Petitgrand and G. Collin, *Physica* **C153-155**, 192 (1988).

[38] P. Burlet, C. Vettier, M. J. G. M. Jurgens, J. Y. Henry, J. Rossat-Mignod, H. Noel, M. Potel, P. Gougeon, J. C. Levet, *Physica* **C153-155**, 1115 (1988).

[39] J. W. Lynn and W-H. Li, *J. Appl. Phys.* **64**, 6065 (1988).

[40] W-H. Li, J. W. Lynn, Z. Fisk, H. A. Mook, and B. C. Sales, *Phys. Rev.* B (to be published).

[41] B. Gillon, D. Petitgrand, A. Delapalme, P. Radhakrishna, and G. Collin, *J. de Physique* (ICM conf., preprint).

[42] A. H. Moudden, G. Shirane, J. M. Tranquada, R. J. Birgeneau, Y. Endoh, K. Yamada, Y. Hidaka and T. Murakami, *Phys. Rev.* **B38**, 8893 (1988).

[43] J. W. Lynn, W-H. Li, S. Trevino and Z. Fisk, *Phys. Rev* **B40**

(9/1, 1989).

[44] D. C. Johnston, A. J. Jacobson, J. M. Newsam, J. T. Lewandowski, D. P. Goshorn, D. Xie and W. B. Yelon, *Chemistry of High-Temperature Superconductors*, ed. by D. L. Nelson, M. S. Whittingham, and T. F. George, *Am. Chem. Soc. Symposium Series* **351** (ACS, Washington, DC 1987). D. C. Johnston, S. K. Sinha, A. J. Jacobson and J. M. Newsam, *Physica* **C153-155**, 572 (1988).

[45] R. J. Cava, B. Batlogg, C. H. Chen, E. A. Tietman, S. M. Zahurak, and D. Werder, *Phys. Rev.* **B36**, 5719 (1987).

[46] P. Zolliker, D. E. Cox, J. M. Tranquada, and G. Shirane, *Phys. Rev.* **B38**, 6575 (1988).

[47] P. F. Miceli, J. M. Tarascon. L. H. Greene, P. Barboux, M. Giroud, D. A. Neumann, J. J. Rhyne, L. F. Schneemeyer and J. V. Waszczak, *Phys. Rev.* **B38**, 9209 (1988).

[48] H. Tang, Z. Q. Qiu, Y.-W Du, G. Xiao, C. L. Chien and J. C. Walker, *Phys. Rev.* **B36**, 4018 (1987). G. Xiao, F. H. Streitz, A. Garvin, Y. W. Du, and C. L. Chien, *Phys. Rev.* **B35**, 8782 (1987).

[49] A. Simopoulos and D. Niarchos, *Phys. Rev.* **B38**, 8931 (1988).

[50] K. Yamada, M. Matsuda, Y. Endoh, B. Keimer, R. J. Birgeneau, S. Onodera, J. Mizusaki, T. Matsuura, and G. Shirane (preprint).

[51] R. J. Birgeneau, D. R. Gabbe, H. P. Jenssen, M. A. Kastner, P. J. Picone, T. R. Thurston, G. Shirane, Y. Endoh, M. Sato, K. Yamada, Y. Hidaka, M. Oda, Y. Enomoto, M. Suzuki, and T. Murakami, *Phys. Rev.* **B38**, 6614 (1988).

[52] C. J. Peters, R. J. Birgeneau, M. A. Kastner, H. Yoshizawa, Y. Endoh, J. Tranquada, G. Shirane, Y. Hidaka, M. Oda, M. Suzuki, and T. Murakami, *Phys. Rev.* **B37**, 9761 (1988).

[53] M. Sato, S. Shamoto, J. M. Tranquada, G. Shirane, and B. Keimer, *Phys. Rev. Lett.* **61**, 1317 (1988).

[54] F. Mezei, B. Farago, C. Pappas, Gy. Hutiray, L. Rosta, and L. Mihaly, *Physica* **C153-155**, 1669 (1988).

[55] G. Aeppli and D. J. Buttrey, *Phys. Rev. Lett.* **61**, 203 (1988).

[55a] G. Aeppli, S. M. Hayden, H. A. Mook, Z. Fisk, S-W. Cheong, D. Rytz, J. P. Remeika, G. P. Espinosa, and A. S. Cooper, *Phys. Rev. Lett.* **62**, 2052 (1989).

[56] J. W. Lynn, W-H. Li, Q. Li, H. C. Ku, H. D. Yang and R. N. Shelton, *Phys. Rev.* **B36**, 2374, (1987).

[57] K. B. Lyons, P. A. Fleury, J. P. Remeika, A. S. Cooper, and T. J. Negran, *Phys. Rev.* **B37**, 7443 (1988).

[58] K. B. Lyons, P. A. Fleury, L. F. Schneemeyer and J. V. Waszczak, *Phys. Rev. Lett.* **60**, 732 (1988).

[59] S. Sugai, S. Shamoto, and M. Sato, *Phys. Rev.* **B38**, 6436 (1988).

[60] K. B. Lyons and P. A. Fleury, *J. Appl. Phys.* **64**, 6075 (1988).

[61] K. B. Lyons (private communications).

[62] B. X. Yang, R. F. Kiefl, J. H. Brewer, J. F. Carolan, W. N.

Hardy, R. Kadono, J. R. Kempton, S. R. Kreitzman, G. M. Luke, T. M. Riseman, D. Ll. Williams, Y. J. Uemura, B. Sternlieb, M. A. Subramanian, A. R. Strzelecki, J. Gopalakrishnan, and A. W. Sleight (preprint). J. Mizuki, Y. Kubo, T. Manako, Y. Shimakawa, H. Igarashi, J. M. Tranquada, Y. Fujii, L. Refelsky, and G. Shirane, *Physica* C (tbp).

[63] For a review see P. Fulde and J. Keller in *Topics in Current Physics,* ed. by Ø. Fischer and M. B. Maple (Springer-Verlag, New York, 1983), Vol. 34, Chapter 9.

[64] R. Joynt and T. M. Rice, *Phys. Rev.* B38, 2345 (1988).

[65] A. A. Abrikosov and G. P. Gor'kov, *Sov. Phys. JETP* 12, 1243 (1961).

[66] For a review see *Topics in Current Physics,* ed. by Ø. Fischer and M. B. Maple (Springer-Verlag, New York, 1983), Vol. 32 and 34.

[67] For a review of crystal field theory see M. T. Hutchings, *Sol. St. Physics* 16, 227 (1964); K. R. Lea, M. J. M. Leask, and W. P. Wolf, *J. Phys. Chem. Sol.* 23, 1381 (1962).

[68] U. Walter, S. Fahy, A. Zettl, S. G. Louie, M. L. Cohen, P. Tejedor, and A. M. Stacy, *Phys. Rev.* B36, 8899 (1987).

[69] A. Furrer, P. Brüesch, and P. Unternähreh, *Phys. Rev.* B38, 4616 (1988).

[70] A. I. Goldman, Y. Gao, S. T. Ting, J. E. Crow, W-H. Li, and J. W. Lynn, *J. Mag. Mag. Matr.* 76-77, 607 (1988).

[71] U. Walter, E. Holland-Moritz, A. Severing, A. Erie, H. Schmidt, and E. Zirngiebl, *Physica* C153-155, 170 (1988).

[72] E. Gering. B. Renker, F. Gompf, D. Ewert, H. Schmidt. R. Ahrens, M. Bonnet, and A. Dianoux, *Physica* C153-155, 184 (1988).

[73] J. W. Lynn, W-H. Li, Q. Li, H. C. Ku, H. D. Yang, and R. N. Shelton, *Rev. Sol. St. Science* 1, 357 (1987).

[74] T. Chattopadhyay, P. J. Brown, D. Bonnenberg, S. Ewert, and H. Maletta, *Europhys. Lett.* 6, 363 (1988). T. Chattopadhyay, P. J. Brown, B. C. Sales, L. A. Boatner, H. A. Mook, and H. Maletta, *Phys. Rev.* B40, 2624 (1989).

[75] J. W. Lynn, T. W. Clinton, W-H. Li, R. W. Erwin, J. Z. Liu, R. N. Shelton, and P. Klavins (preprint).

[76] P. Fischer, K. Kakurai, M. Steiner, K. N. Clausen, B. Lebech, F. Hulliger, H. R. Ott, P. Brüesch, and P. Unternährer, *Physica* C152, 145 (1988).

[77] A. I. Goldman, B. X. Yang, J. Tranquada, J. E. Crow, and C-S Jee, *Phys. Rev.* B36, 7234 (1987).

[78] D. McK. Paul, H. A. Mook, A. W. Hewit, B. C. Sales, L. A. Boatner, J. R. Thompson, and M. Mostoller, *Phys. Rev.* B37, 2341 (1988).

[79] T. Chattopadhyay, H. Maletta, W. Wirgen, K. Fischer, and P. J. Brown, *Phys. Rev.* B38, 838 (1988).

[80] H. A. Mook, D. McK. Paul, B. C. Sales, L. A. Boatner and L. Cussen, *Phys. Rev.* **B38**, 12008 (1988).

[81] D. McK. Paul, H. A. Mook, L. A. Boatner, B. C. Sales, J. O. Ramey and L. Cussen, *Phys. Rev.* **B39**, 4291 (1989).

[82] K. N. Yang, J. M. Ferreira, B. W. Lee, M. B. Maple, W-H. Li, J. W. Lynn, and R. W. Erwin, *Phys. Rev.* B (tbp).

[83] W-H. Li, J. W. Lynn, S. Skanthakumar, T. W. Clinton, A. Kebede, C-S. Jee, J. E. Crow, ant T. Mihalisin, *Phys. Rev.* **B40** (9/1, 1989).

[84] J. C. Ho, P. H. Hor, R. L. Meng, C. W. Chu, and C. Y. Huang. *Sol. St. Commun.* **63**, 711 (1987).

[85] Kadowaki, H. P. van der Meulen, J. C. P. Klaasse, M. van Sprang, J. Q. A. Koster, L. W. Roeland. F. R. de Boer, Y. K. Huang, A. A. Menovsky, and J. J. M. Franse, *Physica* **B145**, 260 (1987).

[86] G. Wortmann, C. T. Simmons, and G. Kaindl, *Sol. St. Commun.* **64**, 1057 (1987).

[87] C. Lin, G. Lu, Z. Liu, Y. Sun, J. Lan, S. Feng, C. Wei, X. Zhu, G. Li, Z. Shen, Z. Gan, F. Chen, J. Chen N. Li, and J. Liu, *Sol. St. Commun.* **64**, 691 (1987).

[88] H. H. A. Smit, M. W. Dirken, R. C. Thiel, and L. J. de Jongh, *Sol. St. Commun.* **64**, 695 (1987).

[89] J. van den Berg, C. J. van der Beek, P. H. Kes, J. A. Mydosh, G. J. Nieuwenhuys, and L. J. de Jongh, *Sol. St. Commun.* **64**, 699 (1987). M. W. Dirken and L. J. de Jongh, *Sol. St. Commun.* **64**, 1201 (1987).

[90] M. E. Reeves, D. S. Citrin, B. G. Pazol, T. A. Friedman, and D. M. Ginsberg, *Phys. Rev.* **B36**, 6915 (1987).

[91] S. Simizu, S. A. Friedberg, E. A. Hayri, and M. Greenblatt, *Phys. Rev.* **B36**, 7129 (1988).

[92] A. P. Ramirez, L. F. Schneemeyer, and J. V. Waszczak, *Phys. Rev.* **B36**, 7145 (1987).

[93] H. D. Yang, H. C. Ku, P. Klavins, and R. N. Shelton, *Phys. Rev.* **B36**, 8791 (1987).

[94] J. A. Hodges, P. Imbert and G. Jehanno, *Sol. St. Commun.* **64**, 1209 (1987).

[95] B. D. Dunlap, M. Slaski, D. G. Hinks, L. Soderholm, M. Beno, K. Zhang, C. Segre, G. W. Crabtree, W. K. Kwok, S. K. Malik, I. K. Schuller, J. D. Jorgensen, and Z. Sungaila, *J. Mag. Mag. Matr.* **68**, L139 (1987).

[96] M. B. Maple, Y. Dalichaouch, J. M. Ferreira, R. R. Hake, B. W. Lee, J. J. Neumeier, M. S. Torikachvili, K. N. Yang, and H. Zhou, *Physica* **B148**, 155 (1987).

[97] J. M. Ferreira, B. W. Lee, Y. Dalichaouch, M. S. Torikachvili, K. N. Yang, and M. B. Maple, *Phys. Rev.* **B37**, 1580 (1988).

[98] B. W. Lee, J. M. Ferreira, Y. Dalichaouch, M. S. Torikachvili,

K. N. Yang, and M. B. Maple, *Phys. Rev.* **B37**, 2368 (1988).

[99] H. Zhou, C. L. Seaman, Y. Dalichaouch, B. W. Lee, K. N. Yang, R. R. Hake, M. B. Maple, R. P. Guertin,and M. V. Kuric, *Physica* **C152**, 321 (1988).

[100] B. D. Dunlap, M. Slaski, Z. Sungaila, D. G. Hinks, K. Zhang, C. Segre, S. K. Malik, and E. E. Alp, *Phys. Rev.* **B37**, 592 (1988).

[101] H. P. Van der Meulen, J. J. M. Franse, Z. Tarnawski, K. Kadowaki, J. C. P. Klaasse, and A. A. Menovsky, *Physica* **C152**, 65 (1988).

[102] F. Nakamura, T. Fukuda, Y. Ochiai, A. Tominaga, and Y. Narahara, *Physica* **C153-155**, 178 (1988).

[103] J. R. Thompson, B. C. Sales, Y. C. Kim, S. T. Sekula, L. A. Boatner, J. Brynestad, and D. K. Christen, *Phys. Rev.* **B37**, 9395 (1988).

[104] C.-S. Jee, A. Kebede, D. Nichols, J. E. Crow, T. Mihalisin, G. H. Myer, I. Perez, R. E. Salomon, and P. Schlottmann, *Solid State Commun.* (in press).

[105] Y. Dalichaouch, M. S. Torikackvili, E. A. Early, B. W. Lee, C. L. Seaman, K. N. Yang, H. Zhou, and M. B. Maple, *Solid State Commun.* **65**, 1001 (1988).

[106] Ce and Tb also affect the superconducting properties, and have a tendency to exhibit mixed valent behavior. However, Ce and Tb may not form the 1-2-3 structure. See K. N. Yang, Y. Dalichaouch, J. M. Ferreira, B. W. Lee, J. J. Neumeier, M. S. Torikackvili, H. Zhou, M. B. Maple, and R. R. Hake, *Solid State Commun.* **63**, 515 (1987); Y. Le Page, T. Siegrist, S. A. Sunshine, L. F. Schneemeyer, D. W. Murphy, S. M. Zahurak, J. V. Waszczak, W. R. McKinnon, J. M. Tarascon, G. W. Hull, and L. H. Greene, *Phys. Rev.* **B36**, 3617 (1987).

[107] L. Soderholm, K. Zhang, D. G. Hinks, M. A. Beno, J. D. Jorgensen, C. U. Segre and I. K. Schuller, *Nature* **328**, 604 (1987).

[108] J. K. Liang, X. T. Xu, S.S. Xie, G. H. Rao, X. Y. Shao, and Z. G. Duan, *Z. Phys.* B, **69**, 137 (1987).

[109] A. P. Goncalves, I. C. Santos, E. B. Lopes, R. T. Henriques, M. Almeida, and M. O. Figueiredo, *Phys. Rev.* **B37**, 7476 (1988).

[110] A. Kebede, C.-S. Jee, D. Nichols,. M. V. Kuric, . E. Crow, R. P. Guertin, T. Mihalisin, G. H. Myer, I. Perez, R. E. Salomon, and P. Schlottmann, *Physica* (tbp).

CHAPTER 9

Electron Pairing: How and Why?

Philip B. Allen

9.1. Introduction

As this chapter is being written, a year has elapsed since high T_c superconductors became an accepted reality. The mechanism for the high T_c is still mysterious. It is not even certain that the superconducting state is BCS-like. This ripe situation is forcing theorists to rethink.

It is beyond my ability to summarize fairly all serious theoretical papers of the last 12 months; instead of trying, my aim will be to explain in my own way as many aspects of the problem as I can. Alternate views can be found in several review papers [1-3] or by scanning conference proceedings such as the June 1987 Berkeley conference [4a] and the March 1988 Interlaken meeting [4b].

This chapter tries to emphasize the basics - Cooper pairing (Sec. 2), Bose Condensation (Sec. 3), and BCS Theory (Secs. 4,5,6). Yang's conjecture that off-diagonal long-range order (ODLRO) is the minimal essence of superconductivity is described in Sec. 7. In Secs. 8,9,10 the popular mechanisms which fit into conventional BCS theory are mentioned. The alternative possibility of superconductivity arising from something other than a conventional Fermi liquid is discussed in Sec. 11, where the terms "weak" and "strong" are introduced for conventional and unconventional theories. Two "strong" theories, bipolarons and resonating valence bonds (RVB), are the subject of Secs. 12 and 14. In Sec. 13 a short explanation is given of the Mott insulator phase of the Hubbard model in the large U, half-filled band case. This is intended to motivate and clarify the notions of RVB theory which are attracting attention from many theorists. In Sec. 15 several models are described in which BCS superconductivity arises via magnetic interactions in a doped Mott insulator.

One can roughly distinguish four classes of superconductors. (1) Ordinary metallic superconductors, such as Pb, Al, Nb, NbN, V_3Si, Nb_3Sn, etc. These all appear to be based on BCS singlet pairing, caused by the electron-phonon interaction. T_c ranges up to 23 K (so far) [5]. (2) "Organic" superconductors such as $(BEDT\text{-}TTF)_2I_3$. T_c is mostly low, but increasing [6]. (3) "Heavy Fermion" superconductors. Here $T_c < 1$ K is usual [7]. (4) High T_c copper oxide-based superconductors, with T_c up to 125 K and possibly higher.

The last three categories share some surprising similarities: (a) the mechanism of superconductivity is not clear; (b) the nature of the pairing scheme is also not clear - BCS triplet or d-wave has not been ruled out; (c) all show a proximity between superconductivity and antiferromagnetic order. In addition, the organics and the CuO superconductors show metal-insulator transitions (no doubt related to antiferromagnetism). A fifth category could be added: (5) ^3He. This is definitely established to be a BCS triplet state, with proximity to ferromagnetic spin ordering, and a low T_c, $\sim 10^{-3}$ K, driven at least partly by spin fluctuations.

It is logical to believe two things about the CuO-based superconductors. First, their behavior should be closely related to that of $Ba(Pb_{1-x}Bi_x)O_3$, a structurally related system showing a similar interplay of superconductivity and a metal/insulator transition [8]. Second, their superconductivity should be closely related to their antiferromagnetic properties. This would help explain the high T_c, because the magnetic energy scale appears to be $\sim 10^3$ K [9]. It is well understood how antiferromagnetism arises from strong Coulomb repulsion (Sec. 13) but is as yet a matter of conjecture and controversy how superconductivity, especially with high T_c, can arise from repulsive interactions. This may require an unconventional, highly-correlated metallic state (such as RVB) rather than the normal state of BCS theory which is a Fermi liquid. A possible phase diagram for $La_{2-x}Sr_xCuO_4$ is shown in Fig. 1. It illustrates the interplay of antiferromagnetism and superconductivity.

Interestingly, the two logical beliefs mentioned above are in apparent conflict. There is no evidence yet for antiferromagnetism in $Ba(Pb_{1-x}Bi_x)O_3$, and the strong correlations which create its insulating state may originate more from electron-phonon than from Coulomb interactions [11]. Therefore it is premature to rule out anything.

Before commencing, a few preliminary remarks. The reader who knows BCS theory can skip to Sec. 8. However, I give in Sec. 4 a derivation of T_c in BCS theory using perturbation theory to find the divergence of the pair susceptibility, and this may be worth reading because it is used later on. I assume the structure of CuO superconductors is familiar (see Chapter 4) and that the electronic states near ε_F are located in the square-planar sheets and on the chains, as is explained in Chapter 5.

Finally I want to clarify what I mean by "BCS theory" [12]. Some take this to mean the electron-phonon interaction, which causes a pairing instability. In this sense the CuO superconductors are probably *not* BCS. However, to others (including myself) the central core of BCS theory is the pairing instability of the Fermi liquid in the presence of an attractive interaction, with the source of this interaction a peripheral aspect. In this sense the new materials may be BCS-like. If future experiments should confirm this, then the remaining issue is to identify the source of attraction.

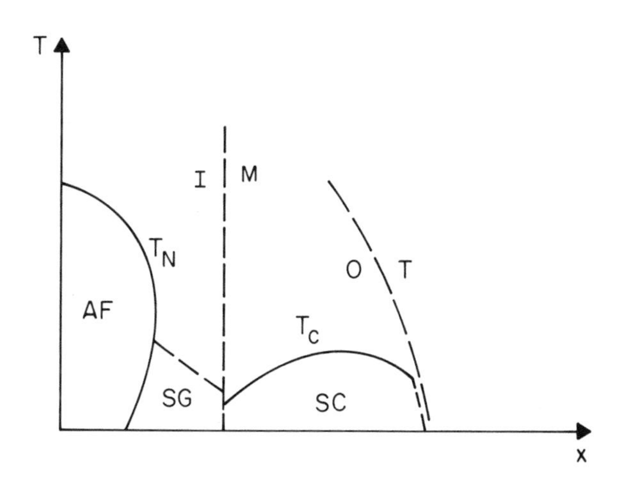

Fig. 9.1. A possible (T, x) phase diagram for $La_{2-x}Sr_xCuO_4$, after Aharony *et al.* [10]. The solid curves are the Neel (T_N) and superconducting (T_c) temperatures separating antiferromagnetic (AF) from non-magnetic insulating (I) and superconducting (SC) from normal metallic (M) regions. Also shown is a possible spin glass (SG) phase and the orthorhombic (O) to tetragonal (T) phase boundary which occurs near $x = 0.2$.

9.2. Cooper Pairs

Cooper [13] discovered that if two electrons, in an otherwise non-interacting gas of electrons, are allowed to interact via an attractive force, they always bind together, with a binding energy

$$E_B = 2\Theta e^{-2/\lambda} \quad . \tag{9.1}$$

A derivation of this formula will be sketched below; λ is $N(\varepsilon_F)V$ where V is the interaction strength and $N(\varepsilon_F)$ is the density of states at the Fermi energy. Generalizing to the case where all electrons attract each other, one should expect an instability of the normal Fermi liquid, and this is the essence of BCS theory. The cause of the instability is the Pauli principle and the sharpness of the Fermi surface. At higher temperatures the Fermi surface is fuzzy over a range k_BT and the instability goes away when $k_BT \approx E_B$.

To derive Eq. (9.1), let the wavefunction for the interacting electron pair be

$$\psi(\mathbf{r}_1,\mathbf{r}_2) = \sum_{\mathbf{k}}' g(k)e^{i\mathbf{k}\cdot\mathbf{r}_1} e^{-i\mathbf{k}\cdot\mathbf{r}_2} \quad . \tag{9.2}$$

To satisfy the Pauli principle, we choose $g(-k) = g(k)$ and a singlet spin arrangement, and we restrict the sum in (9.2) to wave vectors $\mathbf{k}$ *outside* the occupied Fermi sea, $\varepsilon_F < \varepsilon_k = \hbar^2k^2/2m$, which is the meaning of the prime on the sum. Translation symmetry makes the center of mass momentum $\mathbf{Q} = \mathbf{k}_1 + \mathbf{k}_2$ a constant of the motion, and we have chosen $\mathbf{Q} = 0$ or $\mathbf{k}_1 = -\mathbf{k}_2 = \mathbf{k}$ because this gives the lowest energy.

We now operate on $\psi(r_1,r_2)$ with the Hamiltonian $p_1^2/2m + p_2^2/2m + V(\mathbf{r}_1,\mathbf{r}_2)$ and seek the lowest eigenvalue E which will be written as $2\varepsilon_F - E_B$, so that E_B is the binding energy. This gives an equation

$$[2(\varepsilon_k-\varepsilon_F) + E_B]g(k) = \sum_{k'}' |V(k,k')|g(k') \quad . \tag{9.3}$$

Here we are assuming that $V(k,k')$, the Fourier transform of $V(r_1,r_2)$, is negative, or attractive.

Now it helps to consider one specific type of attractive interaction, the Bardeen-Pines [14] form of the phonon-induced interaction.

$$V_{BP}(k,k') = \frac{2\hbar\omega_q |M_{k,k'}|^2}{(\varepsilon_k - \varepsilon_{k'})^2-(\hbar\omega_q)^2} \quad . \tag{9.4}$$

Here $q = k - k'$, ω_q is a phonon frequency, and $M_{k,k'}$ is an electron-phonon matrix element. This equation is derived by making a canonical transformation to eliminate the electron-phonon interaction in lowest order. It is not exact, and is replaced in the more sophisticated Eliashberg [15] version of BCS theory by a time-dependent interaction. After Fourier transforming time to frequency the Eliashberg interaction looks like (9.4) with $\varepsilon_k - \varepsilon_{k'}$, replaced by $\hbar\omega$. The interaction (9.4) is attractive if $|\varepsilon_k - \varepsilon_{k'}|$ is less than $\hbar\omega_q$. A reasonable caricature of Eq. (9.4), which is convenient for use in Eq. (9.3), is

$$V_{BP}(k,k') \approx -V\theta\left(k_B\Theta - |\varepsilon_k - \varepsilon_F|\right)\,\theta\left(k_B\Theta - |\varepsilon_{k'} - \varepsilon_F|\right) \quad , \quad (9.5)$$

where $k_B\Theta \approx \hbar\omega_D$, and θ is the unit step function. Notice that V is a positive number and denotes an attractive interaction. With a reinterpretation of V and Θ, Eq. (9.5) might be used for other types of interactions, such as excitons, plasmons, etc.

Now insert Eq. (9.5) into Eq. (9.3). Then $g(k)$ will depend only on $\varepsilon_k = \varepsilon$, and vanish unless $\varepsilon_F < \varepsilon < \varepsilon_F + k_B\Theta$. In this interval, $g(\varepsilon)$ obeys

$$[2(\varepsilon - \varepsilon_F) + E_B]g(\varepsilon) = V \int_{\varepsilon_F}^{\varepsilon_F + k_B\Theta} d\varepsilon' N(\varepsilon')g(\varepsilon') \quad . \quad (9.6)$$

This equation has a solution, $g(\varepsilon) = c/[2(\varepsilon - \varepsilon_F) + E_B]$, provided E_B obeys the consistency equation

$$1 = V \int_{\varepsilon_F}^{\varepsilon_F + k_B\Theta} d\varepsilon' N(\varepsilon')/[2(\varepsilon' - \varepsilon_F) + E_B] \quad . \quad (9.7)$$

Cooper's answer for E_B, Eq. (9.1), follows immediately from (9.7) upon two further approximations: (1) $k_B\Theta \ll \varepsilon_F$ so that variations of $N(\varepsilon')$ can be neglected and $N(\varepsilon')$ can be replaced by $N(\varepsilon_F)$; (2) $E_B \ll k_B\Theta$, as is justified by Eq. (9.1) in the weak coupling limit $\lambda < 1$. These are reasonable approximations in most metals. A formula for λ also emerges from Eqs. (9.4, 9.5), namely

$$\lambda = N(\varepsilon_F)\,\langle 2|M_{k,k'}|^2/\hbar\omega_Q\rangle \quad (9.8)$$

where the bracket in Eq. (9.8) denotes averaging k and k' over the Fermi surface. The same parameter λ appears in Eliashberg theory for T_c.

It is interesting to examine Eq. (9.7) in a different regime, namely the very dilute limit where $\varepsilon_F \to 0$ and electron pairs are well separated from other electrons. Now $N(\varepsilon)$ is needed near the bottom of the band and cannot be replaced by a constant value. The result depends sensitively on the dimensionality d of the electronic system. For the non-interacting electron gas,

$$N(\varepsilon) = N(c_d/\varepsilon_o)(\varepsilon/\varepsilon_o)^{(d-2)/2} \tag{9.9}$$

where in d = 1, 2, 3, we have $c_1 = 1/2\pi$, $c_2 = 1/4\pi$, and $c_3 = 1/4\pi^2$. The energy scale ε_o is $\hbar^2/2ma^2$, where a is a microscopic length defined by $Na^d = \Omega$, where N = number of atoms and Ω = d-dimensional sample volume. Then, setting $\varepsilon_F = 0$, Eq. (9.7) becomes

$$\frac{1}{\lambda} = \int_0^\theta \frac{dx\; x^{(d-2)/2}}{x+\delta} \tag{9.10}$$

where the dimensionless variables are $x = \varepsilon/\varepsilon_o$, $\delta = E_B/2\varepsilon_o$, $\theta = k_B\Theta/\varepsilon_o$, and $\lambda = c_dV/2\varepsilon_o$. If we further assume weak binding ($\delta \ll \theta$), the binding energies found from Eq. (9.10) are

$$\delta = (\pi\lambda)^2 \qquad (d=1) \tag{9.11a}$$

$$\delta = \theta e^{-1/\lambda} \qquad (d=2) \tag{9.11b}$$

$$\delta = (4\theta/\pi^2)(1-\lambda_c/\lambda)^2 \quad (d=3) \quad [\text{provided } \lambda > \lambda_c = 1/2\theta^{\frac{1}{2}}] \tag{9.11c}$$

In 1d, there is always a bound state with binding scaling as V^2 and independent of the cutoff energy Θ. In 2d, there is also always a bound state, but for small V, E_B is exponentially small, with a scale set by Θ. In 3d, no binding occurs until V exceeds a critical strength $V_c = 4\pi^2\varepsilon_o \sqrt{\varepsilon_o/k_B\Theta}$.

The results for the dimension-dependence of binding are essentially the same as the well-known single-particle quantum results for particles in wells. These results contrast sharply with Cooper's result (9.1), which is independent of dimension and reminiscent of the 2d result (9.11b). Because of the Pauli principle, the electrons in Cooper's problem explore a section of k-space with a density of states analogous to Eq. (9.9) with d=2.

The role of the sharpness of the Fermi surface is to cause a logarithmic divergence of the integral in Eq. (9.7) as E_B goes to zero. This is what permits a bound state $E_B > 0$ no matter how small V is. Suppose the background Fermi sea had a non-zero temperature. This could be modeled by using in Eq. (9.3) a factor $1 - f(\varepsilon_k')$ instead of a sharp cutoff $\theta(\varepsilon_k' - \varepsilon_F)$ in the sum on the right. The logarithmic divergence in Eq. (9.7) would cut off at k_BT and there would be no binding for k_BT greater than the zero temperature binding energy. For the same reason the superconducting instability goes away as T increases.

As a final note, one can imagine solutions of Eq. (9.3) with other symmetries; for example. a triplet state with $g(-k) = -g(k)$. For an interaction like Eq. (9.4) one expects to find E_B largest for the most symmetric state, namely the singlet. Other bound states of smaller E_B are irrelevant to superconductivity. For special kinds of potentials $V(k,k')$, a triplet state (or any other state of different symmetry) might become the most strongly bound, leading to a more exotic BCS-like state. The superfluid phase of ^{3}He at millikelvin temperatures is the only proven example, having triplet spin pairing and $\ell = 1$ orbital symmetry. The $\ell = 0$ channel presumably has a repulsive interaction because of the hard core. It is speculated that certain "organic" superconductors and "heavy Fermion" superconductors may also involve triplet pairs or possibly singlet $\ell = 2$ ("d-wave") pairs. These all have low T_c's, and it seems to me unlikely that exotic pairing could lead to a high T_c.

9.3. Bose Condensation

If we could regard a Cooper pair as a good composite particle (as we can a ^{4}He atom, for example, when probed at energies of a few eV or less) then a Cooper pair would be a Boson, having spin 0 in the singlet $\ell = 0$ case and spin $\mathbf{J} = \mathbf{L} + \mathbf{S}$ with $J = 0$, 1, or 2 in the $\ell = 1$ triplet case. In ordinary metallic superconductors, Cooper pairs actually have a large radius $\xi \sim 10^2$-10^4 Å, and are therefore strongly overlapping, and poorly defined composite particles. Nevertheless there is a sense in which the BCS ground state is a Bose condensate of Cooper pairs. Therefore it is appropriate to review Bose condensation here. It should be born in mind that unlike the ideal gas Bose condensate, in the BCS case the Boson pairs *do not exist except when condensed*. However, it is possible to imagine an alternate scenario where a non-BCS (or at least non-conventional BCS) superconductor is formed out of preexisting pairs. Consider for example, ^{4}He atoms, which are good Bose particles up to $T > 10^4$ K. The superfluid transition occurs at $T_\lambda = 2.18\ K$ which is quite close to the condensation temperature $3.2\ K$ of the corresponding gas of non-interacting He atoms. If these particles were endowed with a special charge which enabled them to couple to external fields without altering their interaction with each other, then the superfluid phase of ^{4}He would be superconducting. Each ^{4}He atom is made of two electrons and an alpha particle (two Fermions and a Boson). Thus this imaginary charged gas of ^{4}He atoms would be a non-BCS superconductor in which Bosons based on electron pairing would be formed at very high temperatures (almost 10^5 K) but would condense only at low temperatures (2.18 K).

The idea of Bose condensation was put forward by Einstein, who realized that the formula which fixes the chemical potential μ_b for Bosons,

$$N = \sum_k [\exp(\beta(\varepsilon_k-\mu_b)) - 1]^{-1} \quad , \tag{9.12}$$

is not necessarily correctly expressed by the corresponding integral

$$N = \int_0^\infty d\varepsilon N(\varepsilon) \; [\exp(\beta(\varepsilon-\mu_b)) - 1]^{-1} \tag{9.13}$$

Let us assume that the Bosons are free particles of mass m. Equation (9.9) for $N(\varepsilon)$ applies to Bosons as well as Fermions. Then Eq. (9.13) becomes

$$(\beta\varepsilon_o)^{d/2} = c_d \int_0^\infty dx \; x^{d/2-1}[\exp(x-\beta\mu_b) - 1]^{-1} \quad . \tag{9.14}$$

The quantity $\beta\mu_b$ must be negative in order for the Bose-Einstein occupation function to be positive. The difficulty is that for $d >$ 2, Eq. (9.14) has no solution for $\beta\mu_b$ if $\beta\varepsilon_o$ is too large, because the integral in (9.14) is bounded above by the value it attains when $\mu_b = 0$,

$$c_d \int_0^\infty \frac{dx \; x^{d/2-1}}{e^x - 1} = q(d) \quad . \tag{9.15}$$

For $d = 1$ or 2, q(d) is infinite which means that a solution $\beta\mu_b < 0$ can be found for any value of $\beta\varepsilon_o$. However, q(3) is ~0.0586, so that there is a minimum value of $\beta\varepsilon_o$, corresponding to $1/\beta_c = k_B T_c \cong 6.62\ \varepsilon_o$, and for lower T, no positive value of $-\beta\mu_b$ exists which satisfies Eq. (9.14). Einstein resolved the dilemma by observing that the process of converting (9.13) to (9.14) does not do justice to the discrete nature of the sum at small k. In particular, the minimum energy state, $k = 0$, must be treated separately, and can develop a macroscopic occupancy below T_c. Anderson [16] has discussed how this macroscopic occupancy requires phase coherence, and Yang [17] has shown how this state exhibits what he calls "off-diagonal long range order" (ODLRO), a property which would lead to flux quantization, the Meissner effect, and superconductivity if the particles had charge.

It is interesting to notice that dimensionality plays a large role in Bose condensation, in a fashion closely related but reciprocal to the role of dimensionality discussed in Sec. 2 for pair binding. In low dimensions, there are proportionately many states near $\varepsilon = 0$ to accommodate the Bosons (because $N(\varepsilon)$ is large) whereas for $d > 2$ there are fewer states near $\varepsilon = 0$ which forces condensation into the $\varepsilon = 0$ state. In the jargon of phase transitions (of which this is the simplest example) "enhanced quantum fluctuations" in low dimensionality destroy the condensed or ordered state. But exactly this availability of low energy states in low d is what promotes pair binding. This creates a dilemma for Boson scenarios of superconductivity: If d is low, Fermions can more easily bind to form Bosons but can't Bose condense, whereas if d is high, condensation is permitted but binding of Fermions into Bosons is suppressed (except in degenerate Fermi systems where the Fermi surface promotes binding).

9.4. BCS Theory for T_C

At high T, the thermal broadening of the Fermi surface eliminates the binding of Cooper pairs. As T is lowered to T_c, a collective rearrangement occurs in which all electrons near ε_F bind into a pair condensate. Below $T = T_c$, energy 2Δ is required to unbind a pair of electrons from this condensate. This "gap" Δ is temperature dependent, and goes to zero as T increases to T_c. This is analogous to a ferromagnet, where the energy required to flip a spin goes to 0 as T approaches the Curie temperature T_c from below. In fact, the BCS theory is completely analogous to a mean field theory of a magnetic transition, so it is appropriate to review the simplest such theory, the Curie-Weiss theory of ferromagnetism.

Consider $\mathcal{N}$ permanent moments μ_i which experience an external field **B** but not each other.

$$\mathrm{H}_{ext} = - \sum_i \mu_i \cdot \mathbf{B} \tag{9.16}$$

Elementary statistical mechanics shows that the susceptibility $\chi = \partial M / \partial B$ is

$$\chi_o = \frac{\mathcal{N}\mu^2}{k_B T} \tag{9.17}$$

(this is for $s = \frac{1}{2}$; classical spins have an extra factor 1/3). Now suppose there is an interaction between moments

$$\mathrm{H}_{int} = - \mathrm{J} \sum_{<ij>} \mu_i \cdot \mu_j \tag{9.18}$$

where the sum runs over nearest-neighbor pairs. Mean field theory replaces μ_i by $<\mu> + \delta_i$ where the fluctuation δ_i is $\mu_i - <\mu>$ and is regarded as small. Omitting quadratic terms $\delta_i \delta_j$, (9.16) and (9.18) can be combined

$$H_{MF} = - \sum_i \mu_i \cdot (\mathbf{B} + zJ <\mu>) \tag{9.19}$$

where $<\mu> = M/\mathcal{N}$ and z is the coordination number. The magnetization of the system described by (9.19) obeys

$$M = \chi_o(B + zJM/\mathcal{N}) = \chi B \tag{9.20}$$

$$\chi = \chi_o/(1 - zJ\chi_o/\mathcal{N}) \tag{9.21}$$

Using Eq. (9.17) for χ_o, the mean field susceptibility χ diverges as $c/(T-T_c)$ where T_c is the solution of $zJ\chi_o/\mathcal{N} = 1$, or $T_c = zJ\mu^2/k_B$.

In the case of superconductivity, there is also a susceptibility which diverges as T_c is approached from above, but it does not couple to any readily available field and thus is hidden from measurement. The key question is, what plays the role of the "order parameter", analogous to $M = \sum_i <\mu_i>$ for the magnet. The BCS answer is that a nonclassical quantity, the "pair amplitude", plays this role. Playing the roles of the moment operators μ_i are the pair creation and destruction operators

$$b_k = c_{k\uparrow}\, c_{-k\downarrow}$$

$$b_k^+ = c_{-k\downarrow}^+\, c_{k\uparrow}^+ \quad . \tag{9.22}$$

Suppose we imagine a field F which coupled to such an operator

$$H_{ext} = - \sum_k b_k^+ \, F\, e^{-i\omega t} + hc \tag{9.23}$$

where *hc* means the Hermitian conjugate term. Then in response there would be a pair field

$$\phi = \sum_k \langle b_k \rangle \tag{9.24}$$

and the pair susceptibility would be

$$\chi_p \equiv \frac{\partial \phi}{\partial F} \quad . \tag{9.25}$$

To compute χ_p, we can use a Kubo-type formula

$$\chi_p(\omega) = - (i/\hbar) \sum_{kk'} \int_0^\infty dt e^{+i\omega t} \langle [b_k^+(0),\, b_{k'}(t)] \rangle \tag{9.26}$$

completely analogous to a well known Kubo formula for the magnetic case [18]. For non-interacting electrons it is easy to evaluate Eq. (9.26). The t-dependence of $b_{k'}(t)$ is $b_{k'}(0)\exp(2i(\varepsilon_{k'} - \mu)t/\hbar)$, and

(9.26) becomes

$$\chi_p^o(\omega) = \sum_k \frac{1-2f_k}{2(\varepsilon_k-\mu)-\hbar\omega} = \int d\varepsilon N(\varepsilon) \frac{\tanh\beta(\varepsilon-\mu)/2}{2(\varepsilon-\mu)-\hbar\omega} \quad . \tag{9.27}$$

At $T = 0$, this integral has a logarithmic divergence arising from the sharpness of the Fermi surface. It is convenient to introduce a cutoff $\theta(k_B\Theta - |\varepsilon_k-\mu|)$ into the k-sum in (9.23), to avoid an uninteresting ultraviolet divergence in Eq. (9.27). The susceptibility then becomes

$$\chi_p^o(0) = \tfrac{1}{2} N(\varepsilon_F) \int_{-k_B\Theta}^{k_B\Theta} d\varepsilon \frac{\tanh\beta\varepsilon/2}{\varepsilon}$$

$$= N(\varepsilon_F) \log \frac{1.13\Theta}{T} \quad . \tag{9.28}$$

The last step above is a familiar manipulation in BCS theory and is explained, for example, by Rickayzen [19].

Notice that both the Curie susceptibility χ_o for free spins, and the pair susceptibility, χ_p^o for free electrons, diverge as $T \to 0$. These systems become infinitely polarizable and an arbitrarily weak interaction favoring alignment will induce a spontaneous polarization at low T. The analog for superconductivity of the magnetic interaction (9.18) would be a term of the form

$$H_{red} = -V \sum_{kk'}^{c.o.} b_k^+ b_{k'} \tag{9.29}$$

where c.o. denotes that a cutoff is used as in Eq. (9.5). The subscript "*red*" stands for "*reduced*" as will be explained below. Mean field theory now replaces b_k by $\langle b_k \rangle$ plus the fluctuation $\delta_k = b_k - \langle b_k \rangle$. The same is done for b_k^+, and the term $\delta_k^+ \delta_{k'}$ is dropped. Then the external field in (9.23) and the mean field version of (9.29) combine to give

$$H_{MF} = -\sum_k b_k^+ (Fe^{-i\omega t} + V\phi) \tag{9.30}$$

where Eq. (9.24) is used for ϕ. This equation is the analog of (9.19). Finally the pair field ϕ is computed.

$$\phi = \chi_p^o (Fe^{-i\omega t} + V\phi) \equiv \chi_p Fe^{-i\omega t} \tag{9.31}$$

$$\chi_p = \frac{\chi_p^o}{1 - V\chi_p^o} \quad . \tag{9.32}$$

These are the analogs of (9.20) and (9.21). Equation (9.32) is the pair susceptibility in mean field approximation. It diverges at ω = 0 when $T = T_c$, where T_c is the solution of $1 = V\overset{o}{\chi}_p(T_c)$. Using Eq. (9.28), we find the famous BCS result,

$$T_c = 1.13\ \Theta \exp(-1/N(\varepsilon_F)V) \quad . \tag{9.33}$$

This is the analog of the magnetic result $T_c = zJ\mu^2/k_B$. The exponential dependence of T_c on $1/V$ tends to make T_c small, and is a consequence of the logarithmic divergence (9.28) of the non-interacting pair susceptibility.

In the case of magnetic systems, mean field theory is not usually a very good approximation; especially in lower dimensions $d < 3$, the neglected fluctuations qualitatively change the behavior of the system. For superconductivity this is usually not true. Unlike Eq. (9.18) where only z nearest neighbor spins j interact with spin i, in Eq. (9.29) the number of wave vectors k' interacting with k grows as the size of the system, making fluctuations relatively unimportant. The BCS solution is exact in the thermodynamic limit for the interaction (9.29). However, unlike the magnetic case, the interaction (9.29) contains only a small piece of the complete interaction. If a pairwise interaction $V(r_1,r_2)$ exists between electrons, then, expressed in a Bloch-state basis, the interaction is

$$H_{int} = \sum_{kk'q\sigma\sigma'} V(k,k',q)\, c^{+}_{k+q\sigma} c^{+}_{k'-q\sigma'} c_{k'\sigma'} c_{k\sigma} \ . \tag{9.34}$$

The interaction (9.29) has taken only a small subset of the terms of Eq. (9.34), namely those with $\sigma' = -\sigma$ and $k' = -k$, and is called the BCS "*reduced*" Hamiltonian. The coupling constant $V(-k,k,q)$ is then relabeled as $-V(k,k+q)$; the sign change is a peculiar BCS convention which allows attractive V's to be positive. The final replacement of $V(k,k+q)$ by V in Eq. (9.29) is only a convenience which permits a closed form solution.

There is no proof that the terms omitted from Eq. (9.34) are innocuous and cause no further new phenomena, but the success of BCS theory provides empirical proof for many materials. BCS theory is built on Landau theory which asserts that all terms of (9.34) are absorbed in building a ground state and quasiparticle excitations which interact weakly. BCS theory adds to Landau theory the hypothesis that a small subset of terms of (9.34), namely (9.29), have an important additional effect in altering the ground state and excitation spectrum. All remaining terms are believed to define the parameters of the Landau theory, but not to alter it qualitatively or destroy it.

9.5. The Interaction in BCS Theory

A rigorous procedure for locating T_c within the philosophy of BCS theory is to find the divergence of the pair susceptibility, Eq. (9.26). The mean field theory of Sec. 9.4 is exact (in the thermodynamic limit) for the special "*reduced*" Hamiltonian (9.29), but for a more general interaction, such as (9.34), or if electron-phonon interactions are present, a more powerful scheme for evaluating Eq. (9.26) is needed. This is provided by graphical (Feynman-Dyson) perturbation theory. In order to use this theory, a related correlation function is needed [20],

$$\chi_p(i\omega_\lambda) = \tfrac{1}{2} \sum_{k,k'} \int_{-\beta}^{\beta} d\sigma \, e^{i\omega_\lambda \sigma} \langle T_\sigma [b_{k'}(\sigma) b_k^{\dagger}(0)] \rangle \qquad (9.35)$$

Here σ is an imaginary time, ω_λ is a Matsubara frequency, and T_σ is a "time-ordering" operator. From the function (9.35) the susceptibility (9.26) can be retrieved, but (9.35) is better suited to the development of a systematic perturbation theory.

First consider the case where the only interaction is the reduced interaction (9.29). Then we should be able to recover the BCS results (9.32, 9.33) from perturbation theory. The terms in the perturbation theory for χ_p correspond to diagrams:

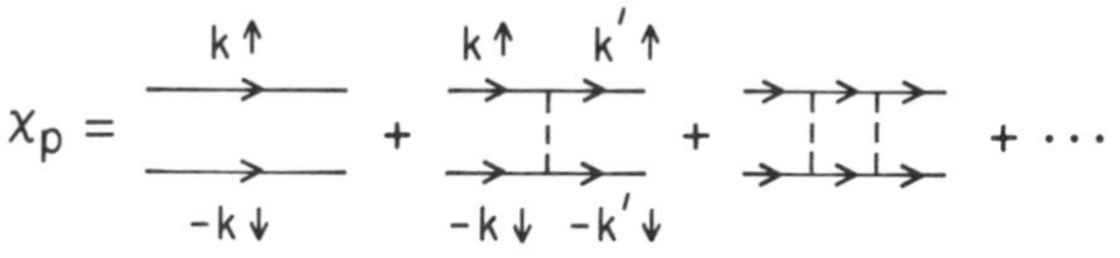

k↑ k↑ k′↑
$\chi_p =$ + + + ··· (9.36)
−k↓ −k↓ −k′↓

After integrating over k and k', a simple infinite series results: $\chi_p = \overset{o}{\chi}_p + \overset{o}{\chi}_p V \overset{o}{\chi}_p + \cdots$ The nth order term is just $\overset{o}{\chi}_p (V \overset{o}{\chi}_p)^n$ and the sum of the resulting geometric series is just Eq. (9.32). Notice that because of the form of H_{red}, there can be no terms corresponding to graphs like

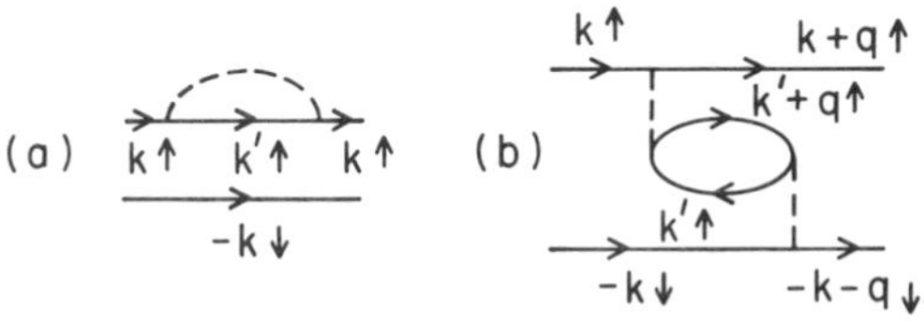

(a) k↑ k′↑ k↑ −k↓ (b) k↑ k+q↑ k′+q↑ k′↑ −k↓ −k−q↓ (9.37)

although any physical interaction, such as (9.34), permits them. These graphs are important in a complete theory. Graph (a) renormalizes the one particle properties and graph (b) screens the interaction.

Next consider the case where there is only an electron-phonon interaction

$$H_{int} = \sum_{kQ\sigma} M_{k,k+Q} c^{+}_{k+Q\sigma} c_{k\sigma} (a_Q + a^{+}_Q) \tag{9.38}$$

$$H_o = \sum_{k\sigma} \varepsilon_k c^{+}_{k\sigma} c_{k\sigma} + \sum_{Q} \hbar\omega_Q a^{+}_Q a_Q \ . \tag{9.39}$$

It should be understood that both phonons and electrons have band indices and the symbols k, Q can be considered short for $(\mathbf{k},n)$ and $(\mathbf{Q},j)$. Many new kinds of graphs are possible. Fortunately, only a simple subset is important:

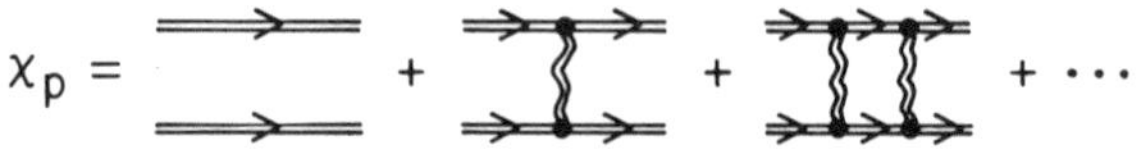

χ_p = + + + ··· (9.40)

This looks exactly like Eq. (9.36), but it is actually more complicated. The double lines are shorthand for other infinite series:

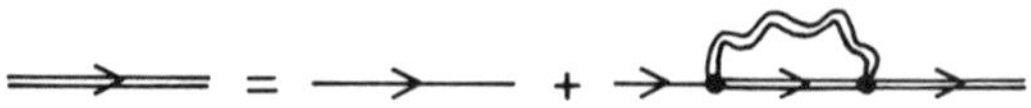

= + (9.41)

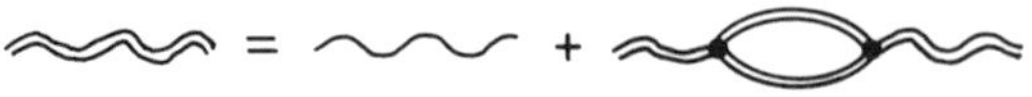

= + (9.42)

Equation (9.42) renormalizes the phonon frequencies. Fortunately, if the frequencies are known from experiment, these are already renormalized so we can bypass this step. Unfortunately, if the frequencies have not been measured, then evaluating (9.42) properly (including Coulomb effects not shown in (9.42)) is very difficult. Weber [21] and Weber and Mattheiss [22] have made a good preliminary attempt on oxide superconductors, but their work has adjustable constants and is not quantitatively predictive. Equation (9.41) contains the "mass renormalization" of the electronic states which causes a significant correction to T_c.

When the terms of Eq. (9.40) are summed, an integral equation results which is different from the original BCS equation using the Bardeen-Pines interaction. The reason for the difference is that there is a time-delay in phonon propagation. This results in an ω-dependence of the gap $\Delta(k,\omega)$ and of the integral equation. Thus Eq. (9.40) corresponds to a time-dependent generalization of BCS theory. It is called Eliashberg theory, and has been thoroughly studied because it is the basis for understanding conventional metallic superconductors. The explicit construction of Eliashberg theory by this route was done by Allen [20] including Coulomb and impurity effects.

Besides the graphs shown in Eq. (9.40-9.42), it is appropriate to consider graphs like

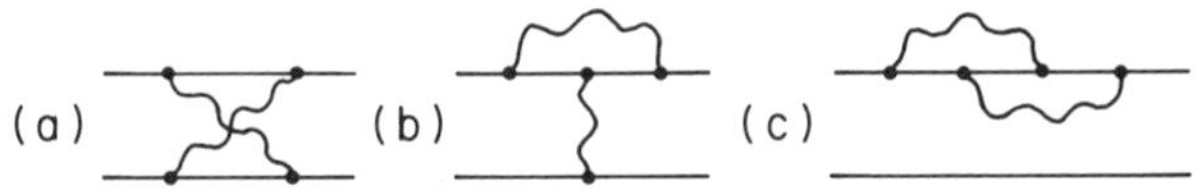

(a) (b) (c) (9.43)

These have phonon lines crossing each other. Migdal [23] has shown that each graph of this kind is negligible in comparison to a similar but uncrossed graph which is already contained in Eqs. (9.40-9.42). This situation, known as Migdal's "theorem", is special to the electron-phonon problem, and is related to the smallness of a parameter sometimes denoted $(m/M)^{1/2}$, but more properly denoted as $N(\varepsilon_F)\hbar\omega_D$.

Next consider the case of an ordinary pairwise Coulomb interaction as in Eq. (9.34). There is an important set of terms which is the analog of Eqs. (9.40-9.42).

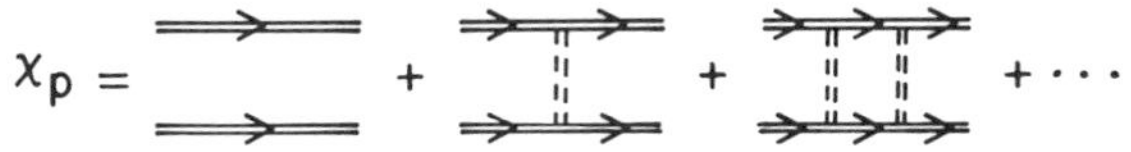

χ_p = + + + ⋯ (9.44)

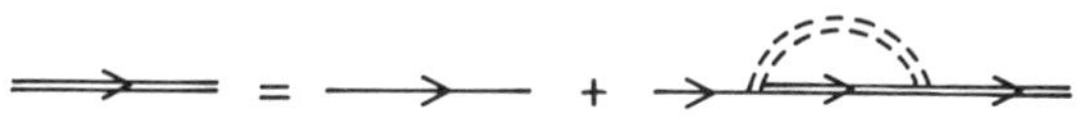

= + (9.45)

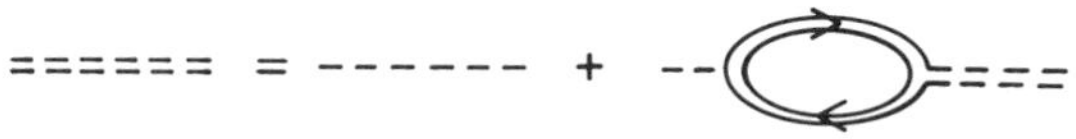

= + (9.46)

Equation (9.46) is the Coulomb interaction screened in random phase approximation (RPA), and Eq. (9.45) represents the "screened exchange" approximation to the electron quasiparticle states. Formal answers can be written for the theory at this level, but evaluation of these formal expressions for real materials is very difficult. To make matters worse, none of the graphs analogous to Eq. (9.43) are *a priori* negligible. Formally one includes many of these by inserting "vertex functions" into Eqs. (9.45) and (9.46), making them "formally exact". Like Eq. (9.42), Eq. (9.46) and to some extent Eq. (9.45), correspond to measurable spectral properties, but such measurements are normally done only in limited spectral ranges. Thus we often must guess at the appropriate screening function and quasiparticle energies, using LDA (local density approximation) band theory as an imperfect guide.

Finally, consider a generalization of the graphs like (9.43a), that is, "maximally-crossed" graphs.

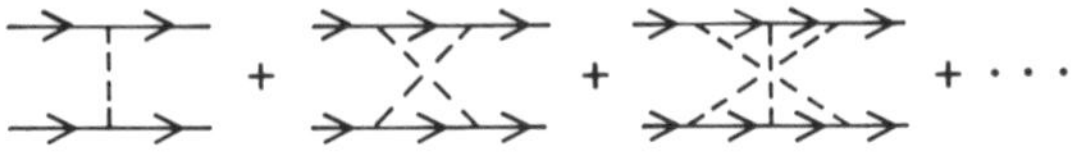

$$+ \quad + \quad + \cdots \tag{9.47}$$

If we redraw the graphs, they simplify as in

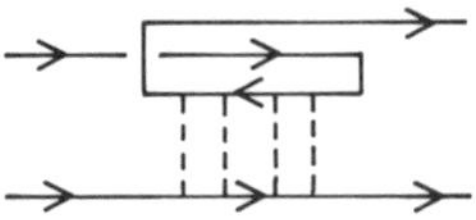

$$\tag{9.48}$$

There is a piece of this graph which is very closely akin to the transverse magnetic susceptibility χ_{+-},

$$\chi_{+-}(Q,\omega) = -i\sum_{k,k'}\int_0^\infty dt e^{i\omega t}\left\langle [c^{+}_{k+Q\uparrow}(t)c_{k\downarrow}(t), c^{+}_{k'-Q\downarrow}(o)c_{k'\uparrow}(o)]\right\rangle \tag{9.49}$$

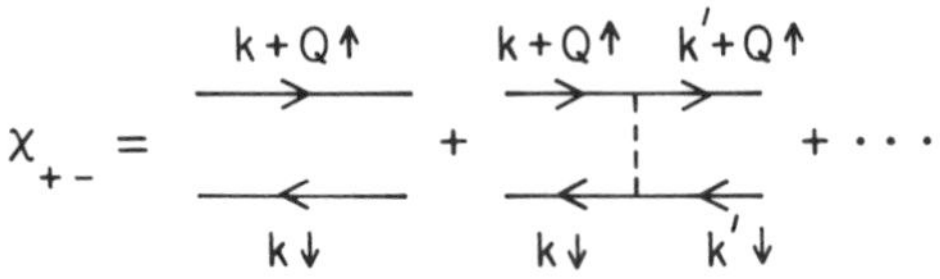

k+Q↑ k+Q↑ k'+Q↑

$$\chi_{+-} = \quad + \quad + \cdots \tag{9.50}$$

k↓ k↓ k'↓

In the paramagnetic state, χ_{+-} is equivalent to the longitudinal susceptibility χ_{zz}, which for $Q \to 0$, $\omega \to 0$ is the measured $\chi = \partial M/\partial B$. Experiment shows that some metals such as Pd have large values of χ corresponding to incipient ferromagnetism. The ladder series (9.50) contains a mechanism for the enhancement, but it is not possible to prove that other graphs are insignificant. In simple models [24] this series (9.50) can be evaluated, and $\text{Im}\chi_{+-}(Q,\omega)$ shows peaks which are well-enough defined to be given a name, "paramagnons". These "particles" have not been easy to find by neutron scattering experiments. If we bend the graph (9.48) yet again, it looks like

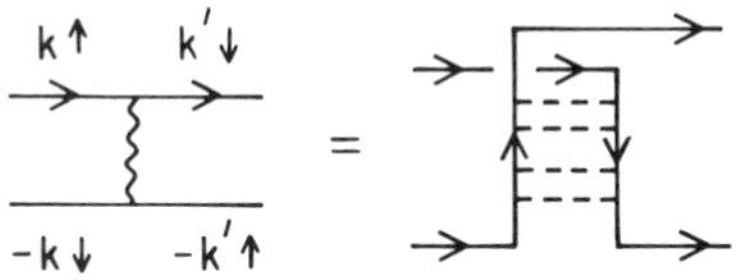

k↑ k′↓ -k↓ -k′↑ (9.51)

This graph is interpreted physically as a pair of electrons exchanging a paramagnon and flipping their spins. This is a well-defined event in metallic ordered magnets such as rare earth elements where itinerant nonmagnetic electrons interact with the ordered spin array of the localized f electron system, but it is a fairly fuzzy event in a paramagnet. Nevertheless, paramagnons provide a potentially useful physical concept. Unfortunately there are many other high-order terms in perturbation series related to (9.51) but not easily summable. Paramagnon theory is not especially well-defined. At some point we are forced to stop looking for more graphs, sum up what we have, and hope to explain T_c. Spin fluctuation graphs of the type in Eq. (9.51) are harmful to singlet superconductivity, and this explains the absence of superconductivity in Pd. The basis for phonon mechanisms as well as plasmon, exciton, and weak-coupling spin-fluctuation mechanisms are all in Eq. (9.35) and its graphical expansion. Only for the phonon part does the perturbation series converge rapidly enough to be treated rigorously.

9.6. The BCS ground state

A collective pair condensate should have a wave function of the kind

$$\Psi_o = \mathscr{A}\{\psi(\mathbf{r}_1,\mathbf{r}_2)\psi(\mathbf{r}_3,\mathbf{r}_4)\ \dots\ \psi(\mathbf{r}_{N-1},\mathbf{r}_N)\} \tag{9.52}$$

where ψ is a pair function like Eq. (9.2) and $\mathscr{A}$ denotes antisymmetrization. In fact, the BCS wavefunction has this form [25] but it is usually written in a very different way. Notice that the electronic wavefunction for $N/2$ ^{4}He atoms would have the same form. Equation (9.52) is even under exchange of pairs of indices such as $(1,2)\leftrightarrow(3,4)$. Thus Eq. (9.52) can describe composite Bosons, but in the BCS case, different ψ's so strongly overlap that such an interpretation is not very appropriate.

By analogy with Eq. (9.2), (9.52) can be rewritten as

$$|\Psi_o\rangle = (\sum_k g(k)\ c^+_{k\uparrow}\ c^+_{-k\downarrow})^{N/2}|0\rangle \tag{9.53}$$

where $|0\rangle$ is the vacuum. Because of the anticommutation relations of the c_k^+'s, this state is fully antisymmetric, but unfortunately very awkward for calculations. Therefore a trick is used:

$$|\Psi_{BCS}\rangle = \exp(\sum_k g(k)\, c_{k\uparrow}^+ c_{-k\downarrow}^+)|0\rangle \tag{9.54}$$

The series expansion of the exponential in (9.54) contains the term (9.53) and many others with all possible exponents n. Only if Eq. (9.54) is dominated by terms with $n \approx N/2$ does it make a sensible approximation to Eq. (9.53).

The next trick is to rearrange (9.54)

$$|\Psi_{BCS}\rangle = \prod_k \exp(g(k)\, c_{k\uparrow}^+ c_{-k\downarrow}^+)|0\rangle$$

$$= \prod_k (1 + g(k)\, c_{k\uparrow}^+ c_{-k\downarrow}^+)|0\rangle \tag{9.55}$$

where the last line of (9.55) is an exact consequence of the fact that $(c_{k\uparrow}^+ c_{-k\downarrow}^+)^2 = 0$. To normalize Eq. (9.55), each factor $(1+g(k)\, c_{k\uparrow}^+ c_{-k\downarrow}^+)$ should be divided by $(1+|g(k)|^2)^{1/2}$. To retrieve the ground state of a normal metal, choose $g(k) \to \infty$ for $\varepsilon_k < \varepsilon_F$ and $g(k) = 0$ for $\varepsilon_k > \varepsilon_F$. The probability that a state k is occupied is $n_k = |g_K|^2/(1+|g_k|^2)$, the mean number of electrons is $2\Sigma_k n_k = \langle N\rangle$ and the rms fluctuation $[\langle N^2\rangle - \langle N\rangle^2]^{1/2}$ is $2[\Sigma_k n_k(1-n_k)]^{1/2}$ which is less than $\sqrt{2N}$, i.e. small enough to justify using (9.54) in place of (9.53).

Equation (9.55) is the standard form of the BCS wavefunction. It is very convenient for calculations. The $g(k)$'s are treated as variational parameters, and turn out all to have the same phase, $g(k) = |g(k)|e^{i\phi}$. The amplitude for having exactly $2n$ electrons has phase $n\phi$. The phase difference between the component of Ψ with n pairs and the component with $n + 1$ is always ϕ. This phase coherence is an essential property, leading directly to the possibility of supercurrents $\mathbf{j}_s \propto \nabla\phi$.

Further details on the ground state and excited states are available in many texts, e.g., Rickayzen [19].

9.7. Off-Diagonal Long Range Order (ODLRO)

The essence of BCS theory is the pairing instability and the formation of a coherent pair condensate. Yang [17] attempted to answer a deeper version of this question: What is the essence of superconductivity which has somehow been correctly captured by BCS theory? His answer is that it is ODLRO, a property of the reduced density matrix. The functions are defined

$$f_1(r,r') \equiv \mathrm{tr}[\hat{\rho}\ \hat{\psi}^+(r)\ \hat{\psi}(r')] \tag{9.56}$$

$$f_2(r,r') = \mathrm{tr}[\hat{\rho}\ \hat{\psi}_\downarrow^+(r)\ \hat{\psi}_\uparrow^+(r)\ \hat{\psi}_\uparrow(r')\ \hat{\psi}_\downarrow(r')] \tag{9.57}$$

where $\hat{\rho}$ is the density matrix, equal to $e^{-\beta H}/Z$ in equilibrium, where $Z = \mathrm{tr}e^{-\beta H}$. The operator $\hat{\psi}_\sigma^+(r)$ creates a particle of spin σ at position r. By way of contrast, consider a correlation function

$$g(r,r') = \mathrm{tr}[\hat{\rho}\ \delta\hat{n}(r)\ \delta\hat{n}(r')] \tag{9.58}$$

where $\delta\hat{n}$ is the deviation of the particle density from it average value

$$\delta\hat{n} = \sum_\sigma \hat{\psi}_\sigma^+(r)\hat{\psi}_\sigma(r) - \bar{n} \quad . \tag{9.59}$$

In a liquid, $g(r,r') \to 0$ as $|\mathbf{r} - \mathbf{r}'|$ gets large, but in a solid, $g(r,r') \to \langle\delta n(r)\rangle\langle\delta n(r')\rangle$ which remains non-zero for large $|\mathbf{r} - \mathbf{r}'|$. This is an example of long-range order. If f_1 or f_2 [Eqs. (9.56), (9.57)] should remain finite as $|\mathbf{r}-\mathbf{r}'|$ gets large, this would be an expression of ODLRO, a purely quantum effect involving off-diagonal elements of the density operator which have no classical analog. Yang showed that $f_1(r,r')$ cannot remain finite at large $|\mathbf{r} - \mathbf{r}'|$ in a system of Fermions, but in a Bose condensate, f_1 becomes $\langle\psi^+(r)\rangle\langle\psi(r')\rangle$ which is nonzero at large separations. Similarly for a normal metal, f_2 goes to zero at large $|\mathbf{r} - \mathbf{r}'|$ whereas in a BCS superconductor, f_2 becomes $\langle\psi_\downarrow^+(r)\ \psi_\uparrow^+(r)\rangle\langle\psi_\downarrow(r')\ \psi_\uparrow(r')\rangle$ which is not zero. To be specific, since

$$\psi_\uparrow(r) = \sum_k \psi_k(r)\ c_{k\uparrow} \tag{9.60}$$

the pair amplitude has expectation value

$$\langle\psi_\uparrow(r)\ \psi_\downarrow(r)\rangle = \sum_k |\psi_k(r)|^2\ g(k)/(1+|g(k)|^2) \quad . \tag{9.61}$$

This has phase ϕ, and vanishes in the normal state.

Yang made the following statements based on either proofs or plausible conjectures: (a) There is only one route by which f_1 or f_2 exhibit ODLRO and this route is chosen by the ideal Bose gas (for f_1) and by the BCS superconductor (for f_2). (b) The existence of ODLRO is a sufficient condition for quantization of magnetic flux, the quantum being hc/e or $hc/2e$ depending on whether f_1 or f_2 exhibits ODLRO. (c) From flux quantization, the Meissner effect and superconductivity are almost certainly derivable. From these statements it is expected that ODLRO will be a property not only of BCS superconductors, but of any alternate type of superconductivity which may be found. In the "RVB" picture to be described later, Anderson conjectures a ground state with ODLRO. As a counter-example, Fröhlich [26] made a very interesting theory of a one-dimensional "superconductor" based on an incommensurate charge-density-wave (CDW) distortion which introduced a gap at the Fermi surface. Because of incommensurability, the CDW is free to slide in a perfect crystal, carrying current in a collective way. The effect is realized in such materials as $NbSe_3$. Unfortunately, perfect conductivity is destroyed by even a small number of impurities (as it is also in a pure non-superconducting metal at T = 0). In fact the impurities pin the CDW giving zero conductivity unless the field is big enough to de-pin the CDW. The missing ingredient in Fröhlich's theory (and also in a perfect normal metal at T = 0) is ODLRO.

9.8. Eliashberg Theory of Electron-Phonon Superconductors

This is a much studied subject with a recent review [5] so this section will be a series of comments.

(a) The basic equations can be derived from Eqs. (9.40-9.42) and (9.44-9.46). The interaction is both **r** and ω dependent. The Coulomb interaction in first approximation is

$$V_c = \langle k'\uparrow,-k'\downarrow | \int d\mathrm{r}''\ \varepsilon^{-1}(\mathrm{r},\mathrm{r}'',\omega)\ e^2/|\mathbf{r}''-\mathbf{r}'|\ |k\uparrow,-k\downarrow\rangle \quad (9.62)$$

and is repulsive over a large fraction of (r,r',ω)-space. The electron-phonon interaction is

$$V_{ph} = 2|M_{kk'}|^2\ \hbar\omega_{k-k'} / \hbar^2[\omega^2-\omega^2_{k-k'}] \quad (9.63)$$

and is attractive when $\omega^2 < \omega^2_{k-k'}$. The dimensionless parameters λ and μ are defined by Fermi surface averages

$$-\lambda = N(\varepsilon_F)\ \langle V_{ph}(k,k',\omega=0)\rangle \quad , \quad (9.64)$$

$$\mu = N(\varepsilon_F)\ \langle V_c(k,k',\omega=0)\rangle \quad . \quad (9.65)$$

Note that (9.64) is identical to Eq. (9.8).

(b) In a naive BCS approach one would expect

$$T_c(\text{BCS}) = \Theta \exp(-1/(\lambda-\mu)) \quad . \tag{9.66}$$

In an approximate treatment of Eliashberg theory [27] one gets

$$T_c(\text{McMillan}) = \Theta \exp\left[-1/\left(\frac{\lambda}{1+\lambda} - \mu^*\right)\right] \quad , \tag{9.67}$$

$$\mu^* = \mu/(1+\mu \log(\varepsilon_F/\hbar\omega_D)) \quad . \tag{9.68}$$

The parameter λ is "renormalized" by the factor $(1+\lambda)^{-1}$ which enters via Eq. (9.41). The parameter $\mu \approx 1$ is "renormalized" to $\mu^* \approx 0.1$ by an entirely different effect: The slow propagation of phonons (retardation) allows electrons to interact without being close at the same time, thus reducing the Coulomb repulsion.

(c) The parameter λ ranges from 0.1 (Cu,Na) to 1.5 (Pb) or even larger ($\gtrsim 2$ in amorphous Pb/Bi alloys). When $\lambda \gtrsim 1.2$, the form of Eq. (9.67) starts to fail; after a wide crossover region, the asymptotic regime with $T_c \propto \sqrt{\lambda}$ occurs (with $\lambda \gtrsim 10$, an apparently unrealistic case) [28].

(d) Detailed, first principles calculations of T_c are still quite unreliable because of the difficulty of calculating ω_Q. If ω_Q is known from experiment, the reliability of theory is not bad.

(e) T_c is bounded above by $0.18\sqrt{\lambda\langle\omega^2\rangle}$. This can also be expressed as $0.18\sqrt{\Sigma_a \eta_a/M_a}$, where a runs over atoms of mass M_a and η_a is the "McMillan-Hopfield parameter". Although values of T_c as large as 100 K are allowed, this would require fairly extreme choices of η_a and M_a.

(f) A practical limit to T_c is apparently set by problems of structural stability [29]. As Eq. (9.42) suggests, strong electron-phonon coupling drives phonon frequencies down, risking a second order instability. First order instabilities are probably more serious, but very difficult to anticipate theoretically.

(g) Although most high T_c superconductors are quite anharmonic for reasons given in (f), there is little evidence to suggest that anharmonicity *per se* affects T_c beneficially or adversely relative to a corresponding harmonic solid with equivalent vibrational frequencies. Anharmonic effects to a large degree are absorbed into the structure of Eliashberg theory with no experimentally detectable change [30].

(h) The new oxide superconductors probably can not be explained on the basis of electron-phonon theory. This is suggested by isotope effect measurements, attempts to calculate T_c, and past experience on the difficulty of getting high T_c by this mechanism.

(i) On the other hand, a very small isotope shift by no means proves anything. A nice analysis is by Rainer and Culetto [31].

PdD has a higher T_c than PdH (reverse isotope effect) and yet is believed to be electron-phonon induced with a healthy $T_c \sim 12$ K.

9.9. Excitons and Plasmons

Given that known electron-phonon superconductors all have $T_c < 23$ K, it is logical to look for new mechanisms if higher T_c's are wanted. It is also logical to blame low T_c's on the low energy $\hbar\omega_D$ of the phonon. This immediately suggests that a higher-energy Boson might act in place of (or in addition to) phonons, yielding a higher T_c. The subject has been nurtured by Little [32] and Ginzburg [33]; recent reviews are by Ginzburg and Kirzhnits [34], Ruvalds [35], and Little [36]. Since they have been the target of some ridicule (Matthias [37]), it is appropriate to acknowledge that their optimism (although not yet their specific mechanisms) has now been validated.

When a test charge is inserted into a solid, three kinds of distortion occur: The lattice distorts (virtual phonons), the electrons can repopulate any partly filled band (virtual plasmons), or they can mix in components from higher unfilled bands (virtual excitons). The words "plasmon" and "exciton" are used quite loosely here. To a spectroscopist, an exciton is a narrow resonance in Imε, and a plasmon is a narrow peak in -Imε^{-1}. Most of the spectral weight is not in narrow peaks, but broadly distributed in ω. It is necessary to use all this broad structure to get a significant interaction, and the usage of the terms "exciton" and "plasmon" is correspondingly broadened. If the test charge is time-dependent (for example, moving) then these distortions follow in time, with a lag which is larger for a lattice distortion than electronic distortions. A second test charge will feel both the direct instantaneous repulsion of the first test charge and a time-dependent attraction from these dynamical distortions. The usual Eliashberg picture assumes that electronic polarization serves only to cancel part of the direct Coulomb repulsion, and that only phonon polarization is sufficiently strong and retarded to contribute to pair binding. Thus a reasonable question is whether, in the absence of phonons, the other polarization mechanisms can (in principle) cause superconductivity.

9.9.a. Plasmons

Rietschel and Sham [38] have answered this question affirmatively for the free electron gas using the RPA (Lindhard) version of ε; i.e. for the plasmon mechanism. Their numerical calculations use the full (**k** and ω-dependent) Eliashberg equations with no other interaction besides $\varepsilon^{-1}(Q,\omega)v(Q)$. This is always repulsive (positive) for $\omega = 0$, but has an attractive region for $\omega_{min} < \omega < \omega_p$ and Q/k_F not too large. Their results depended sensitivity on the density of the electrons, expressed by the parameter $r_s = 1.92/a_B k_F$. For $r_s < 3$, superconductivity does not occur; at $r_s = 4$, T_c is ~ 0.3 mK, and at $r_s \simeq 5$, T_c is ~ 0.22 K. This result is surprising for two reasons. First, ω_p is so large ($\hbar\omega_p/\varepsilon_F = 0.94\sqrt{r_s}$) that no simple scenario based on retardation can make this seem intuitively necessary. It is a source of optimism that electronic pairing is sufficiently versatile to take successful advantage of the limited attractive part of $(\mathbf{q},\omega)$ space. Second, alkali metals and noble metals are not superconducting. It is difficult to see how the less dense alkalis with $4 < r_s < 6$ could fail to superconduct given both plasmon and phonon attraction, unless something else occurs to cancel the plasmon attraction.

This question has been further addressed by Grabowski and Sham [39], who approximately include the effects of the lowest order vertex corrections and the graph with crossed Coulomb interactions. The solution of Rietschel and Sham is totally altered by these corrections, and the net effective interaction becomes again repulsive. One lesson to be drawn from this is that superconductivity induced by high energy Bosons is difficult to predict because results are very model-sensitive. We are unable to calculate interactions to the necessary accuracy. Grabowski and Sham conclude with the remark, "One or two tens of degrees Kelvin in the increase of T_c above that due to phonon attraction appears possible, but a T_c higher than the liquid nitrogen temperature appears impossible by purely raising the Boson energies." Of course, now that T_c exceeds 100 K, it might be appropriate to rethink this conclusion.

For metals more complicated than alkalis, band effects become significant and the free-electron model becomes less relevant and possibly misleading. One measure of this is the *f*-sum rule, an exact relation for any system:

$$\frac{2}{\pi}\int_0^\infty d\omega\, \omega\varepsilon_2(\omega) = \frac{4\pi n e^2}{m} \quad . \tag{9.69}$$

On the right of (9.69) n is the total number of electrons (including core electrons) and m is the free electron mass. In many cases it is reasonable to integrate the left hand side up to an energy less than the minimum for core electron excitation, and to use for n on the right the valence electron density. In band theory, $\varepsilon_2(\omega)$ has two distinct parts, a Drude region representing acceleration of electrons within partially filled bands ($\hbar\dot{k} = -eE$) and an interband term. These can be separately calculated; results are given in Table 9.1 for representative cases. In the first column, Ω_p^2 is the contribution to the integral (9.69) from the Drude part of ε_2. The last column gives the fraction of the spectral weight which is in the intraband (Drude) channel. As materials become more complicated there are more bands at a given **k** and less room for Drude-like behavior. The result for La_2CuO_4 is not pathological, but typical of a complicated metal; the calculation assumes La_2CuO_4 is metallic rather than semiconducting and that **E** is polarized in the metallic a-b plane. Table I shows that most of the electronic polarizability of complicated materials is in "exciton" (interband) rather than "plasmon" (intraband) form.

There have been proposals that the nearly *2d* plasmons of quasi-*2d* metals like La_2CuO_4 are especially beneficial for superconductivity, but so far there is no convincing calculation of this, even for an electron gas, let alone a real material with interband dominance.

9.9.b. Excitons

Crudely speaking, two versions of this idea exist. Allender, Bray, and Bardeen (ABB, [43]) proposed getting a metal into such intimate contact with a polarizable, narrow gap semiconductor that the metallic electrons would be able to interact strongly with interband excitations of the semiconductor. Pairing would then occur by exchange of these virtual excitations. It was not proposed that the electron-hole pairs should be required to bind to each other to form Wannier excitons. Instead the word "exciton" is used loosely to describe the sort of virtual excitation involved in the polarization of a valence band.

The exciton mechanism, like the plasmon mechanism, surely

Table I. The Drude plasma frequency Ω_p for Aℓ and Nb [40], Nb_3Ge [41], and La_2CuO_4 [42]. The free electron values ω_p are based on the assigned valences: Aℓ(3), Nb(5), Ge(4), La(3), Cu(11), and O(6).

	$\hbar\Omega_p$(eV)	$\hbar\omega_p$(eV)	$(\Omega_p/\omega_p)^2$
Aℓ	12.4	15.8	0.62
Nb	9.1	19.6	0.22
Nb_3Ge	3.7	19.5	0.04
La_2CuO_4	3.0	17.2	0.03

exists, and helps in the sense that it weakens the direct Coulomb repulsion of electrons. Unlike the plasmon in the free electron gas studied by Sham *et al.*, there is no obvious simple model system which should be subjected to rigorous theoretical investigation. The merit of the ABB paper was that it proposed a testable mechanism involving thin metal layers on semiconductor surfaces. Careful experiments by Miller *et al.* [44] did not see the effect. Inkson and Anderson [45] thought the ABB estimates for T_c were much too high, and that no significant improvement of T_c ought to occur by this scheme.

The other version of this idea is in the original paper by Little [32]. It is almost the same as the ABB mechanism except it is presumed that it is better *not* to have intimate contact between the electrons involved in pairing and the entity undergoing electronic polarization. Spatial separation should reduce the interaction strength, but even at 14 Å, e^2/r is 1eV if unscreened. Separation eliminates exchange interactions between Cooper pairs and polarizing electrons, which otherwise would reduce the attraction. Little also proposed looking in 1*d* organic systems with polarizable side chains. In recent years superconductivity has been found in 1*d* "organic" metals [6] but these do not look much like Little's proposal. Two dimensions is probably more favorable, and Little [36] proposes that the out-of-plane oxygen-copper bonds may be the polarizing entity.

9.9.c. Excitonic Insulator

If a semiconductor has a very narrow indirect gap, then in principle electrons could be promoted across this gap, motivated by the possibility that they could gain more energy by binding to the holes left behind than they lost in promotion energy. This is called the "excitonic insulator" phase [46]. Lo and Wong [47] proposed that additional carriers in such a system might be superconducting. Abrikosov [48] pointed out that a phase related to the excitonic insulator could be a candidate for a high T_c superconductor, namely one where instead of an exciton condensate, the heavy holes crystallized and the light electrons paired by exchanging "phonons" of the hole lattice.

9.10. Spin Fluctuations

In a magnetic medium, a free electron can scatter off the spin system, emitting a spin wave. Superconducting systems are generally not magnetically ordered, except perhaps at temperatures lower than

T_c as in rare-earth-substituted $YBa_2Cu_3O_7$ (see Chapter 8). An exception is URu_2Si_2 [49,50], and superconductivity in this heavy Fermion metal may be caused by exchange of antiferromagnetic spin fluctuations. A more common situation is for a metal to be close to magnetic ordering when becoming superconducting. In this situation one expects to have relatively long-lived local spin fluctuations, for example, locally ferromagnetic regions in a ferromagnet above its Curie temperature, or in a metal like Pd which is an incipient ferromagnet. The scattering of electrons from these fluctuations will alter the tendency to form pairs. Usually this alteration has been assumed to be harmful. Pd is not superconducting even though the value of λ is probably $\sim$ 0.3 - 0.5. The reason almost certainly is that a spin up electron will attract other electrons of up spin, creating a locally ferromagnetic region, and repelling the down spin electron needed for singlet Cooper pairing. This is shown diagrammatically in Eq. (9.51). Clearly this same process should be helpful to triplet pairing, and is believed to be an important source of triplet pairing in liquid ^3He. So far there is no proof that any metal has a triplet pairing state, although UPt_3 and other heavy Fermion systems, as well as some "organic" superconductors, are possible candidates. These candidates all have a low T_c.

Antiferromagnetic spin fluctuations are a more complicated situation [51-53]. It is not so obvious whether an up spin electron will create or destroy local antiferromagnetic polarization, or whether this will attract or repel other electrons of opposite (or parallel) spin. Different answers seem to emerge in different regions of k-space and r-space. These questions were first raised because of the experimental discovery of closely related antiferromagnetic and superconducting states in organic metals, and then in heavy Fermion metals. Recently they have acquired a new urgency because of the discovery of antiferromagnetism in La_2CuO_{4-y} [54] and $YBa_2Cu_3O_{6+y}$ [55] (see Chapter 8).

An interesting simple scenario has been proposed by Schrieffer et al. [56]. The idea is that free carriers in a material wanting commensurate antiferromagnetism will weaken the tendency to order (for example, by changing the Fermi surface "nesting" property in $\chi(Q)$). Thus an up-spin electron repels local spin order. Other free carriers, with either spin orientation, will be attracted to the region of depleted spin, (called a "spin bag") giving a net attractive interaction.

9.11. Weak versus Strong Coupling

All mechanisms discussed so far are derivable by perturbation theory for the pair susceptibility χ_p (Eq. 9.26) starting from a normal

Fermi liquid reference state; in other words, they are in the "weak" regime. It is conventional to call Pb a "strong coupling" superconductor because with $\lambda \sim 1.5$ it exhibits deviations from "normal" BCS-like behavior ($2\Delta/k_BT_c > 3.52$, etc.). But in the sense used in this section, Pb is "weak". An example of "strong" would be ^{3}He viewed as a collection of electrons and nuclei. At T $\sim 10^6$ K, one would have a plasma of electrons and ^{3}He nuclei; at high enough densities this would give two interpenetrating quantum Fermi liquids. This would be the wrong Fermi liquid starting point for a perturbative theory of superfluidity in liquid ^{3}He at 10^{-3} K.

By analogy, exotic forms of superconductivity could emerge in solids from highly correlated electronic states which are not describable as ordinary Fermi liquids. Bipolaronic and RVB superconductors are such proposals, and will be described in subsequent sections. Neither of these yet constitutes a full theory in the way that BCS is for Fermi liquids. One can imagine scenarios in which the final superconducting state looks BCS-like, and others in which it doesn't. For example, the ^{3}He case just described starts at very high T and P as a Fermi liquid, and ends at very low T and P looking BCS like, with a low T Fermi liquid in between, and two crossover regions, one between plasma and ^{3}He vapor, and one between ^{3}He vapor and normal Fermi liquid ^{3}He (at T $\lesssim$ 1 K). But ^{4}He presents an alternate scenario, ending with a Bose condensate which would be superconducting if endowed with an extra hidden charge to couple it to external E-fields. This superconductor would not look BCS like, and at T_c the superconducting state would evolve into a normal Bose liquid.

Presumably all states share the property of ODLRO, and show flux quantization, the Meissner effect, and supercurrents. Even if the oxide superconductors turn out to be in the weak regime, they have enlarged our vision, and a search for strong superconductors will endure.

9.12. Bipolarons

In Sec. 9.2 the Cooper problem was addressed and solved in two limits: The usual Cooper limit [$\varepsilon_F \gg E_B$, $\varepsilon_F \gg \Theta$] and the dilute limit [$\varepsilon_F \ll E_B$, $\varepsilon_F \ll \Theta$]. In the latter case electrons form mobile bound pairs provided the attractive interaction is strong enough to support a bound state (which requires a critical strength in three dimensions). The "bipolaron" is just such a mobile bound pair of electrons, bound by electron-phonon interactions. There is no unambiguous evidence that bipolarons exist in any known solid, but a plausible case has been made [57] that the metal-insulator transitions seen in Ti_4O_7 at 140 K and 150 K could be described as bipolaronic. Specifically it is assumed that as T is reduced, Ti atoms form dimers of Ti^{3+} (one dimer, on average, per cell, with two undimerized Ti^{4+} atoms per cell). At 150 K the dimers form in a disorganized, liquid form, and at 140 K these pre-formed dimers crystallize into a relatively normal insulating phase. In the "liquid" phase between 140 and 150 K one has pairs of electrons bound to pairs of Ti^{3+} ions and free to move by breaking bonds and forming new ones neighboring undimerized Ti^{4+} atoms which then form new dimer bonds. Later Chakraverty [58] proposed that in the liquid state the bipolarons could Bose condense to form a new kind of superconductor.

Actually this kind of superconductivity is not a new idea. Ogg [59] mistakenly believed that he had seen high T_c superconductivity in metal-ammonia solutions. His explanation was that free electrons might prefer to pair in cavities in NH_3, and could then Bose-condense. The idea is identical to more modern proposals. Another precursor of BCS theory [60] involved Bose-condensed electron pairs and subsequently it has been shown that much of BCS theory can be re-cast in this form. Leggett [61] worked out a model showing the crossover from BCS condensation to Bose condensation of molecular pairs. Other oxides besides Ti_4O_7 have provoked theoretical speculation about bipolarons. $SrTiO_3$ supports superconductivity when doped n-type with n as low as $10^{19}/cm^3$, possibly the most dilute superconductor known, and thus a natural suspect for bipolaronic effects.

A case closely related both to Ti_4O_7 and to CuO-based superconductors is $Ba(Pb_{1-x}Bi_x)O_3$ which is insulating when $x > 0.35$ and superconducting for $x < 0.35$, with a surprisingly high $T_c \sim 15$ K occurring close to the phase boundary. Rice and Sneddon [11] proposed that the superconducting phase was normal BCS and electron-phonon driven, but with increasing x and increasing electron-phonon coupling, "real-space pairing" took over and created an insulator. The real-space pairs are hypothesized to consist of pairs of electrons bound to pairs of Bi atoms making Bi^{3+}-Bi^{5+} frozen dimers, i.e. another kind of bipolaronic crystal.

It is possibly accidental but nevertheless interesting that the idea of bipolaronic superconductivity, specifically in Jahn-Teller distorted systems, is part of what motivated Bednorz and Müller [62] to search for superconductivity in CuO-based systems. Scalapino *et al.* [63] argue that CuO-based superconductivity may be electron-phonon driven, and on the border between bipolaronic and BCS-like.

Finally, a related notion is the Anderson "negative-U" center, an isolated but immobile bipolaron. An example is a vacancy in crystalline Si, which has a stable neutral configuration; when a single additional electron localizes at the vacancy site, a local atomic rearrangement occurs which makes it energetically favorable for a second electron (of opposite spin; a singlet pair) to bind to the same site. The lattice relaxation overcomes the Coulomb repulsion, as if there were a negative value for the on-site "Hubbard" Coulomb repulsion U. This idea was originally proposed by Anderson [64]. It was suggested [65,66] that if such sites occurred in a superconducting metal, T_c would be enhanced by the additional binding available if Cooper pairs temporarily fell into negative-U centers. This creates a difficult theoretical problem, being a hybrid of a perturbative BCS problem and a local non-perturbative problem with an internal dynamical degree of freedom (the lattice distortion). Schüttler *et al.* [67] claim to have made some progress by numerical methods and to verify that the enhancement is significant.

9.13. The Hubbard Model in Strong Coupling

The best evidence that superconductivity in CuO-based superconductors has something to do with the Hubbard model comes from diffraction experiments showing ordered antiferromagnetism in La_2CuO_4 [54] and $YBa_2Cu_3O_{6+x}$ [55] (see Chapter 8). In this section the Hubbard model will be discussed for large U and nearly half-filling. In particular, the relation to a Heisenberg antiferromagnet will be explained. Anderson [1] has recently reviewed this subject.

The Hubbard model is defined by three specifications: (1) a Hamiltonian

$$H = t \sum_{<ij>,\sigma} c^{+}_{i\sigma} c_{j\sigma} + U \sum_{i} n_{i\uparrow} n_{i\downarrow} \quad , \tag{9.70}$$

(2) a lattice which defines the pattern of bonds $<ij>$ on which the electrons can hop with matrix element t, and (3) a filling factor x, which I define such that $x = 0$ is the half-filled case (a standard reference point) and x counts the number of holes as in $La_{2-x}Sr_xCuO_4$. The Coulomb energy U inhibits double occupancy. I now state what can be called the "Hubbard" hypothesis:

> **"Hubbard" Hypothesis: The fundamental physics of the oxide superconductors is contained in the Hamiltonian (9.70) on a 2-*d* square lattice for small numbers of holes $0 \leq x \leq 0.2$.**

Since the properties of the Hubbard model in this limit are not yet under good theoretical control, there is no proof or disproof, and testing the hypothesis is regarded as a key issue. Many people suspect that it is wrong, but that a modified Hubbard hypothesis (with added features such as electron-phonon effects, two bands, or next neighbor interactions) is right, with superconductivity still driven by the Coulomb U term. This does not diminish the urgency of testing the simple hypothesis because it may be impossible to work out the consequences of modifications without first having a better understanding of the unmodified model.

There are various versions of the Hubbard Hypothesis, some of which claim ultimately to give "weak" superconductivity via spin fluctuations. In this section and the next I will focus on versions which hope to generate "strong" superconductivity, by first condensing into a new state (claimed by Anderson [68] to be a "resonating valence-bond" (RVB) state). This new state is not a conventional Fermi liquid, so the low T superconducting condensate is not conventional BCS.

One approach to understanding the strong-coupling Hubbard model is exact diagonalization of finite-size systems. Let us consider the $x = 0$ case. With one site and one electron there are two states, $\uparrow$ and $\downarrow$, both having zero energy. With two sites and two electrons there are six states, each described by a Slater determinant: $|1\rangle = (1\uparrow,1\downarrow)$, $|2\rangle = (2\uparrow,2\downarrow)$, $|3\rangle = (1\uparrow,2\downarrow)$, $|4\rangle = (1\downarrow,2\uparrow)$, $|5\rangle = (1\uparrow,2\uparrow)$, and $|6\rangle = (1\downarrow,2\downarrow)$. If t is neglected compared to U, H is diagonal in this basis. States 1 and 2 have energy U, and 3-6 have energy 0. After t is turned on, a triplet of states of energy 0 remains, ($|5\rangle$, $|6\rangle$, and $2^{-1/2}(|3\rangle + |4\rangle)$) and one state, $2^{-1/2}(|1\rangle - |2\rangle)$, has energy U. The singlet state plays a special role:

$$|\mathrm{VB}\rangle = 2^{-\frac{1}{2}}(1\uparrow 2\downarrow - 1\downarrow 2\uparrow) = 2^{-\frac{1}{2}}(|3\rangle - |4\rangle) \quad . \tag{9.71}$$

It is a Heitler-London "valence bond" state. Two eigenstates are mixtures of $|\mathrm{VB}\rangle$ and ionized states $2^{-1/2}(1\uparrow 1\downarrow + 2\uparrow 2\downarrow)$, with eigenvalues $U/2 \pm \sqrt{(U/2)^2+4t^2}$. For t $\ll U$, the ground state is the $|\mathrm{VB}\rangle$ state with a weak admixture of ionized states, and the energy is $-4t^2/U$.

As the number of "atoms" N increases, the size of the space grows rapidly, as $(2N)!/(N!)^2 \sim 4^N$. Of these, exactly 2^N correspond to states with only one electron per site, giving a 2^N-fold degenerate ground state of energy 0 when t = 0. These states (denoted $|\Sigma\rangle$) are characterized by the spin orientation on each site. It is this very large ground state degeneracy which makes the problem so interesting and difficult.

For even a modest nine atom (3x3) $2d$ array, direct diagonalization of the 48,620-order matrix is not easy. A considerable simplification occurs if t is small, and the problem is truncated to the 2^N-order space (512 for N = 3x3) of singly occupied states. This truncated problem has spins at each site but no "charge" degrees of freedom. Hence it describes an insulator, as is reasonable for small enough t/U. The H-matrix is 0 in this truncated basis, and it is necessary to do first order degenerate perturbation theory in the parameter t/U. This introduces matrix elements

$$\langle \Sigma' | H_{eff} | \Sigma \rangle = \sum_n \frac{\langle \Sigma' | H | n \rangle \langle n | H | \Sigma \rangle}{E_o - E_n} \tag{9.72}$$

between the states $|\Sigma\rangle$, $|\Sigma'\rangle$ of the singly occupied subspace $E(\Sigma) = E_o = 0$. The intermediate states $|n\rangle$ always have one site doubly occupied ($E_n = U$). The resulting matrix (9.72) turns out to be the same as the matrix of

$$H_{eff} = J \sum_{<ij>} (\sigma_i \cdot \sigma_j - \tfrac{1}{4}) \tag{9.73}$$

where $\boldsymbol{\sigma} = (\sigma_x, \sigma_y, \sigma_z)$ are the Pauli matrices. When applied to the two-site problem, (9.73) has eigenvalues 0 (triply degenerate) and -J (singly degenerate) leading to the identification $J = 4t^2/U$. The additive constant -J/4 per bond is then dropped from (9.73), which is the antiferromagnetic Heisenberg Hamiltonian. In a three-dimensional simple cubic lattice, (9.73) has an antiferromagnetically ordered low temperature phase (the "Neel" state), a phase transition at $T_N \sim J$, and a high temperature paramagnetic phase with disordered local moments. This is believed to provide a correct qualitative description for systems such as NiO and CuO which are antiferromagnetic insulators at low T and remain insulating even above T_N irrespective of whether the number of electrons per unit cell is even or odd. (Alternative descriptions also exist for NiO, etc; spin-density-functional band theory captures at least some of the physics correctly (see [69])).

In one dimension, the Heisenberg antiferromagnet has no long range spin order, even at $T = 0$. The ground state is given by the famous "Bethe *ansatz*" [70,71], and is very complicated. It is easy to see why ordered spins are not favored, by making variational estimates of ground state energies using simple trial wavefunctions. The simple antiferromagnetic state $|\text{Neel}\rangle$ is $|1_\uparrow, 2_\downarrow, 3_\uparrow, 4_\uparrow, \ldots\rangle$ and has energy per site -J/4, which is also the energy per bond, and comes entirely from the $s_{iz}s_{jz}$ part of (9.73). A "valence bond" state $|\text{VB}\rangle$ of the form $|\text{VB}(1,2), \text{VB}(3,4), \ldots\rangle$ does better. Each valence bond has energy -3J/4, with equal contributions from the three Cartesian components $s_{i\alpha}s_{j\alpha'}$. Since there are two sites per bond, the energy per site is -(3/8)J, 50% lower than the Neel state. The "quantum fluctuations" carried by the operator

$$\tfrac{1}{2}(s_i^+ s_j^- + s_i^- s_j^+) = s_{ix}s_{iy} + s_{jx}s_{jy} \tag{9.74}$$

are responsible for destroying the ordered state for small spin ($s = \tfrac{1}{2}$) and low dimensionality. In a two dimensional square lattice, the ground state of Eq. (9.73) is not known. It may be ordered antiferromagnetically, but it is quite certain that no long-range magnetic order persists above $T = 0$.

For a *2d* triangular lattice, the best "Neel" state has three spin sublattices oriented at 120°. It is not likely that much long range magnetic order occurs in the ground state. The triangular geometry causes "frustration". Anderson [72,73] argued that in this case the ground state would be a "resonating valence bond" (RVB) state

$$|\mathrm{RVB}\rangle = \sum_{\mathrm{P}} \mathrm{c(P)}\,|\mathrm{VB},(1,2),\ \mathrm{VB}(3,4),\ \ldots\rangle \tag{9.75}$$

where the sum runs over all permutations P of bond arrangements. It is possible that such a state is also the ground state of the *2d* square lattice. Even if this is not true, still when second-neighbor exchange coupling is introduced or else a finite hole-concentration x, these sources of frustration could stabilize an RVB state. Anderson [68] has proposed that an RVB state with finite x would be a novel kind of superconductor. The scale of T_c would be related to J $\sim$ 1000 K rather than $\omega_D \sim$ 200 K.

9.14. RVB Theory

This theory starts with the hypothesis that for insulating La_2CuO_4, the ground state is of the RVB form (9.75). Next it is necessary to add holes, or remove electrons. With a single hole, the wavefunction becomes

$$|\mathrm{RVB},\ 1\ \mathrm{hole}\rangle = \sum_{\mathrm{P}} \mathrm{c}'(\mathrm{P})\,|\mathrm{H}(1)N\alpha(2)\ \mathrm{VB}(3,4)\ \mathrm{VB}(5,6)\ \ldots\rangle \tag{9.76}$$

Note that the missing electron alters two sites: H(1) means a missing electron on site 1, $N\alpha(2)$ means that site 2 is neutral but not valence-bonded and thus has a free spin $\alpha = \pm \tfrac{1}{2}$. The sum runs over all permutations of sites, and c′(P) are variational.

Actually, the starting hypothesis has been overstated and can be weakened. In the pure undoped case, La_2CuO_4 may have a Neel ground state (strongly assisted by weak *3d* interplanar spin coupling). Beyond a critical doping level (e.g. a few percent Sr, Ba, or Ca atoms in place of La) the Neel state disappears. The hypothesis says that states like (9.76) take over. It has not been proved that such a state occurs in the large U, small x Hubbard model. It may be necessary to admit some next-neighbor interactions or some electron-phonon effects to stabilize the state (9.76).

It was pointed out by Kivelson, Rokhsar, and Sethna (KRS, [74]) that the state (9.76) can be thought of as embodying two "topological defects" (or solitons): A site of charge $+|e|$ which behaves like a Boson, and a neutral site of unpaired spin which behaves like a Fermion. Although necessarily created together, these two "particles" wander separately through the crystal. Their statistical (Fermi/Bose) properties were explicitly verified by checking the sign of wavefunctions like (9.76) with two identical defects present, as the locations of defects were moved adiabatically until they switched places.

At this point the scenario of Anderson *et al.* [75,76] and of KRS diverge from each other. In the latter picture the neutral Fermions ("spinons") have a finite creation energy (of order J or t^2/U) and thus are not present except as virtual pairs in the ground state; a second hole added to the state (9.76) will bind to the "spinon" creating a second charged Boson and leaving no Fermion degrees of freedom. Superconductivity is then expected as a Bose condensation of the charged Bosons.

The picture of Anderson involves a gapless spinon spectrum with a ground state of N occupied spinon states and a spinon Fermi surface. The nesting property of this "pseudo-Fermi surface" accounts for the antiferromagnetism. Doping destroys the antiferromagnetism but does not shrink the pseudo Fermi surface; charged Bosons are added which Bose condense. The superconducting transition is supposed to be a pairing transition at still lower temperature within the Bose condensate [77].

These scenarios are changing rapidly, as is the experimental situation, so it is too early to perceive the shape into which the final picture may evolve.

9.15. Oxygen Holes and Copper Spins

Various experiments suggest that carriers in Cu-O superconductors are located primarily on oxygen atoms, rather than in hybridized Cu(3d) Oxygen (2p) antibonding bands as suggested by band theory. For example, Cu-K edge X-ray absorption (1s → 4p transitions) seems to give a clear fingerprint of the valence state of Cu. Tranquada [78] find that pure La_2CuO_4 has the fingerprint of the Cu^{2+} ($3d^9$) ion, and does *not* evolve toward Cu^{3+} as carriers are introduced by Sr doping ($La_{2-x}Sr_xCuO_4$ with $0 \leq x \leq 0.2$). The oxygen 1s to 2p transitions have been explored by Nücker *et al.* [79] using electron energy-loss spectroscopy in $La_{2-x}Sr_xCuO_4$ and $YBa_2Cu_3O_{7-y}$. Their spectra seem to show a filled O(2p)-shell for values of x,y in the insulating region, but holes in this shell, roughly proportional to x and y, for conducting samples. This leads to the view that the Cu sublattice remains insulating with $s = \frac{1}{2}$ Cu(d^9) correlations even in samples which are metallic. Long range antiferromagnetic (AF) order is lost, short range spin order of some kind is presumed to remain.

Two interesting models have been proposed which treat the oxygen holes as a conventional Fermi liquid and achieve BCS pairing by an attractive interaction caused by Cu spins. Both models, like the "spin-bag" model mentioned in Sec. 9.10, derive the attraction from the fact that the oxygen hole alters the magnetic energy of the Cu sublattice in such a way that it is preferable to bunch the carriers spatially. The specific mechanisms, however, are quite different.

Birgeneau, Kastner, and Aharony (BKA, [80]) require that superconducting samples should exhibit fluctuating AF order with a correlation length of several planar Cu-Cu spacings; this is quite different in principle from the RVB spin organization, but available neutron scattering data [81] do not seem to distinguish these pictures unambiguously. The BKA model says that the oxygen hole (carrying $s = \frac{1}{2}$) tends strongly to orient its two neighboring Cu spins into a $S = 3/2$ ferromagnetic cluster. (See Fig. 2a). A second such cluster of opposite S will then prefer to be close because the pair of clusters is less disruptive to fluctuating AF order than either one individually. The source of the ferromagnetic alignment energy of Fig. 2a is Hund's rule energy (from ordinary

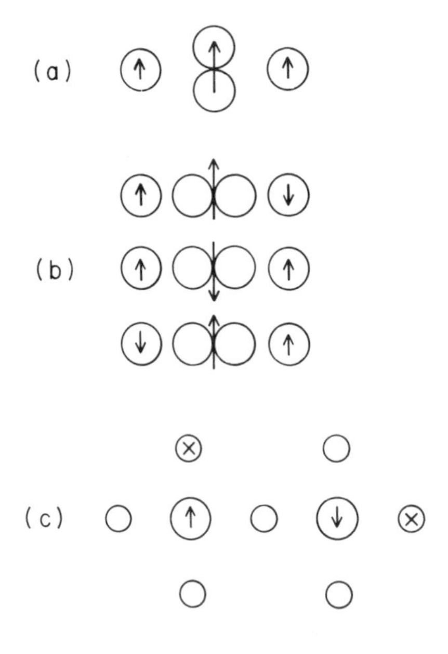

Fig. 9.2. A Cu-O-Cu group in the CuO_2 plane, showing possible arrangements of the spin of a hole on oxygen coupling to its two Cu neighbor spins. (a) model of Birgeneau, Kastner and Aharony (1988); (b) model of Emery and Reiter (1988); (c) possible sites of oxygen holes denoted by x which experience attraction due to enhanced superexchange.

Coulomb exchange energy) which dominates over AF exchange (t^2/U effects) because the oxygen hole is assumed to lie in a p_y-orbital (where x is the Cu-O-Cu bond axis and x-y is the CuO_2 plane). This orbital is orthogonal to Cu ($d_{x^2-y^2}$) and the hopping matrix element t vanishes by symmetry. The p_y-orbital is assigned the highest energy based on an electrostatic calculation. (See also Guo *et al.*, [82]).

Emery and Reiter (ER, [83]) assign the oxygen hole to the p_x-orbital (see Fig. 2b) which is favored by antibonding covalency as in band theory. Then one expects an antiferromagnetic interaction J′ between Cu spins and O spins, which should be larger than the previous Cu-Cu coupling J. Spin configurations shown in Fig. 2b are expected for a Cu-O-Cu complex, dominated by the $\uparrow\downarrow\uparrow$ state where Cu atoms are ferromagnetically arranged. Just as in the BKA model, this complex disrupts local AF order. A second such complex, dominantly $\downarrow\uparrow\downarrow$, would be attracted in order to minimize the disruption to AF energy. However, ER identify a separate mechanism which causes attraction and makes no specific demand on the nature of the fluctuating magnetic order. The $2d$ $s = \frac{1}{2}$ Heisenberg antiferromagnet is apparently a marginal case, so theory provides little guidance as to the type of magnetic order expected in doped weakly $3d$ systems, and the ER theory is safer for making fewer assumptions. Their new mechanism is "enhanced superexchange". The Cu-Cu exchange J is only partly given by t^2/U, because the hopping parameter t to go from Cu to Cu is small. Instead, the principal route is to hop indirectly via the intervening oxygen ion using a larger matrix element t′. But then the process is fourth order, and complicated expressions result, such as

$$J' \sim \frac{t'^4}{(\varepsilon_p+V)^2}\left[\frac{2}{U_d}+\frac{4}{U_p+2\varepsilon_p}\right] \tag{9.77}$$

where ε_p is the O(2p) upper valence orbital energy, relative to the copper ε_d orbital energy chosen to be zero; U_p and U_d are onsite (Hubbard) Coulomb repulsions on O and Cu respectively, and V is a first neighbor Coulomb repulsion ("extended" Hubbard model). Now suppose that there is an oxygen 2p hole on one of the six oxygens adjacent to one (but not both) of the copper atoms in question (see Fig. 2c). Then by the nearest neighbor repulsion V, this raises the energy ε_d of a Cu-hole and correspondingly diminishes some of the denominators in Eq. (9.77). If a second O(2p) hole occurs in another of the six atoms (see Fig. 2c), this will further diminish denominators in (9.76). The effect is bigger than the additive energy changes of two single oxygen satellite holes, and causes attraction of oxygen holes located at second neighbor sites. For reasons mentioned before, this interaction is more effective for antiparallel spins and this favors singlet BCS pairing.

9.16. Postscript

In the last few months there has been an interesting controversy concerning the statistics (i.e. the behavior under interchange) of charge-carrying and spin-carrying excitations of the RVB state. The situation is summarized in table II.

Authors	Ref	holon	spinon
KRS	74	+	-
KL	84	$\pm i$	$\pm i$
HL; RC	86,87	-	+
K	88	+	?

Table II. Phase change $e^{i\phi}$ of the RVB wavefunction after interchange of two identical quasiparticles.

As mentioned in sec. 9.14, Kivelson, Rokhsar, and Sethna (KRS [74]) were the first to notice that spin and charge degrees of freedom were likely to decouple in an RVB state. Their assignment of Bose statistics to $S = 0$, $q = |e|$ "holons" and Fermi statistics to $S = \frac{1}{2}$, $q = 0$ "spinons" agrees with naive expectations, but has been challenged. Kalmeyer and Laughlin (KL [84]) have proposed "fractional" statistics, which can be constructed in two dimensions [85]. Haldane and Levine (HL) [86] and Read and Chakraborty (RC) [87] have argued for Bose and Fermi, but in reverse order to KRS. Finally, Kivelson (K) [88] has reexamined these arguments. He agrees with HL and RC when the hole hopping term is ignored, but finds that the holon turns back into a boson once it is given a reasonable hopping matrix element. The spinon can go either way. This controversy seems to have sharpened the understanding of the meaning of statistics in two dimensions. Three dimensions apparently is different and lacks the rich possibilities found in $d = 2$. Kivelson's [88] argument also clarifies the subtle issue of flux quantization in units of $h/2e$ even when the object condensing is a charge $|e|$ boson rather than charge $-2e$ Cooper pair. Experiments clearly indicates $h/2e$.

An experimental discovery [89] worth noting is superconductivity at $T_c \sim 30K$ in the system $Ba_{0.6}K_{0.4}BiO_3$. Like the $Ba(Pb_{1-x}Bi_x)O_3$ system, this is an approximately cubic perovskite with a metal-insulator transition but no sign of local magnetic moments. The lack of a unifying theory for these diverse but evidently related high T_c systems is an embarrassment. It is safe to say that the discovery of high T_c superconductors has already altered our view of physics, and that this process will continue.

Acknowledgements

I thank V. Emery, S. Kivelson, W. E. Pickett, J. Serene and C. N. Yang for helpful discussions. This work was supported in part by NSF grant no. DMR 84-20308.

References

[1] P. W. Anderson in *Frontiers and Borderlines in Many Particle Physics* (Varenna Summer School, to be published).
[2] T. M. Rice *Z. Phys.* B**67**, 141 (1987).
[3] P. Fulde *Physica C* **153-155**, 1769 (1988).
[4a] *Novel Superconductivity*, ed. by S. A. Wolf and V. Z. Kresin (Plenum, New York, 1987).
[4b] *Physica C* **153-155** (1988).
[5] P. B. Allen and B. Mitrovic in *Solid State Physics*, v.**37** pp. 1 (1982).
[6] D. Jerome and F. Creuzet in *Novel Superconductivity*, ed. by S. A. Wolf and V. Z. Kresin (Plenum, New York, 1987) p. 103.
[7] Z. Fisk, D. W. Hess, C. J. Pethick, D. Pines, J.L. Smith, J. D. Thompson and J. O. Willis *Science* **239**, 33 (1988).
[8] S. Uchida, K. Kitazawa and S. Tanaka, *Phase Transitions* **8**, 95 (1987).
[9] G. Shirane *et al.* *Phys. Rev. Lett* **59**, 1613 (1987).
[10] A. Aharony, R. J. Birgeneau, A. Coniglio, M. A. Kastner and H. E. Stanley *Phys. Rev. Lett.* **60**, 1330 (1988).
[11] T. M. Rice and L. Sneddon, *Phys. Rev. Lett.* **47**, 689 (1981).
[12] J. Bardeen, L.N. Cooper and J. R. Schrieffer *Phys. Rev.* **108**, 1175 (1957).
[13] L. N. Cooper *Phys. Rev.* **104**, 1189 (1956).
[14] J. Bardeen and D. Pines *Phys. Rev.* **99**, 1140 (1955).
[15] G. M. Eliashberg *Zh. Eksp. Teor. Fiz.* **38**, 966; **39**, 1437 [Sov. Physics JETP **11**, 696; **12**, 1000] (1960).
[16] P. W. Anderson *Basic Notions of Condensed Matter Physics*, (Benjamin/Cummings, Menlo Park, CA, 1984) p. 16.
[17] C. N. Yang, *Rev. Mod. Phys.* **42**, 1 (1962).
[18] D. Forster *Hydrodynamic Fluctuations, Broken Symmetry, and Correlation Functions* (W.A. Benjamin, Reading, MA, 1975).

[19] G. Rickayzen in *Superconductivity*, ed. by R. D. Parks (Dekker, New York) p. 51, (1969).

[20] P. B. Allen in *Modern Trends in the Theory of Condensed Matter*, ed. by A. Pekalski and J. Przystawa (Lect. Notes in Physics **115**, Springer, Berlin, 1980) p. 388.

[21] W. Weber *Phys. Rev. Lett.* **58**, 1371 (1987).

[22] W. Weber and L. F. Mattheiss *Phys. Rev.* **B37**, 599 (1988).

[23] A. B. Migdal *Zh. Eksp. Theor. Fiz.* **34**, 1438 [Sov. Phys. JETP **7**, 996 (1958)].

[24] S. Doniach and E. H. Sondheimer *Green's Functions for Solid State Physicists* (W.A. Benjamin, Reading, MA, 1974).

[25] P. G. de Gennes *Superconductivity of Metals and Alloys* (W.A. Benjamin, New York, 1966).

[26] H. Fröhlich *Proc. Roy. Soc. (London)* A **223**, 296 (1954).

[27] W. L. McMillan *Phys. Rev.* **167**, 331 (1968).

[28] P. B. Allen and R. C. Dynes *Phys. Rev.* **B12**, 905 (1975).

[29] P. B. Allen in *Dynamical Properties of Solids*, ed. by G. K. Horton and A. A. Maradudin (North-Holland, Amsterdam, 1980) v.3 pp. 95-196.

[30] A. E. Karakozov and E. G. Maksimov *Zh. Eksp. Teor. Fiz.* **74**, 781 [Sov. Phys. JETP **47**, 358 (1978)].

[31] D. Rainer and F. J. Culetto *Phys. Rev.* **B19**, 2540 (1979).

[32] W. A. Little *Phys. Rev.* **134**, A1416 (1964).

[33] V. L. Ginzburg *Zh. Eksp. Theor. Fiz.* **47**, 2318 (1964) [Sov. Phys. JETP **20**, 1549 (1965)].

[34] V. L. Ginzburg and D. A. Kirzhnits eds., *High-Temperature Superconductivity* (Consultants Bureau, New York, 1982).

[35] J. Ruvalds *Adv. Phys.* **30**, 677 (1981).

[36] W. A. Little in *Novel Superconductivity*, ed. by S. A. Wolf and V. Z. Kresin (Plenum, New York, 1987) p. 341.

[37] B. T. Matthias *Comments on Sol. State Phys.* **3**, 93 (1970).

[38] H. Rietschel and L. J. Sham *Phys. Rev.* **B28**, 5100 (1983).

[39] M. Grabowski and L. J. Sham *Phys. Rev.* **B29**, 6132 (1984).

[40] D.A. Papaconstantopoulos *Handbook of the Band Structure of Elemental Solids* (Plenum, New York, 1986).

[41] P. B. Allen, W. E. Pickett, K. M. Ho and M. L. Cohen *Phys. Rev. Lett.* **40**, 1532 (1978).

[42] P. B. Allen, W. E. Pickett and H. Krakauer *Phys. Rev.* **B36**, 3926 (1987).

[43] D. Allender, J. Bray and J. Bardeen *Phys. Rev.* **B7**, 1020 (1973).

[44] D. L. Miller, M. Strongin, O. F. Kammerer and B. G. Streetman *Phys. Rev.* **B13**, 4834 (1976).

[45] J. C. Inkson and P. W. Anderson *Phys. Rev.* **B8**, 4429 (1973).

[46] B. I. Halperin and T. M. Rice *Rev. Mod. Phys.* **40**, 755 (1968).

[47] S. C. Lo and K. W. Wong *Nuovo Cim.* **10B**, 361 and 383 (1972).

[48] A. A. Abrikosov (1978) *Pisma. Zh. Eksp. Teor. Fiz.***27**, 235 [*JETP*

Letters **27**, 219 (1978)].

[49] T. T. M. Palstra, A. A. Menovsky, J. van den Berg, A. J. Dirkmaat, P. H. Kes, G. J. Nienwenhuys and J. A. Mydosh *Phys. Rev. Lett.* **55**, 2727 (1985).

[50] M. B. Maple, J. W. Chen, Y. Dalichaouch, T. Kohara, C. Rossel, M. Torikachvili, M. W. Elfresh and J. D. Thompson *Phys. Rev. Lett.* **56**, 185 (1986).

[51] M. T. Béal-Monod, C. Bourbonnais and V. J. Emery *Phys. Rev.* **B34**, 7716 (1986).

[52] D. J. Scalapino, E. Loh, Jr. and J. E. Hirsch Phys. Rev. **34**, 8190 (1986).

[53] K. Miyake, S. Schmitt-Rink, and C. M. Varma *Phys. Rev.* **B34**, 6554 (1986).

[54] D. Vaknin *et al. Phys. Rev. Lett.* **58**, 2802 (1987).

[55] J.M. Tranquada *et al. Phys. Rev. Lett.* **60**, 156 (1988).

[56] J. R. Schrieffer, X.-G. Wen and S.-C. Zhang *Phys. Rev. Lett.* **60** 944 (1988).

[57] B. K. Chakraverty and C. Schlenker *J. de Physique* **37**, suppl. C4, 353 (1976).

[58] B. K. Chakraverty *J. de Physique Lett.* **40**, L99 (1979).

[59] R. A. Ogg, Jr. *Phys. Rev.* **69**, 243 (1946).

[60] J. M. Blatt (1974) *Theory of Superconductivity*, (Academic, New York, 1974).

[61] A. J. Leggett in *Modern Trends in the Theory of Condensed Matter*, ed. by A. Pekalski and J. Przystawa (Lect. Notes in Physics **115**, Springer-Verlag, Berlin, 1980) p. 13; J. Phys. (Paris) **41**, C7-19 (1980). P. Nozieres and S. Schmitt-Rink, *J. Low Temp. Phys.* **59**, 195 (1985).

[62] J. G. Bednorz and K. A. Müller *Z. Phys.* **B64**, 189 (1986).

[63] D. J. Scalapino, R. T. Scalettar and N. E. Bickers in *Novel Superconductivity*, ed. by S. A. Wolf and V. Z. Kresin (Plenum, NY, 1987) p. 475.

[64] P. W. Anderson *Phys. Rev. Lett.* **34**, 953 (1975).

[65] E. Simanek *Solid State Commun.* **32**, 731 (1979).

[66] C. S. Ting, D. N. Talwar and K. L. Ngai *Phys. Rev. Lett.* **45**, 1213 (1980).

[67] H.-B. Schüttler, M. Jarrell and D. J. Scalapino, in *Novel Superconductivity*, ed. by S. A. Wolf and V. Z. Kresin (Plenum, NY, 1987) p. 481.

[68] P. W. Anderson *Science* **235**, 1196 (1987).

[69] K. Terakura, A. R. Williams, T. Oguchi and J. Kübler, *Phys. Rev. Lett.* **52**, 1830 (1984).

[70] H. Bethe *Z. Phys.* **71**, 205 (1931).

[71] D. C. Mattis *The Theory of Magnetism I* (Springer, Berlin) p. 175 (1981).

[72] P. W. Anderson Mater. *Res. Bull.* **8**, 153 (1973).

[73] P. Fazekas and P. W. Anderson *Philos. Mag.* **30**, 432 (1974).

[74] S. A. Kivelson, D. S. Rokhsar and J. P. Sethna *Phys. Rev.* **B35**, 8865 (1987).

[75] G. Baskaran, Z. Zou and P. W. Anderson *Solid State Commun.* **63**, 973 (1987).

[76] P. W. Anderson, G. Baskaran, Z. Zou and T. Hsu *Phys. Rev. Lett.* **58**, 2790 (1987).

[77] J. M. Wheatley, T. C. Hsu and P. W. Anderson *Phys. Rev.* **B37**, 5897 (1988).

[78] J. M. Tranquada, S. M. Heald and A. R. Moodenbaugh *Phys. Rev.* **B36**, 5263 (1987).

[79] N. Nücker, J. Fink, J. C. Fuggle, P. J. Durham and W. M. Temmerman, *Phys. Rev.* B37, 5158 (1988).

[80] R. J. Birgeneau, M. A. Kastner and A. Aharony *Z. Phys.* B **71**, 57 (1988).

[81] R. J. Birgeneau *et al.*, preprint.

[82] Y. Guo, J.-M. Langois and W. A. Goddard III, *Science* **239**, 896 (1988).

[83] V. J. Emery and G. Reiter (preprint).

[84] V. Kalmeyer and R. B. Laughlin, *Phys. Rev. Lett.* **59**, 2095 (1987); R. B. Laughlin, *Phys. Rev. Lett.* **60**, 2677 (1988).

[85] F. Wilczek, *Phys. Rev. Lett.* **49**, 957 (1982).

[86] F. D. M. Haldane and H. Levine (preprint).

[87] N. Read and B. Chakraborty (preprint).

[88] S. Kivelson (preprint).

[89] L. F. Mattheiss, E. M. Gyorgy, and D. W. Johnson, Jr., *Phys. Rev. B* **37**, 3745 (1988). R. J. Cava, B. Battlogg, J. J. Krajewski, R. C. Farrow, L. W. Rupp Jr., A. E. White, K. T. Short, W. F. Peck Jr., and T. Y. Kometani, *Nature* **332** 814 (1988).

CHAPTER 10

Superconducting Devices

Fernand D. Bedard

10.1. Introduction

All electronics should be done at room temperature. The only reason for going to cryogenic operating temperatures is in order to achieve high performance. Traditionally, this label is characterized by very high switching speed (or bandwidth), small size, low electrical power, and reliability. These attributes have long been ascribed to superconductive electronics and yet with only modest "customer" acceptance over the past 77 years of its history. The first attempt is attributable to Dudley Buck [1] for the invention of the cryotron switching device for digital computing. This device was demonstrated to work, was later modified and developed into a "manufacturable" form, thus becoming an incipient "technology". As it was being pursued, its advantages were gradually overtaken and overwhelmed by semiconductor technology until the cryotron disappeared from view. Its inherent speed limitations were demolished by the discovery of superconductive tunneling, quasi-particle and coherent [2, 3], which launched a major attempt to evaluate devices which exploited these phenomena and then create a technology. Directed also toward digital computing switches, this work still has not gained a market-place, but neither has it been abandoned. Along the way, remarkably high-quality fabrication processes were developed which have allowed very high performance special function electronics to be built: millimeter wave mixers, detectors, amplifiers; picosecond sampling circuits; ultra-precise voltage standards; very sensitive magnetometers; dispersive delay lines and possibly low-power linear A/D converters. Now, with the discovery of "High Temperature Superconductivity" by Bednorz and Muller [4] and the follow-on work by Tanaka, Kitizawa et al. [5], Chu et al. [6], Maeda et al. [7], Herman et al. [8], it appears that there is a new oppor tunity to insert superconductivity into electronic systems, i.e., make it a technology. What has changed?

First, the institutional abhorrence for liquid helium temperature operation may well be overcome by using liquid hydrogen, neon, or nitrogen for cooling or by the opportunity to cryogenically cool a system with more efficient, smaller, more reliable refrigerators which operate at these temperatures. Second, as time has passed, technical requirements have put more pressure on the mainstream semiconductor technologies: higher speeds, lower power, smaller size, greater reliability, lower noise are being demanded. This has often forced designers to consider and use refrigeration and cooling of systems to low temperatures. As a result, it becomes reasonable to suggest and evaluate the enhancement of semiconductor electronics by the use of superconducting components which operate in a common environment (not necessarily the same temperature). Finally, there are some functions which are sought after which neither semiconductors nor superconductors alone can provide, but which a merger of the best features of both materials technologies can deliver.

For superconductivity to become an electronics technology there must be a sufficiently high "payoff" to overcome the acknowledged complication of cryogenic cooling even to "warm" temperatures such as 77 K. Furthermore, in order to intelligently assess where it may fit in and therefore what directions to take, one must know and assess the competing technology solutions, their realistic limitations and their future performance. A newcomer on the block must prove himself in competition.

10.2 Cryotron

The first superconductive electronic switching device for a large application area was introduced by Dudley Buck with his invention of the wire-wound cryotron [1]. In this device, the switching was produced by using the magnetic field suppression of superconductivity; the application was in digital computing. The basic operation is straightforward (Fig. 1). Current inserted into coil A produces a magnetic field within the coil which is sufficient to quench the superconductivity of the "gate" wire, but yet not large enough to cause the driving coil itself to become resistive. This can be accomplished by the proper choice of materials and operating temperature; in the original case the coil was made of niobium (T_c = 9.3 K) and the gate of tantalum (T_c = 4.3 K). This control drive current results in all the gate current being shunted to the path controlled by coil B. If these branched lines are themselves electrically connected to coils, then the current steered can be used to

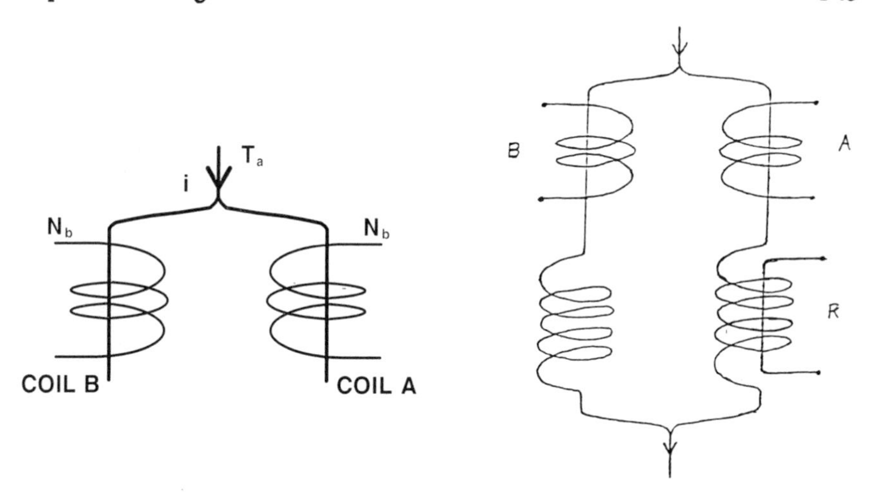

Fig. 10.1. Wire Wound Cryotron. Fig. 10.2. Cryotron Memory Cell.

control other current paths and thereby one can carry out logical operations such as "OR", "AND". In addition, if the devices are connected as in Fig. 2, a memory cell can be made. The input current I_g divides equally between the two legs, since their inductances are the same. Control current into coil A will cause I_g to transfer into the left leg with an L/R time constant. Upon removal of A (first), then I_g, a persistent circulating current is set up, CCW. To "read", one simply reapplies I_g and reads the resistive/superconductive state of sensor R; in this case, with no current through the right leg of the cell, R will be superconductive. Clearly, had coil B been driven at the start, R would be resistive in the read-out state. Notice that there is no power dissipated in the standby state and also that the persistent current will be maintained in the memory loop indefinitely and not be affected by repeated "read-outs". These attractive features are overwhelmed by the low performance character and obvious manufacturing problems for a large system (as viewed today, but not necessarily as seen in the time frame of the 1950s).

The first issues of concern are device and circuit speed, which will be determined by the time required for the niobium coil to generate a magnetic field of sufficient strength to exceed the critical field of the tantalum gate at the operating temperature (Fig. 3) [16]. The field strength H_{cc} generated by the solenoid is:

$$H_{cc} = I_c n \quad , \tag{10.1}$$

where H_{cc} is in ampere-turns/meter, I_c is in amperes, and n is the number of turns/unit length.

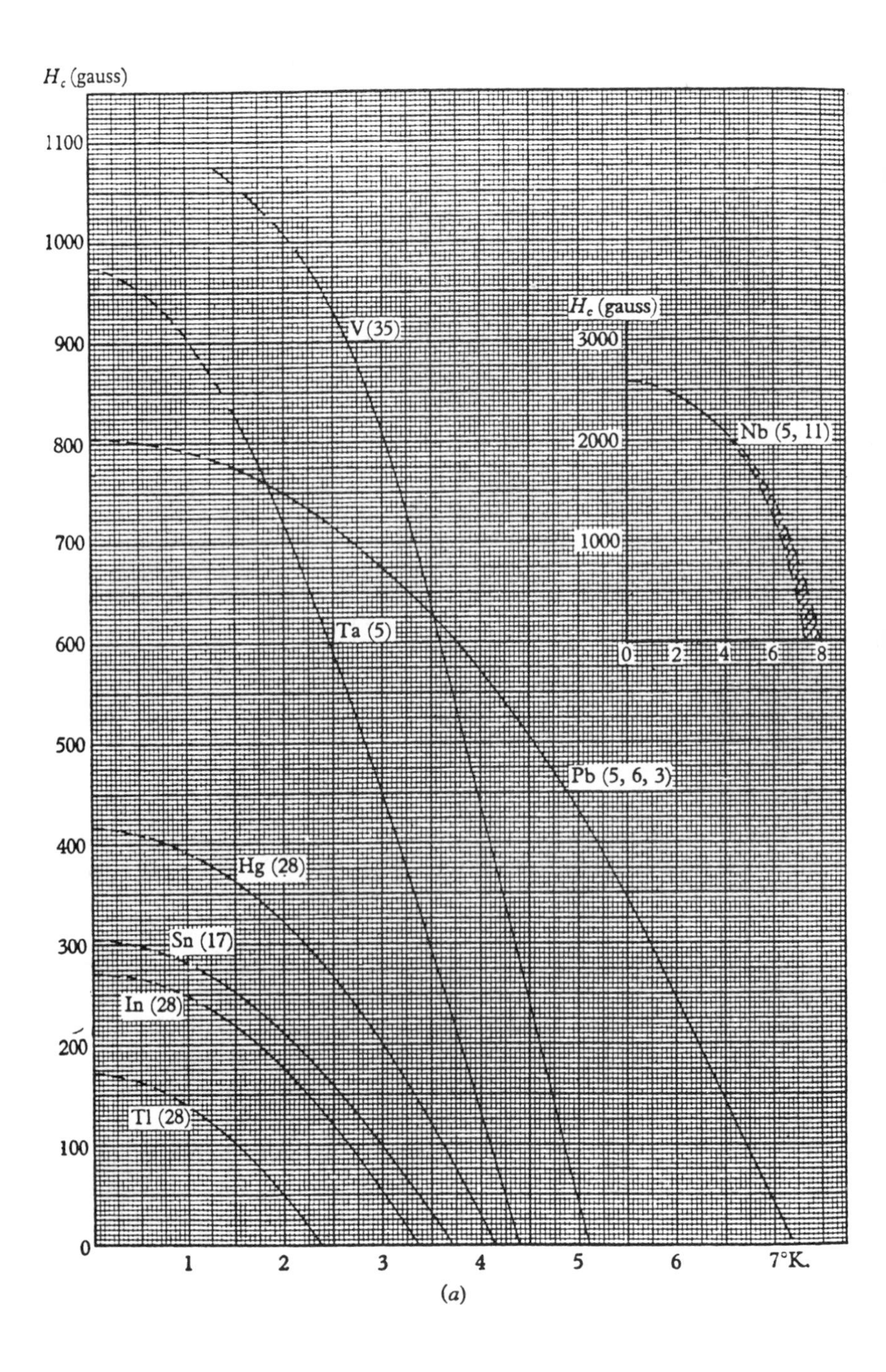

Fig. 10.3. Critical magnetic field H_c versus T_c (from D. Schoenberg [16]).

In any logic or memory circuit configuration this solenoid will be driven by, or will drive, a gate wire whose resistance will determine the switching time, L/R, of the circuit. The inductance L of a single layer solenoid is

$$L = \frac{\mu_o \pi d^2 n^2 \ell}{4} , \qquad (10.2)$$

where d is the average diameter of the coil, and ℓ is the length of the coil (meters).

If the gate diameter is approximately d, then its resistance is

$$R(\text{ohms}) = \frac{\rho \ell}{A} = \frac{4\rho\ell}{\pi d^2} ,$$

where ρ is the resistivity. Thus

$$t = \frac{L}{R} = \mu_o \frac{\pi^2 d^4 n^2}{16\rho} .$$

For n = 100 turns/cm, $\rho \simeq 5\text{x}10^{-7}$ Ω-cm and d = 0.025 cm (0.010″ dia.), we get t $\simeq$ 60 μsec—not very impressive!

Note that the full resistance of the gate is not immediately reached, which adds to the switching time spread of the logic gates. In the case of a wire-wound cryotron, reducing d is not realistic and reducing n encounters another obstacle. In practice, a control solenoid will not only be driven by a gate resistor, it will also be required to drive a gate. Thus, unless one inserts a current amplifier (very unrealistic) then we must demand that the control current required to switch the gate be smaller than the current being switched, i.e., we want current gain;

$$G \equiv \frac{I_g}{I_c} \geq 1 . \qquad (10.3)$$

To appreciate the implications of this, let us assume that the gate current may be increased until it produces a magnetic field at its wire surface equal to the gate's own critical field at the operating temperature, i.e.,

$$H_{cg} \approx \frac{I_g}{\pi d} . \qquad (10.4)$$

Then, comparing H_{cg} with H_{cc}, we get

$$\frac{H_{cc}}{H_{cg}} = \frac{n I_c \pi d}{I_g} \qquad (10.5)$$

and

$$G = \frac{I_g}{I_c} = \pi n d \quad .$$

In practice, Eq. (10.4), Silsbee's hypothesis [9], is never completely satisfied, even in type-I superconductors. That is, the gate current is always less than that calculated from (10.3), so that from our previous device parameters,

$$G = \pi(200)\,(0.025) = 5\pi$$

is not realized, but something in the range of 2–4 is more likely. Finally, and most importantly, the field generated by the control current must exceed the critical field H_c of the gate wire. Even though the gain is greater than 1, no real system will be designed to switch logic devices at the ampere current range for each gate; the power dissipation, particularly at 4 K, would be prohibitive. Since the energy stored in *each cell*, $\frac{1}{2}Li^2$, will be dissipated in the gate resistance every clock cycle, we have from Eq. (10.2):

$$L = \mu_o \frac{\pi d^2 n^2 \ell}{4} \quad .$$

For our previous parameters, and $\ell \simeq 1$ cm, $L \simeq 10^{-8}$ H. Then the power per device, p, for a representative $I_c \simeq 300$ ma, becomes $p = \frac{1}{2}LI_c^2 \simeq 5 \times 10^{-10}$ Joules/cycle/device.

The total power $P = pN/\tau$ for $N = 10^5$ devices (a small size) and $\tau = 10^{-4}$ seconds (very slow) becomes $P \simeq 1$ watt.

Clearly, L and I_c *both* must be reduced for time constant and power dissipation reasons. Reducing I_c demands that the temperature of operation be raised (to allow a smaller H_c), which is generally not convenient.

For all of the above reasons and for credibility in large scale fabrication of these devices, it was clearly necessary to change the structure from a wire-wound device to a thin film one. This proposal was made in 1957 [10] and considerable work took place to choose materials, fabrication techniques, and circuits, and to characterize them in the hope that a much more attractive technology would result. The basic concept of the current switch stayed unchanged, but was implemented with a "crossed film" control line and gate (Fig. 4). The electrical changes were significant, and the performance changes were likewise large.

Once again, a magnetic field created by the control line current is used to make the gate line resistive and transfer the gate current to another path which has zero resistance. In this configuration the device inductance is greatly reduced for two separate reasons. First, the inductance of the control line film now

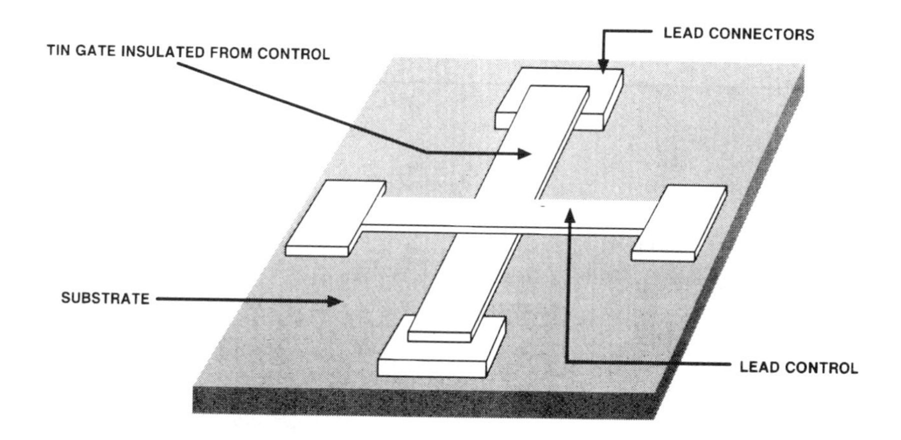

Fig. 10.4. Cross-Film Cryotron.

becomes:

$$L = \frac{\mu_o}{2\pi} \ell \, ln\left[\frac{2\ell}{w}\right] \quad , \tag{10.6}$$

where t is the insulator thickness, ℓ is the line length, and w is the control line width (this neglects the effect of the gate-control film crossing).

If we place a ground plane beneath the gate line we can significantly reduce this inductance while increasing the field strength at the gate (Fig. 5). The magnetic field is now confined to the space between the control line and the ground plane, which produces a "uniform" field with the intensity

$$B = \mu_o H \simeq 4\pi \times 10^{-7} \frac{I_c}{w} \quad , \tag{10.7}$$

where **B** is in teslas and I_c is the control current. Now,

$$L \simeq \frac{\mu_o t \ell}{w} \qquad \text{(when } t \ll w, \ell) \quad .$$

The parameter values d = 0.3 μm, ℓ = 1.5 mm, w = 100 μm, and t = 0.3 μm inserted into Eq. (10.7) produce L = 7 picohenry versus 10 nanohenries for Eq. (10.6).

The magnetic field at our disposal must be determined for "practical" current levels. From Eq. (10.7), with I_c = 100 ma and w = 100 μm, we find

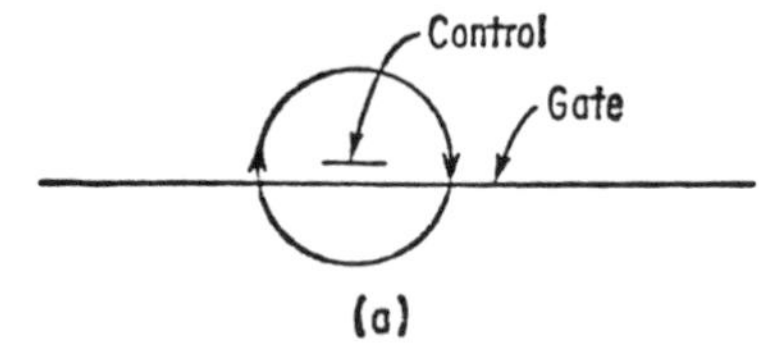

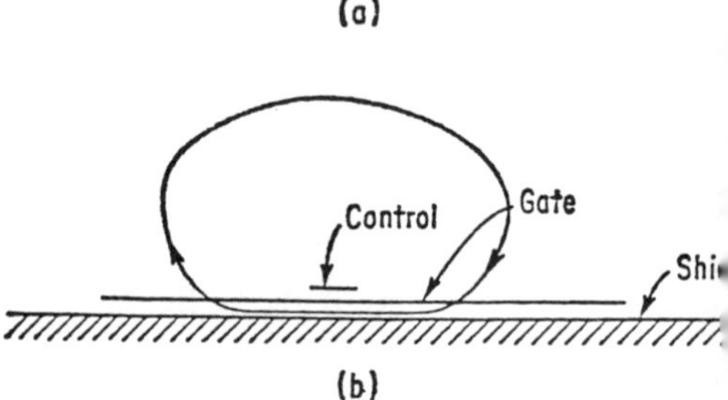

Fig. 5 Magnetic field distribution for thin film line. a) Field without Ground Plane. b) Field with Ground Plane. (From J. W. Bremer [14].)

$$B = \frac{4\pi \times 10^{-7}(0.1)}{10^{-4}} \simeq 10^{-3}\ \mathrm{T}\ (10\ \text{gauss}) \quad . \tag{10.8}$$

This magnetic field strength is acceptable for some gate materials, such as tin, lead, or indium, where again type-I behavior allows a reasonably sharp resistive transition (Fig. 6) [10]. Nevertheless, this transition is broad enough that, for the gate resistance to reach half its normal value, 10% to 100% more current is called for than the $\mathbf{B}_c$ vs T_c curve would indicate.

Our next concern is the gate resistance which may be obtained. The region of the gate line which is made normal has a resistance

$$R = \frac{\rho w}{W d} \quad , \tag{10.9}$$

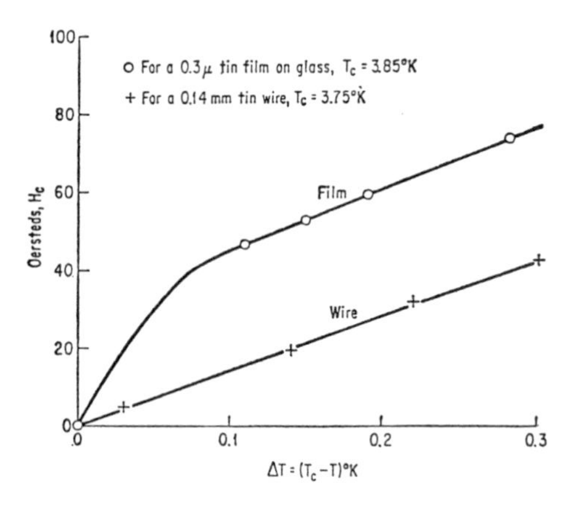

Fig. 10.6. Critical field, H_c, parallel to the sample axis, versus temperature difference (from J. W. Bremer [14]).

where W = gate line width, d = thickness, and ρ = resistivity. The time constant becomes

$$\tau = \frac{L}{R} = \frac{4\times 10^{-7} t d \ell W}{\rho w^2} ,$$

which for ℓ = 1.5 W, reduces to

$$\frac{L}{R} = \frac{6\pi \times 10^{-7} W^2 t d}{\rho w^2} . \quad (10.10)$$

For example, for our previous parameters d = 0.3 μ, W = 1 mm, w = 100 μm, t = 0.3 μm, and $\rho = 6\times 10^{-7}$ Ω -cm, we get

$$\frac{L}{R} = \simeq 3\times 10^{-10} \text{ s} .$$

This result is overly optimistic since the actual gate resistance vs. control current curve is not abrupt (Fig. 7) [15]. Once again, this device also provides current gain for the right design parameters. This gain comes about due to the high field produced by the narrow control line (width w) over the wide gate line (width W). Introducing corrections for sample size (film thickness) relative to the penetration depth (at the operating temperature) one can calculate the current gain, and show that

$$G = \frac{I_{cg}}{I_{cc}} \simeq \frac{2W}{\omega} .$$

This would indicate a gain G $\simeq$ 20 for our example, whereas G $\simeq$ 6 experimentally. Thus L/R becomes more like 5 nanoseconds.

Finally, the power dissipation needs to be estimated. In this case, the situation is much more favorable:

$$p = \tfrac{1}{2}(7\times 10^{-12})(10^{-1})^2 \simeq 4\times 10^{-14} \text{ J} .$$

For N = 10^6 devices and T = 1 μsec (clock period),

$$P \simeq (4\times 10^{-14})(10^6)(10^6) = 4\times 10^{-2} \text{ W} ,$$

an attractively small number except that for these device and system speeds the refrigeration power (1000×) and complexity makes the solution now only competitive with room temperature semiconductors.

It is reasonable to ask whether high temperature superconductivity makes a difference in the conclusion reached. First, the new material resistivity is ~ 200 $\mu\Omega$-cm versus 0.5 $\mu\Omega$-cm for the elemen-

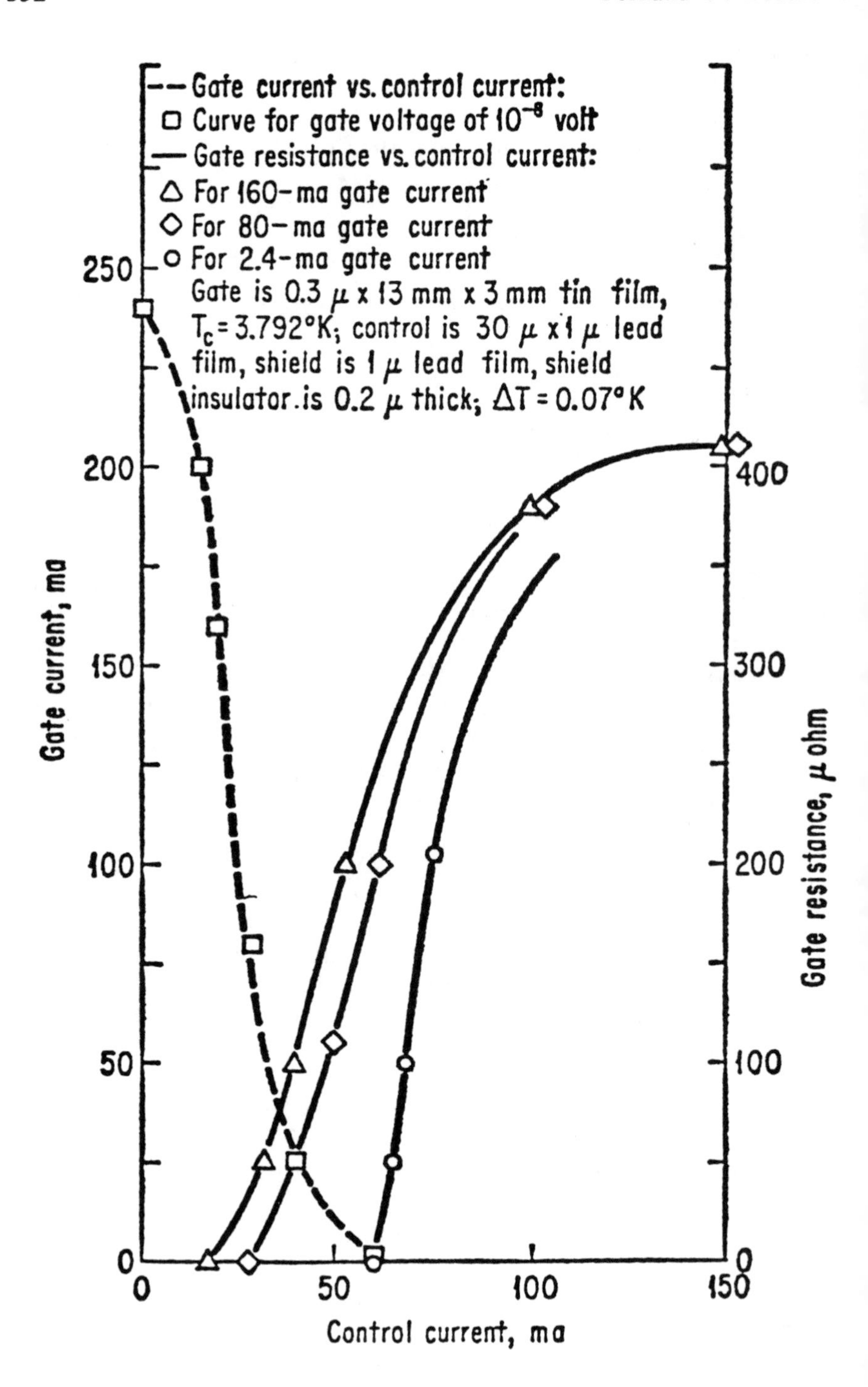

Fig. 10.7. Transfer function for the shielded Cryotron (from J. W. Bremer [14]).

tary superconductors. Second, modern lithography allows linewidths in the micron range and third, the refrigeration complexity is greatly simplified.

From Eq. (10.10), for example, the L/R time constant of the "new" device would improve by $1/\rho$, roughly 200/0.5 = 400. *If* this were achieved, the device would improve from $\tau \simeq 5 \times 10^{-9}$ s to $\tau \simeq 10^{-11}$ s. From Eq. (10.6), and with $\ell = 1.5W$, $W = 15\ \mu m$, $w = 1\ \mu m$, and $t = 0.3\ \mu m$, the inductance would become $L = 5 \times 10^{-12}$ H, very attractive for stored energy.

The first stumbling block is the magnetic field requirement. From Eq. (10.7):

$$B = \frac{4\pi \times 10^{-7} I_c}{w} \simeq 1.2 \times 10^4\ I_c \text{ (gauss)}$$

For the devices to be useful, the power per device needs to be roughly 1 μwatt at system clock speeds approaching 100's of pico-seconds. Thus

$$10^{-6} \text{ watt/device} = \tfrac{1}{2} Li^2 \times \text{(clock rate)}$$

for a 1 nsec clock, requires

$$i^2 = \frac{20 \times 10^{-16}}{5 \times 10^{-12}} \simeq 4 \times 10^{-4} \text{ amp}^2$$

so that $i \leq 2 \times 10^{-2}$ amps.

Note once again that the full resistance of the gate is not immediately achieved, which adds to the spread of switching times on the logic gates. The field B which the control current produces in order to switch the gate will be [Eq. (10.7)] $B \simeq 240$ gauss. This is unrealisticly low for field switching of a type-II material. Finally, the current density demanded (in order to get current gain via the field generated) is very large. Thus, the attractions of current gain, non-latching behavior, and speed "offered" by the cryotron device using high T_c materials (as now known) are not apparently available.

10.3. Josephson Device

The speed performance limitation which the cryotron device encountered was demolished with the discovery of superconductive tunneling [2,3]. The highly non-linear I-V curve of an S-I-S tunneling device which exhibits Josephson tunneling (see Chap. 3) is shown in Fig. 8. When the device is driven by a voltage source V_s through a series resistor R, it can have two stable voltage states determined by the electrical parameters, so that

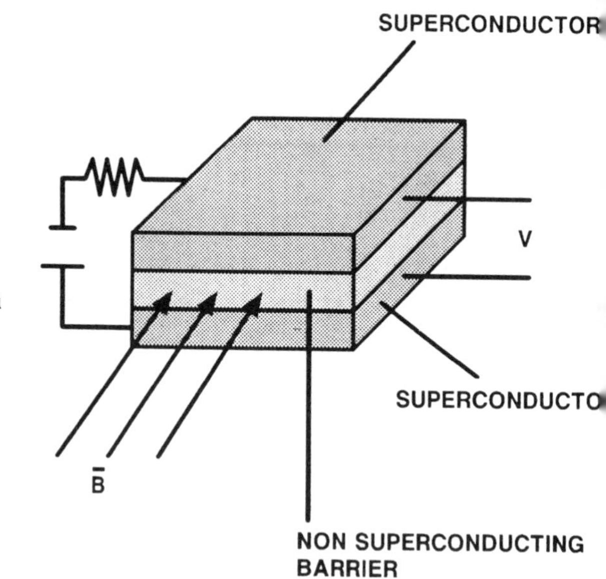

Fig. 10.8. Josephson Junction. (Top) Representative Structure; (Bottom) I-V curve with different load lines.

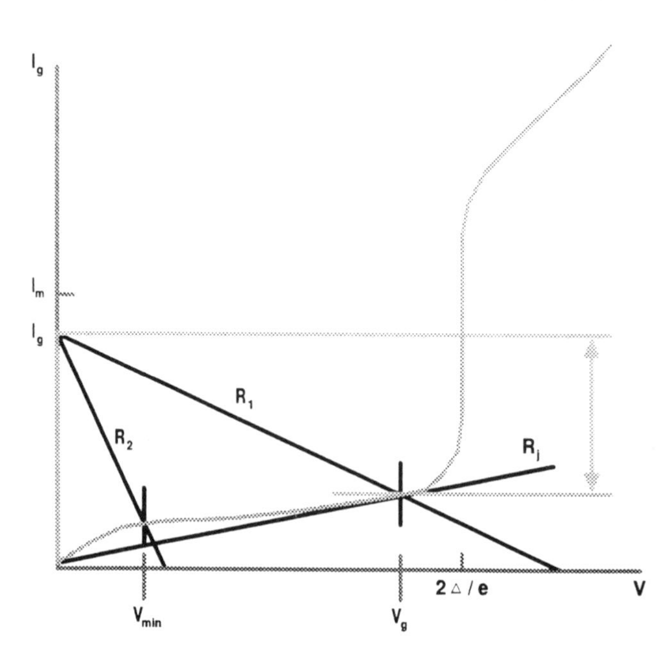

$$V_j = (V_s - iR) \tag{10.11}$$

where V_s is the supply voltage, i is the circuit current, V_j is the junction voltage, and R is the circuit resistance.

The function $(V_s - iR)$ forms the usual "loadline" which intersects the tunneling (I-V) curve at $V = V_j$ and at $V = 0$. For typical currents and tunneling gap voltages, $i \simeq 150\ \mu A$; $2\Delta/e$, twice the gap voltage, is materials dependent and ranges from 2.5 mV for Pb, 3.5 mV for Nb, and 5.5 mV for NbN electrodes. The significance of having a low conductance voltage region is that now, when the device is switched from the zero resistance state (as in the cryotron) to the non-zero voltage state, the drive voltage available to transfer current from the device to a load remains large, thereby providing a higher switching speed; thus we find that for properly designed devices, logic delays of ~2 picoseconds can be achieved (Fig. 9) [46]. On the unfavorable side, however, is the fact that once switched to the voltage state the device remains there until its current is dropped to a sufficiently low value. Hence, this "latching" behavior must be accommodated (or may be exploited) in the circuit design.

The detailed behavior is best described using the model created separately by Stewart and McCumber [11,12]. The total device currents can be represented (Fig. 10) as being composed of terms from the quasiparticle currents through an equivalent resistance R, an AC displacement current through capacitance C, and the "true" Josephson current.

The governing equations then become:

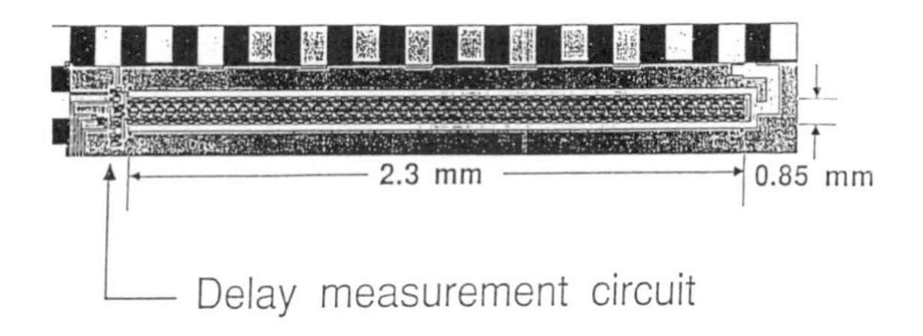

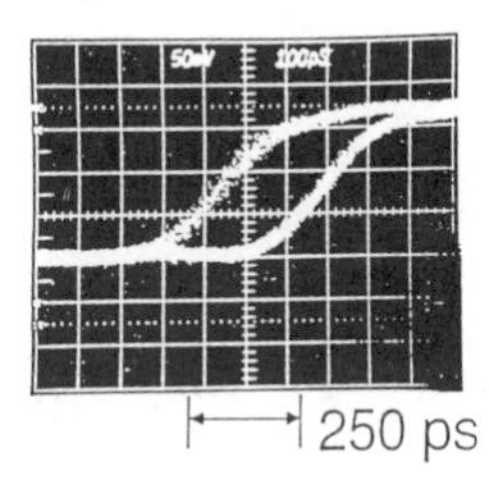

Fig. 10.9 OR Gate delay measurement. (Top) 100 toge chain of OR gates, 2.5 μm diameter Josephson junctions (Nb/A1203/Nb). (Bottom) Delay difference between input and output of 100 gate chain. (From S. Kotani, T. Imamura and S. Hasuo [46].) © 1987 — IEEE

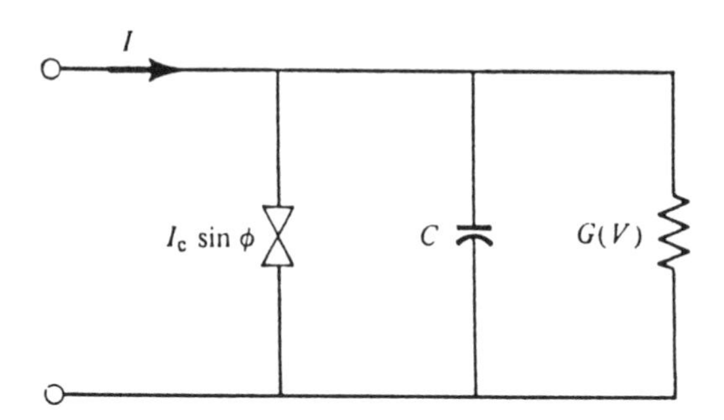

Fig. 10.10. Stewart-McCumber model of a Josephson junction (from W. C. Stewart [11], and D. E. McCumber [12]).

$$V_R = i_R\, R = V \tag{10.12a}$$

$$V_C = \frac{Q}{C} = V \tag{10.12b}$$

$$2eV_j = \frac{h}{2\pi}\frac{d\phi}{dt} = 2eV \tag{10.12c}$$

$$\text{and } i_j = i_0 \sin\phi \quad . \tag{10.13}$$

Since

$$i_s = i_C + i_R + i_j = C\frac{dV}{dt} + \frac{V}{R} + i_o\sin\phi \quad , \tag{10.13a}$$

then from (10.12c) and (10.13)

$$\frac{h}{2e}\frac{1}{2\pi}C\frac{d^2\phi}{dt^2} + \frac{h}{2e}\frac{1}{2\pi}\frac{1}{R}\frac{d\phi}{dt} + i_o\sin\phi = i_s \quad , \tag{10.13b}$$

$$\frac{\Phi_0}{2\pi}C\frac{d^2\phi}{dt^2} + \frac{\Phi_0}{2\pi}\frac{1}{R}\frac{d\phi}{dt} + i_o\sin\phi = i_s \quad , \tag{10.13c}$$

where $\Phi_0 = h/2e$ is the flux quantum.

This result presupposes a junction small enough to allow i_o to be uniform; the magnetic self-field and external field combined are smaller than a flux quantum. Equation (10.13c) describes the motion of a particle of unit mass in a tilted "washboard" potential [13] (see Chap. 3)

$$u(\phi) = \frac{2\pi}{\Phi_0 C}\left[i_s - i_o\sin\phi\right] \quad , \tag{10.13d}$$

i.e.,

$$\frac{d^2\phi}{dt^2} = \frac{1}{RC}\frac{d\phi}{dt} = u(\phi) \tag{10.13e}$$

subject to viscous damping of coefficient 1/RC (cf. Fig. 11).

For $i_s < i_o$, there is a *static* solution of (10.13e) with $\phi < \pi/2$, wherein

$$\frac{d\phi}{dt} = 0 \text{ and } \frac{d^2\phi}{dt^2} = 0, \text{ therefore } u(\phi) = 0 \quad .$$

As i_s is increased so that $|i_s| > |i_o|$, then ϕ increases with time, and

$$V = \frac{\Phi_0}{2\pi}\frac{d\phi}{dt} \neq 0 \quad . \tag{10.12c}$$

In the steady state,

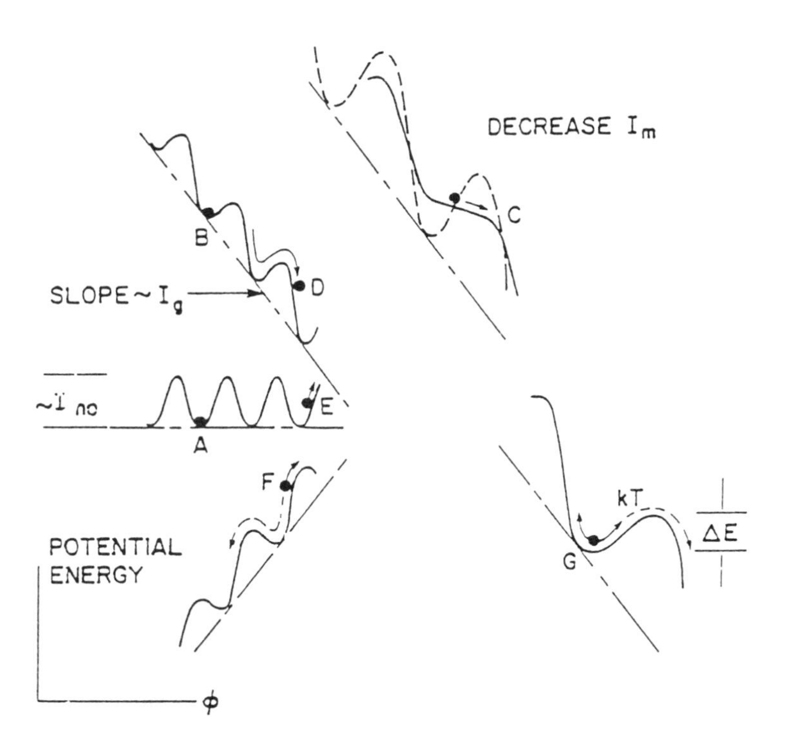

Fig. 10.11. "Washboard" potential model of a single Josephson junction (from H. H. Zappe [13]).

$$\frac{d^2\phi}{dt^2} = 0$$

and thus i_j oscillates with time;

$$i_j = i_o \sin\phi = i_o \sin \frac{2\pi}{\Phi_0} \int V \, dt \quad . \tag{10.13e}$$

For constant V,

$$i_j = i_o \sin \frac{2\pi}{\Phi_0} Vt = i_o \sin 2\pi \left[\frac{2eV}{h} \right] t \quad ,$$

where

$$\frac{2eV}{h} = 483.59 \text{ MHz}/\mu\text{volt} \quad .$$

This current oscillation is based on fundamental constants, is materials independent, and thus makes the Josephson effect a perfect candidate for use as a voltage standard (to be discussed later).

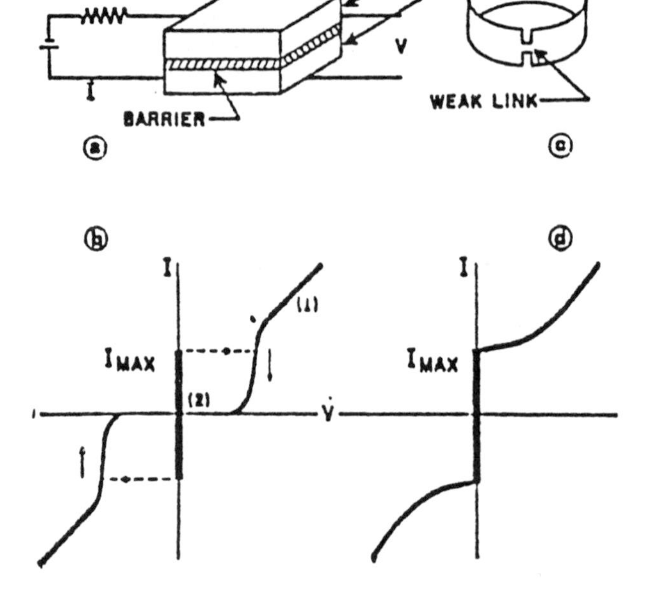

Fig. 10.12. I-V curves of a superconducting Josephson junction; (Left) hysteretic; (Right) non-hysteretic.

For a device with a relatively large RC product, i.e., small damping, it is apparent that once it is in the voltage state, reducing i_s such that $i_s < i_o$ does *not* return the junction to the zero resistance state.

Referring to the washboard potential model, one sees that the device dynamics are indeed complex. Starting from an initial drive current $i_s = 0$, the "particle" stays trapped in a potential well, $V = 0$, since

$$\frac{d\phi}{dt} = 0 \quad .$$

As i_s increases, then ϕ increases until finally the particle "falls" out of the well and proceeds to the subsequent ones, accelerating all the while until damping losses stabilize the average speed. If one now reduces i_s to less than i_o, $d\phi/dt$ does *not* go to zero, i.e., $V \neq 0$, until a sufficiently low i_s is reached whose value is determined by the viscous damping term. Thus the device is hysteretic, which is both an advantage and a problem. The fact that it "latches" provides for a stable two-state system, but it also removes control of the state from the input signal. As one changes the RC product, most commonly by changing R, the I-V curve can be made to transition to a non-hysteretic form, which, in some applications, such as magnetometers, is desirable (Fig. 12).

10.4. A Voltage Standard

The relationship between the Josephson oscillation frequency and the fundamental physics quantities of electron charge e and Planck's constant h led to a series of confirming experiments to assure that there were no materials-dependent factors or correction factors in the relation [17]. These questions were answered conclusively, which led to the proposal and acceptance of the Josephson device as the international standard for defining the volt; $2\,eV = h$, where 1 V (DC) applied to a Josephson junction produces a current oscillation in the junction at a frequency of 483.59 MHz. The existence of these oscillations had been inferred very early in superconductive tunneling research as very prominent steps in the I-V curves of the junctions which were illuminated by microwaves [18]. These "mixing" steps were observed out to the millivolt region and were readily attributed to the intrinsic non-linearity of the Josephson currents. Fundamentally, the Josephson current through a junction is being driven by a voltage

$$V = V_J \sin(\omega_i t) \quad ,$$

where ω_i is the imposed microwave frequency.

Thus, using (10.13e) we find

$$i_j = i_o \sin\left[\frac{2e}{h} V_o \sin(\omega_i t)\right] \quad . \tag{10.14}$$

This is of the form y = a sin (bsin β) and thus can be expanded in a Bessel series to produce

$$i_j = i_o \sum \left[\sin \frac{2e}{h} V_I\right] \sin(n\omega_i t) \quad ,$$

i.e., a spectrum rich in harmonies of ω_i. If the junction is also subjected to a DC voltage V_o, the new expansion will have frequency terms $\omega = (\omega_o + n\omega_i)$, where

$$2eV_o = \hbar\omega_o \quad .$$

Thus for each n there will exist a DC current step. For example, if 10 GHz radiation is impressed upon the junction, the first step will occur at 20.7 μV and successive steps will appear at multiples of that voltage. Very carefully designed electronics is required in order to selectively "lock" the DC voltage supply to a

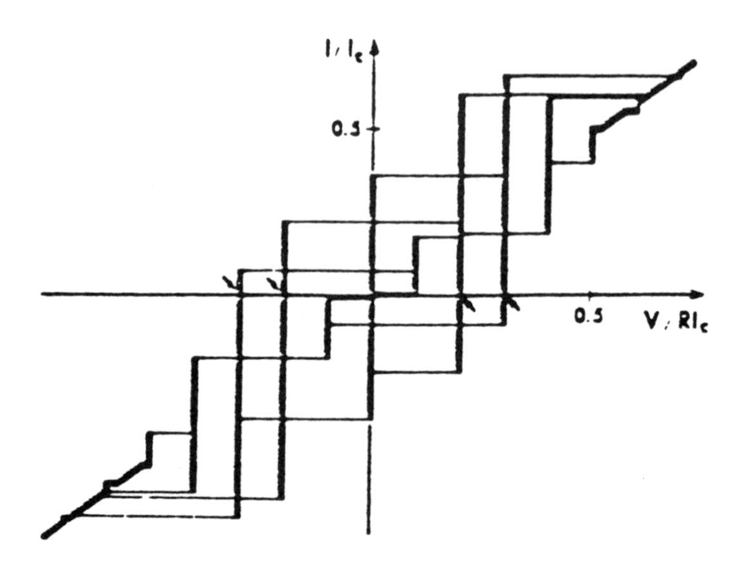

Fig. 10.13. Zero-biased voltage standard. Simulation of I-V curves for Josephson junction at specific drive. The arrows mark voltages at which zero bias current may be maintained. Simulation is for $\omega RC = 50$ ($R = 50X_c$), $\hbar\omega = 0.1$ ($2ei_{oR}$). (From M. T. Levinsen [20].)

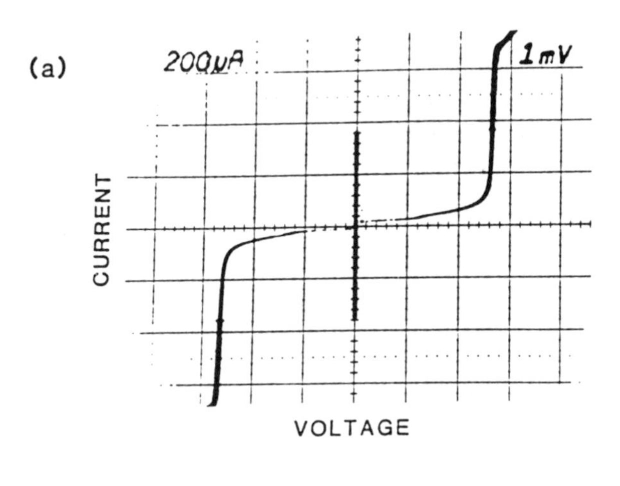

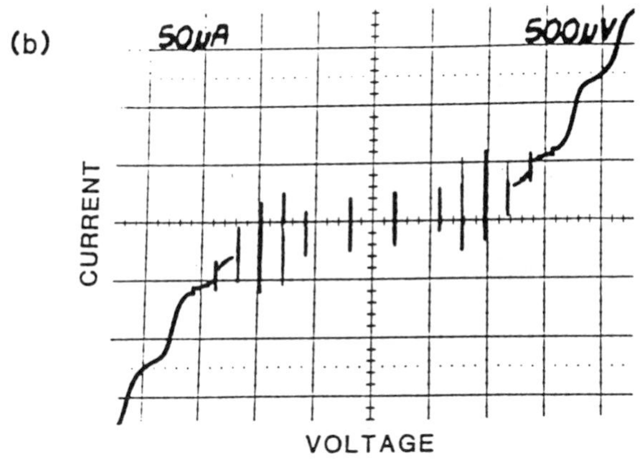

Fig. 10.14. Current-Voltage characteristics of a 15 μm × 30 μm Nb/Nb0x/Pb-In-Au Junction. a) In the absence of microwaves; b) In the presence of 96 GHz microwave radiation. (From R. L. Kautz, C. A. Hamilton, and F. L. Lloyd [21].) © 1987 — IEEE

selected step. The highest voltage that can be used on such a 10 GHz single junction standard is about 5 mV. Because the practical environment is that of the standard chemical cell one would like to cascade an array of such junctions to approach 1 volt. Unfortunately, this means that each junction must be identical (not realistic) or each junction must be separately current biased to be at the same harmonic step. This biasing has been done for up to 20 junctions, which still necessitates a resistive divider to reach the desired 1 volt [19]. A major improvement was made when Levinsen [20] proposed the use of zero-biased Josephson junctions, thereby greatly alleviating the requirements for uniformity [21] (Fig. 13).

Thus a very large series 1500, 6000, 18000 (Fig. 14), driven by 96 GHz microwave radiation, may be operated at output voltages up to 10 V with a precision of $1{:}10^9$.

The frequency response of present low temperature materials easily allows operation of a series of junctions at frequencies well above 96 GHz, thus allowing for an even larger voltage per junction and hence fewer junctions for a given standard output voltage and value of n. If a Josephson junction were available in the new HTS materials, operation at more friendly temperatures would be the principal advantage, rather than a higher voltage per junction.

The reciprocal aspect of the voltage standard equation also lends itself to exploitation for large dynamic range, high speed, very low power analog-to-digital conversion. If the voltage waveform to be digitized is applied to a junction, the resulting oscillations can be made of sufficient amplitude such that they may be counted [22]. In practical applications where a large dynamic range is wanted this voltage-to-frequency conversion technique imposes a requirement to count these pulses at a very high rate. The Nyquist

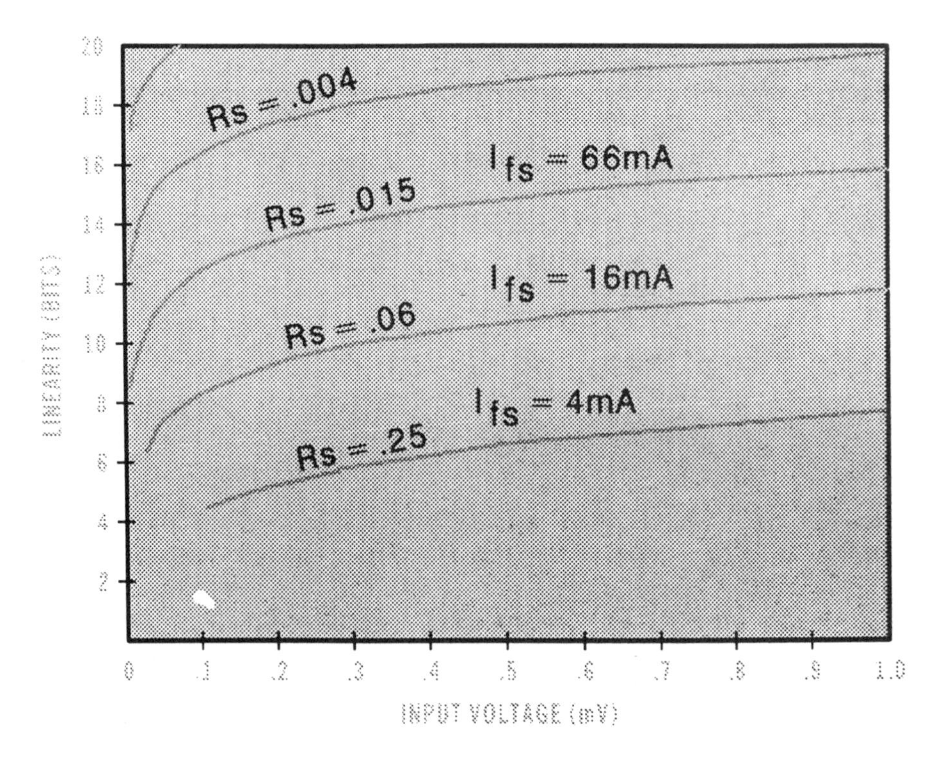

Fig. 10.15. Voltage-to-Frequency converter linearity versus input voltage. Curves are for different junction shunt resistors. (From C. Hamilton [22].)

sampling condition requires that this sampling rate be at least 2× the highest frequency signal to be digitized (Typically 2.5× is used to ease the problem of filtering out the "alias" waveform). For example, a 5 MHz analog signal (200 ns period) digitized to 14 bits (84 dB dynamic range) would call for:

sampling time: 80 ns
maximum counts/window: 2^{14} i.e., 16,000
counting rate: 200 Gb/sec .

This awesome counting rate appears amenable to being solved in the same technology; a two-junction "interferometer" SQUID (to be discussed later) used as a binary counter may be capable of this task. One must note, however, that the intrinsic nonlinear impedance of the junction when used as a voltage-to-frequency converter presents problems for systems in which linearity is demanded [22] (Fig. 15).

10.5. Single-Junction SQUID

For many applications the single Josephson junction is used shunted by an inductor to form a SQUID (Superconducting Quantum Interference Device). An understanding of the behavior of this configuration will provide a solid basis for using Josephson devices in many analog and digital circuits.

Let us consider a Josephson junction in parallel with an inductor L (Fig. 16). We will ignore the effects of the resistance R and the capacitance of the junction and assume quasi-static operation. The input current I divides into the two branches i_L and i_J.

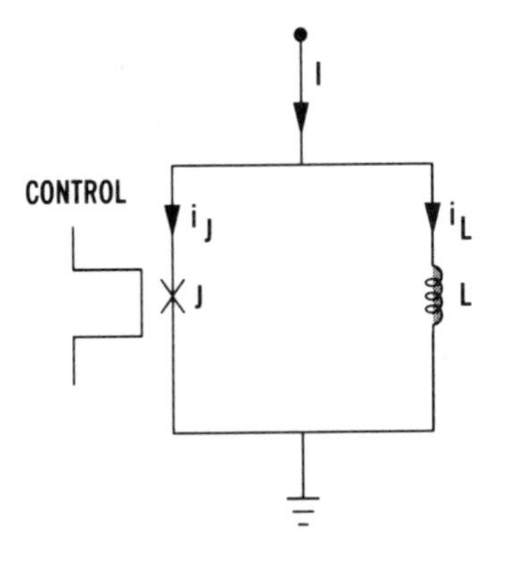

$$I = i_J + i_L$$

1) $i_J = i_0 \sin\phi$

2) $i_L = \frac{h}{2e}\frac{1}{2\pi L}\phi$

$$i_L = \Phi_0 \frac{1}{2\pi L}\phi$$

Fig. 10.16. Single Junction SQUID.

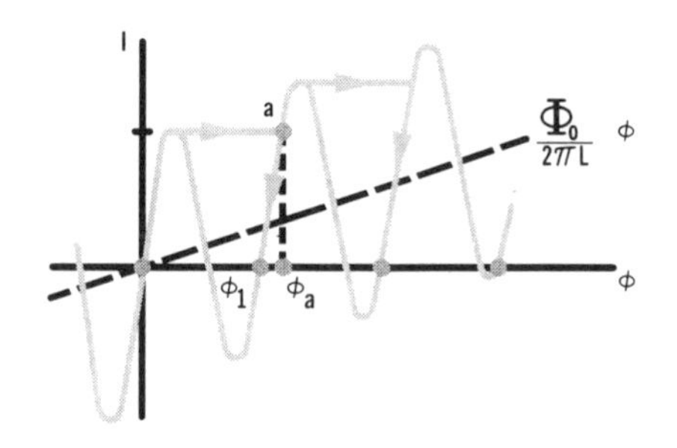

Fig. 10.17. Flux storage. Current vs. phase for a single junction SQUID.

Since the voltage is the same across each device, then

$$V = V_L = V_j \quad , \tag{10.15}$$

$$L \frac{di_L}{dt} = \frac{h}{2e} \frac{1}{2\pi} \frac{d\phi}{dt} \quad . \tag{10.16}$$

Thus,

$$Li_L = \Phi_0 \frac{1}{2\pi} \phi \quad , \tag{10.17}$$

$$I = i_L + i_j \quad ,$$

$$I = \frac{\Phi_0}{2\pi L} \phi + i_o \sin\phi \quad . \tag{10.18}$$

If we plot this current versus the phase ϕ a surprising result ensues (Fig. 17). Suppose that $i_o > \Phi_o$, then, as the input current I is increased we will reach the maximum zero resistance current i_o which the junction will support, at which point the current I will re-partition between i_L and i_J, i.e., the phase ϕ will slip. If now we reduce the input current until I = 0 we will settle at a phase $\phi = \phi_1$, where $i_L \neq 0$, $i_L = i_1$, $i_J = -i_1$, $\phi_1 \simeq 2\pi$, and therefore $i_1 L = \Phi_o$ (approximately). We have captured a quantum of flux; a persistent circulating current i_1 is maintained in the loop. If we increase the input current even further a second such point can be reached and, depending on the ratio of Φ_o/L versus i_o more such stable flux states may be achieved with $n = 0,1,2,3$ ... Given such behavior, very useful and important circuit elements can be designed and used.

In the digital electronics world it is natural to consider exploiting this behavior to create memory cells for high-speed computer use. This has been done by various research groups using different circuit configurations. The simplest such cell is the so-

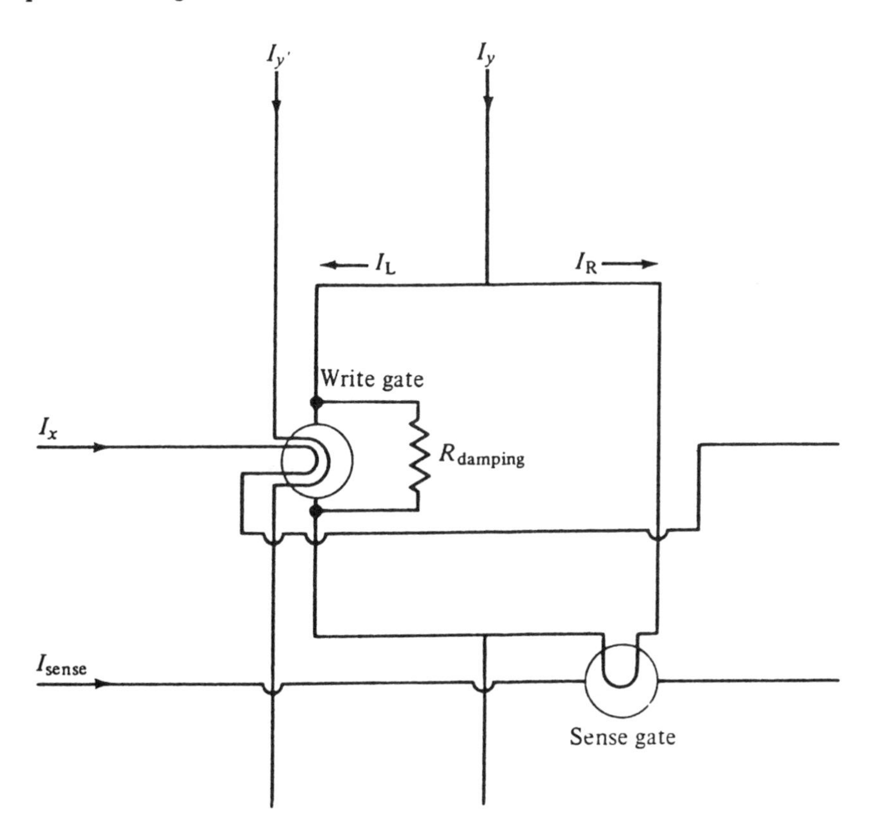

Fig. 10.18. Non-Destructive readout memory cell (from S. M. Faris, et al. [24]).

called Henkels cell [24] (Fig. 18). In this circuit the Josephson device's critical current is itself externally controlled to select whether the input current I will result in stored flux, signifying a "1" state, or no stored flux designating a "0" state. The cell operation is quite simple (Fig. 19) [13]. A current is injected into the "Y" line. If a "1" is to be written the X line is also driven which switches the current to the right leg of the cell. When that current is transferred, the Y current may be removed which will result in a stored persistent current corresponding to one (or several) flux quanta being stored. If the Y control line were not powered, the above process would result in no stored flux. Reading the existence of a stored current is quite straightforward. Driving the Y line when a "1" is stored will cause the full current to pass through the right branch. This right branch line is laid out to pass over, and thus be the control line for, a "sense" gate Josephson device. Thus, a "read" current will transfer the sense gate to the

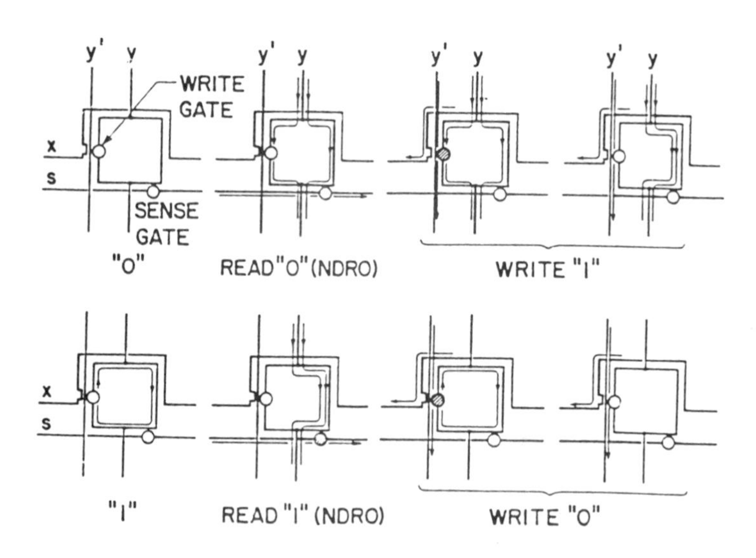

Fig. 10.19. Memory cell operation (from H. H. Zappe [13]).

voltage state. Zero stored flux will produce only half the required current over the sense gate and thus no output voltage will result. The merits of such a memory cell are very significant. First, the energy stored is exceptionally small; with an inductance of 20×10^{-12} Hy the stored energy, $\Phi_o^2/2L$ is 10^{-19} J. Thus at access rates of 10^9 per second one would dissipate on the order of 10^{-10} watts/bit (actually a real memory of 4K bits dissipate much more power than this, ~10 mW, due to the logic decoding circuitry). Second, the memory behaves as a non-volatile one: the stored currents persist indefinitely unless the temperature is raised above T_c.

10.6. SQUID Magnetometer

The single junction SQUID device has been used very effectively for years as an ultra-high sensitivity magnetometer; it has been the standard against which all magnetometers are measured. (Advances in two-junction SQUIDS to produce so-called DC SQUIDS have improved upon the single junction device; this will be discussed later.) When an RF magnetic field of symmetric shape and of such an amplitude that ±1 flux quantum is impressed upon the SQUID, it will produce a mensurable RF power absorption from the driving oscillator [26]. If the device is also subjected to an external magnetic field, the induced opposing circulating current will bias the SQUID preferentially in one current to more greatly exceed the one flux quantum critical current compared with the opposite polarity (Fig. 20). This RF power absorbed reaches a maximum for an external field of $\frac{1}{2}\Phi_o$, then back to a minimum at Φ_o, to a maximum at $\frac{3}{2}\Phi_o$, minimum at $2\Phi_o$, repeating this behavior periodically (Fig. 21). One then amplitude modulates the RF at a lower frequency and synchronously detects the amplitude modulation. If the SQUID is biased at a maximum or a minimum, the synchronous detector will produce an output at twice the modulation frequency. If, however, the bias field is not at the symmetry point, an error signal at the modulation frequency will be produced whose amplitude depends upon the displacement of the field from the symmetry point and of positive or negative phase depending upon which direction the "error" is from the null. This

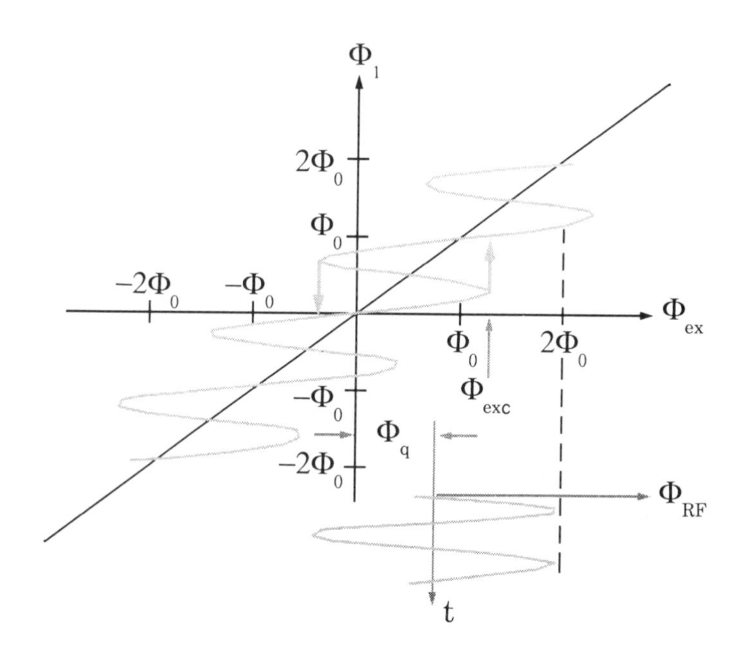

Fig. 10.20. Single junction SQUID magnetometer. An RF field Φ_{RF} is applied to the SQUID in the presence of an external field Φ_q.

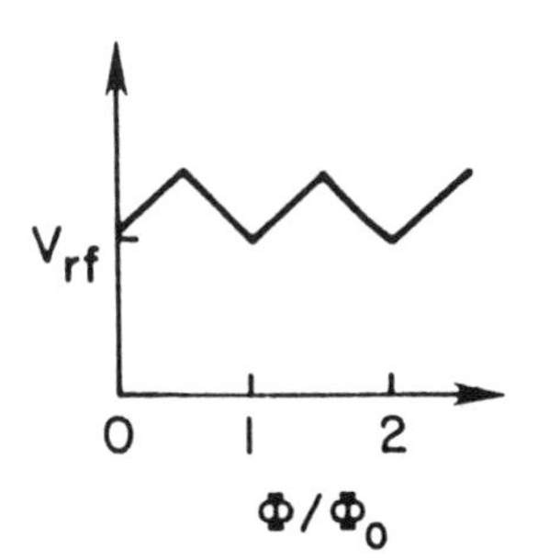

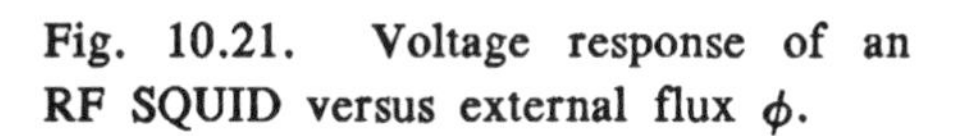

Fig. 10.21. Voltage response of an RF SQUID versus external flux ϕ.

error is used to generate a correction field to return the SQUID back to a null. By this means one can detect externally imposed fields to a small fraction, approximately 10^{-5}, of a flux quantum. In order to improve *field* sensitivity, one commonly uses a large-area superconducting pickup coil to capture the weak external field, and then to transformer-couple the resulting small current into the SQUID. In this fashion field sensitivities of 10^{-11} gauss (10^{-7} tesla) have been achieved.

The new materials present a golden opportunity for the creation of high-sensitivity magnetometers which can now be operated in dewars of greatly extended storage time by use of liquid neon or liquid nitrogen. Since the magnetometer's power dissipation is well below a milliwatt, the cryogen loss is due almost totally to the parasitic heat losses of radiation and conduction. Consequently, a small magnetometer system which holds 20 liters of helium and operates for 20 days should last for ~800 days with liquid neon, or 3^+ *years* with liquid nitrogen. The clear penalty, of course, comes from the added *intrinsic* thermal noise of higher temperature operation [27]. However, for most (but not all) applications the environmental noise greatly overwhelms the sensor noise floor. Since one needs but a single Josephson junction for a magnetometer (or two junctions for a "DC SQUID") it clearly will be the first electronic application, and indeed was the first to be demonstrated [28,29].

10.6.1 A/D Conversion

The SQUID device may also be used very effectively to quantize an analog signal [30]. In this case the pickup coil is replaced by an analog amplifier. As the current in the input transformer increases, the usual counter-current in the SQUID appears until that current reaches the critical value for the junction. This time the i_oL product is chosen to be Φ_o and thus when $i_s = \Phi_o/L$, a single flux quantum is allowed to enter the loop. The dynamics of this event results in a very sharp pulse appearing across the inductor and Josephson junction. This comes about from the stored energy in the inductor being dissipated by discharging through the junction and its parallel resistance and capacitance. If the junction is critically damped, the pulse from this resonant circuit can be of several picoseconds time width. As the transformer input current continues to rise, this event repeats; another flux quantum enters, another voltage spike is created (Fig. 22). By proper design and layout, one expects this sequence to allow as many as 32000 flux quanta to be trapped within the SQUID loop. In addition, and most importantly, one expects the quantized inductance to be field independent, and therefore the input current intervals to be *equally*

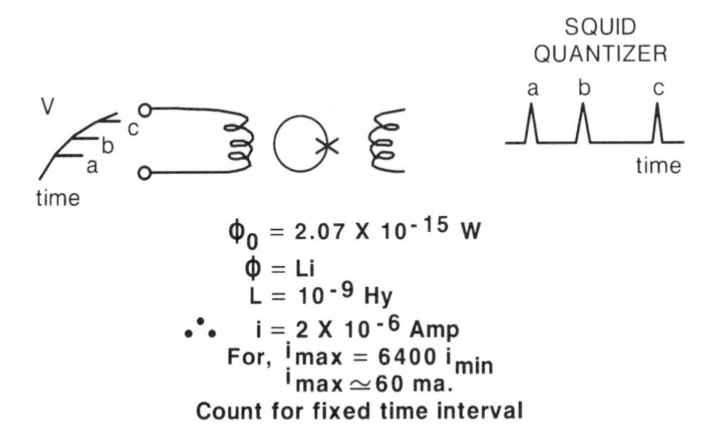

Fig. 10.22. SQUID quantizer. Analog voltage, V, coupled to the SQUID loop, generates a current which exceeds the loop critical at point A, thereby allowing a flux quantum to enter and generate a pulse in the output transformer. A second quantum enters at B, a third at C, etc. thereby producing the pulse train at a, b, c, etc.

spaced. This property of linearity is frequently the most sought-after. In order to complete the conversion process, one must also be able to count these "spikes" in a time commensurate with the highest analog frequency to be converted. This last requirement is itself awesome, demanding counting rates in the neighborhood of 500 GHz. This latter task will be discussed when we address two-junction SQUIDS.

10.7. Two-Junction SQUIDs

An important step in achieving improved device performance and in producing new circuits is taken by putting a second Josephson junction, J_2, in the series loop with L and J_1 (Fig. 23) and inserting the input current at a dividing point on L. That is, divide L into pL and qL, where p + q = 1. The equations for this configuration can be solved by changing variables and the circuit behavior well described [31]. The simplest and probably the most interesting arrangement is to set $p = q = \frac{1}{2}$; then

$$I_{o1} = I_o \sin \phi_1$$

$$I_{o2} = I_o \sin \phi_2$$

and

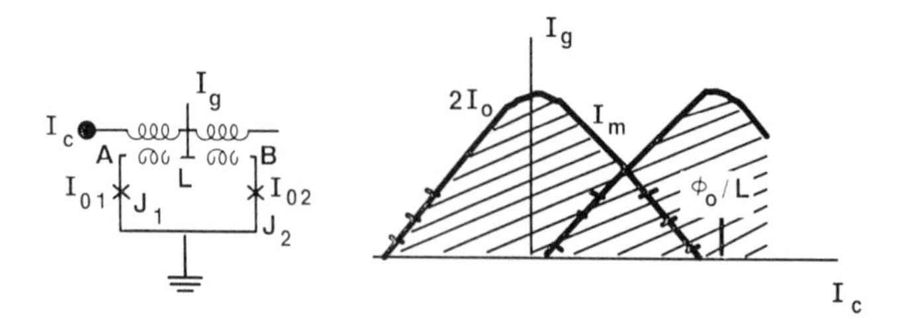

Fig. 10.23. Two-junction interferometer. For a given I_c, zero resistance I_g occurs within the shaded region. The first "lobe" encompasses a zero flux quantum state in the SQUID, the second "lobe" a single flux quantum. In the overlap region the SQUID can be in either flux state.

$$LI_o = \Phi_o \quad .$$

The consequence of also inductively coupling such a SQUID via an input transformer is shown in Fig. 23. For zero current I_c in the transformer primary the SQUID current I_g can be increased to $I_g = I_{01} + I_{02} = 2\ I_o$, at which point *each* junction has reached its critical current. (Note that no flux is yet enclosed by the SQUID loop.) However, in this case, as distinct from the single junction SQUID, the two junctions arrive at the voltage state and *stay* there. The current $2I_o$ cannot flow down one branch alone and thereby allow Φ_o into the loop: That junction's critical current is only I_o. As one applies a current I_c to the input transformer a circulating current, as usual, is set up in the loop and now a smaller gate current I_g will produce the voltage state; for increased I_c the allowed gate current I_g continually decreases. One notes that a competing situation can also be calculated. This is one in which one flux quantum is stored in the loop and the externally imposed current I_c induces a circulating current which *subtracts* from the stored current. This produces the second underlying "lobe" which eventually becomes the unique solution (Fig. 23). In the shaded region where the two states coexist a transition can take place from one lobe to the other, which will allow the insertion or ejection of a flux quantum with the system then returning to the zero voltage state.

The first evaluation and use of this two-junction configuration was in the design of logic circuitry. It was noted by H. Zappe [32] that the conventional Josephson logic device, the magnetically coupled Josephson junction, suffered from being a low impedance device. The junction capacitance presented such a low reactance for high speed pulses (~10 picosecond rise time) that the loading transmis-

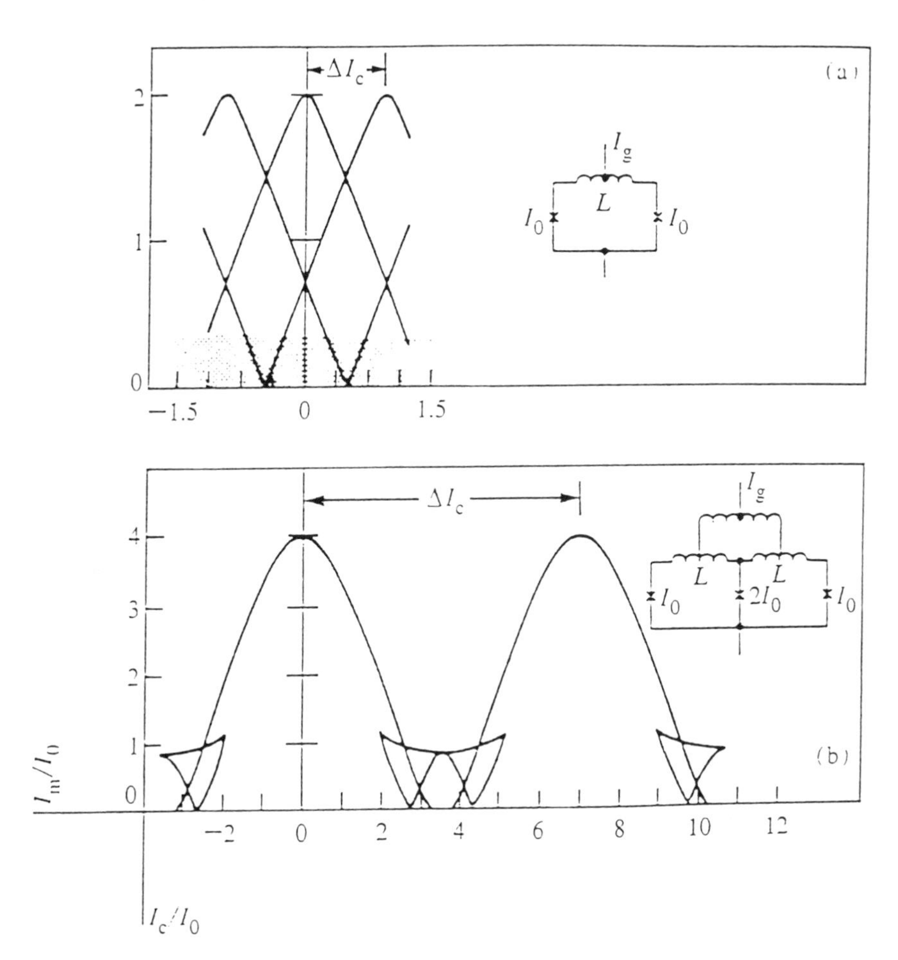

Fig. 10.24. Threshold characteristic for multijunction SQUIDS. a) Two-junction device, $\Phi_o/LI_o = 0.94$. (b) Three-junction device, $\Phi_o/LI_o = 7.0$. (From J. Matisoo [23].)

sion lines would have to be of very low impedance. For example, for lead alloy junction technology the characteristic capacitance was $C_s = 4\ \mu F/cm^2$. Thus for junctions whose diameter is $d = 5\ \mu$, the area $A = 2.2x10^{-7}\ cm^2$ and $C_j \simeq 10^{-12}$ F. For $\tau \simeq 10^{-11}$ sec., $X_c \simeq 2\Omega$. A transmission line of such a low impedance would result in extremely slow signal propagation and be too wide for high density micro-electronics use. Consequently, one needed a higher impedance switch, which the multi-junction SQUID provided (Fig. 24). A complete logic family evolved from this basic element [33] as an example of the versatility of the two-junction SQUID.

An unbalanced "interferometer" SQUID was ingeniously created by

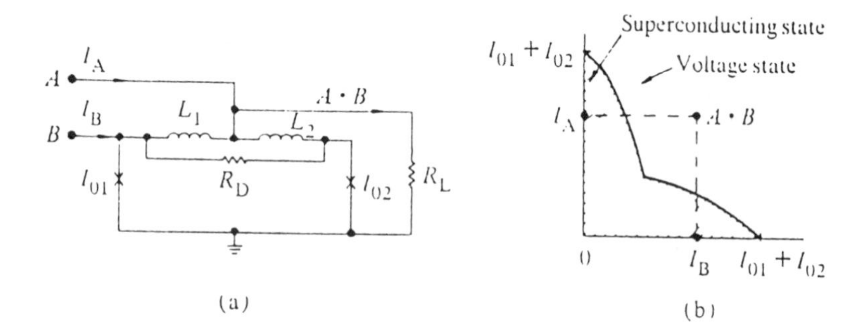

Fig. 10.25. Current injection logic gate. a) Two-junction injection gate; b) Injection gate switches to the voltage state only when both IA and IB are present. (From T. R. Gheewala [33].)

T. Gheewala [33], which produced the operating curve of Fig. 25. This device clearly offered a sizeable transfer of gate current for a given change in control current as well as a generous operating region to allow for process parameter variances. The basic two-junction SQUID configuration was modified and enhanced by Kotani in his so-called Modified Variable Threshold Logic (MVTL) gate design [34] (Table I) (Fig. 26).

10.7.1. DC SQUID

By far the most widespread use of the two-junction SQUID, however, has been the evolution of it into the so-called "DC SQUID" [35,36]. If you put a sufficiently small resistor in parallel with each of

Table 10.1. Performance of the MVTL Gate Family (from S. Hasuo [25]).

Gate \ Min. JJ dia. (μm)	4.0	2.5	2.0	1.5
OR	5.6	4.2	3.3	2.5
2 OR-AND (unit cell)	16.0	11.5		
3 OR-AND (2/3 MAJORITY)	21.0			
Timed Inverter				

FUJITSU AUG 1988 UNIT: PS/GATE OR PS/CELL

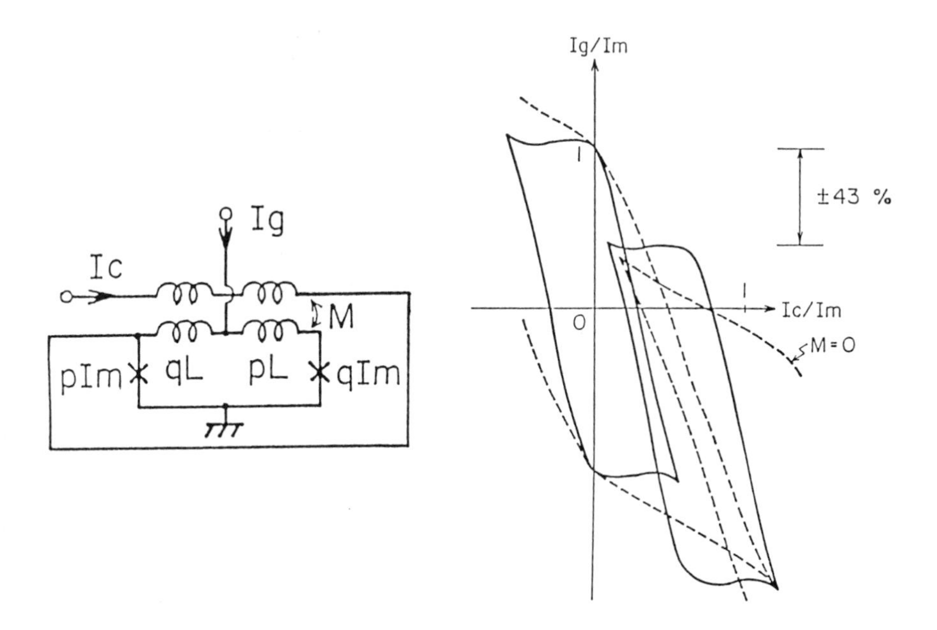

Fig. 10.26. Modified Variable Threshold Logic (MVTL) "OR" Gate. (Left) Equivalent circuit. (Right) Threshold Characteristics. (From S. Kotani, et al. [46].) © 1987 — IEEE

the junctions in the circuit (Fig. 23), where p = q = ½, the resulting I_g vs I_c curve will not be hysteretic [11,12]. As in the single- junction case the allowed zero voltage supply current I_g will decrease as before until you reach $\Phi_0/2$ and then it begins returning to I_g(max). By current biasing this device at its threshold one then detects a voltage change with respect to the applied current I_g which is provided by a superconducting pick-up loop. The device thus offers magnetometer operation without the use of an RF field (a low frequency modulation is used to enable feedback to be employed for linearizing the system). As with the single-junction SQUID, one needs to explore the possible new opportunities presented by higher temperature operation. And, as before, the merits appear the same: Cryogenic convenience and dewar hold time. The disadvantage will be a noisier but very often, more than adequate performance [37].

10.8. Binary Counter

We return to the problem of pulse counting for both the voltage-frequency and flux counting analog/digital converters. It was proposed by Hurrell and Silver [38,39] and demonstrated by Hamilton [40] that the two-junction SQUID could be readily used for that purpose. Referring to Fig. 27, suppose that an input bias current I_{c1} is applied sufficient to set up a clockwise current $i_o/2$ in the loop, where $Li_o = \Phi_o$; i.e., we are biased with $\frac{1}{2}\Phi_o$ externally. If there is, in addition, an input current $I_{g1} = i_o - \delta$, this current will divide equally between the two legs, maintaining the same internal flux state ($\Phi=0$) and producing a current in J_1 less than i_o and zero current in J_2. An input pulse line S is connected to a pulse source such as the flux quantizer or a V/F converter junction. When a positive current pulse appears on the line S it will divide its current equally into the two branches, which will cause J_2 to exceed i_o and switch into the voltage state. The circulating current will now decay to zero, allowing flux to enter the loop, and will build up in the opposite direction until one flux quantum is stored in the loop. This state has a current of $i_o - \delta$ in J_1, zero current in J_2, and a flux of $+\Phi_o$ ($L/2i_o + \Phi_o/2$ from I_{c1}). The proper damping resistance across each junction can provide a very fast current reversal, in picoseconds, into the CCW circulating current state. The switching of J_2 itself produces a very fast voltage pulse which can be fed to an identical succeeding stage. Notice that a second input pulse will trigger J_1, this time to reverse the current back to a CW circulation. However, this second pulse produces a positive spike on J_1 but not on J_2. Consequently, this biased SQUID behaves as a very fast binary counter. Low-speed counting and reading have been verified [41], but to date it has not been possible to directly verify the counting rate and accuracy. However, preliminary measurements which have scaled the output voltage of a series of such binary stages versus the input current have

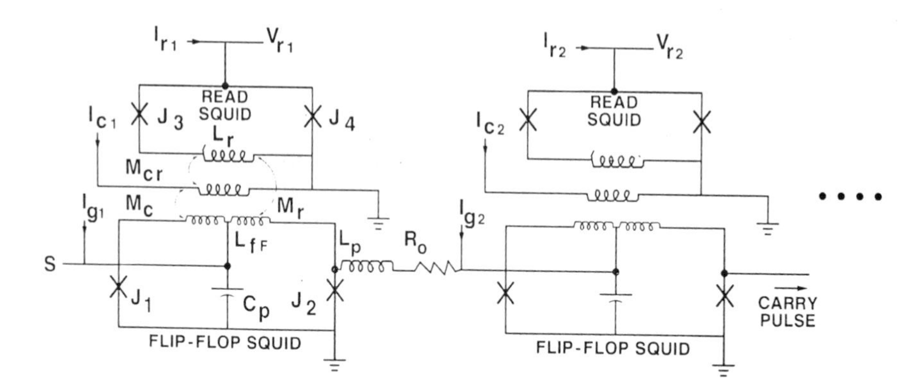

Fig. 10.27. Binary counter (from C. A. Hamilton and F. L. Loyd [47]). © 1982 – IEEE

verified that a factor of two voltage reduction is achieved per stage, consistent with proper operation [40].

In order to use this counter you must be able to read the state of each stage at a prescribed moment in time, with a very small time variation, as usual, on the order of picoseconds. This general requirement is not unique to this particular circuit, but is a direct consequence of trying to sample a fast analog waveform with large dynamic range (Fig. 28).

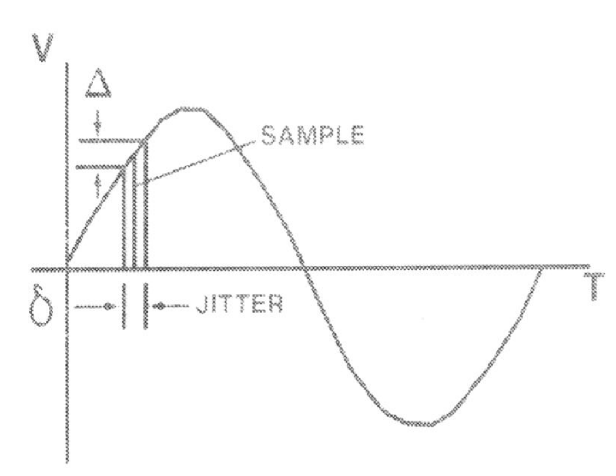

Fig. 10.28. A/D conversion. Aperture jitter requirements for digitizing a 5 MHz analog sine wave to 16 bit precision.

10.9. Sampling Oscilloscope

The most striking demonstration of the speed and timing stability of a Josephson device is given by its use as a gate for a sampling oscilloscope. The original circuit [41,42] was motivated by the need to verify and characterize the operation of Josephson junction logic gates. Connecting room temperature probes not only disturbed the circuits but also presented great difficulty in measuring the low voltage fast rise-time waveforms. Two basic circuit elements are required for the function. The first is a very fast (~ picosecond) pulser, and the second is essentially an "AND" gate which sums the pulse with the analog waveform and an external bias current (Fig. 29) [43]. For a fixed *difference in time* between the input signal and the pulse, the bias current I_b is increased until the summing junction "fires". The experimental signal and the pulse are synchronously, repeatedly, triggered with the bias current automatically adjusted to provide a ~50% firing rate. The pulser consists of J_1, J_2, and R_2. An external trigger pulse drives J_1, an interferometer, into the voltage state which then causes the transfer of current into J_2. The time constant for that rise is determined by

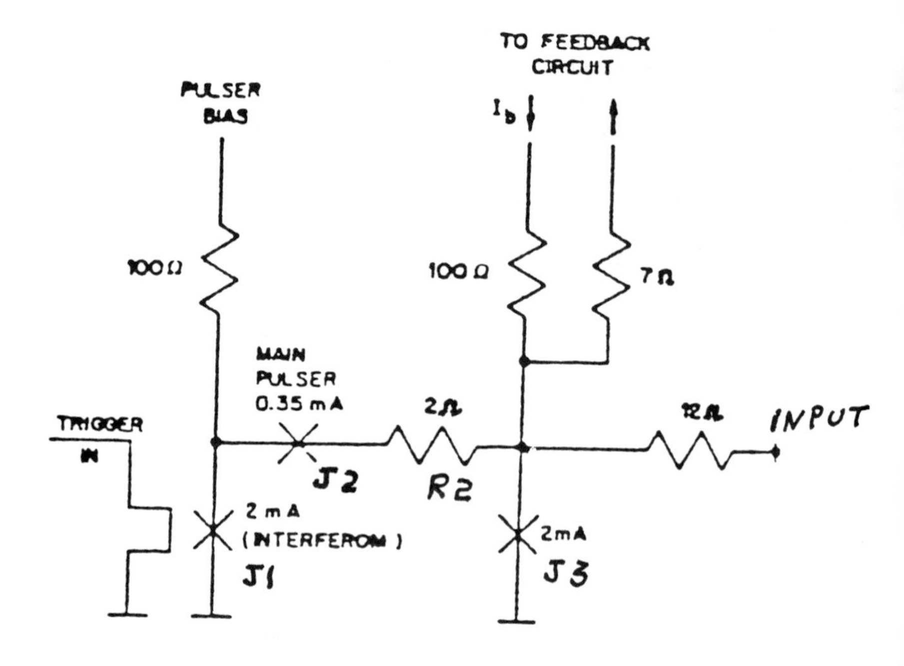

Fig. 10.29. Josephson sampler. Component values are nominal. (From P. Wolf, B. J. Van Zeghbroeck, and U. Deutch [48].) © 1985 — IEEE

both J_1 and L, the inductance in the line, J_2, R_2, J_3 (the summing junction). When the current through J_2 exceeds J_2's critical current, J_2 switches abruptly into the voltage state, thereby reducing the current through J_3 in a time determined by L/R_2. With proper selection of damping resistors across the junctions and of L, R_2, J_1, J_2, a very fast pulse can be created. Note that in the pulser circuit the "tip" of the pulse is used to trigger the summing junction.

The time and amplitude stability of the pulser, stability of the AND gates's current operating point, as well as the time stability of the experiment-pulser time differences, are such that 5 picoseconds sampling resolution is available in a commercial instrument [44] and 2.1 picoseconds has been demonstrated in the laboratory (Fig. 30) [43].

If one used a high temperature superconductor in this sampling circuit you could anticipate several benefits. First, if the room temperature-to-low temperature transmission line could be made with these new materials, greater bandwidths might result. Second, if a Josephson tunnel junction were available with the expected much larger energy gap this would translate into an intrinsically faster device. However, raising the *operating* temperature to, say 50 K,

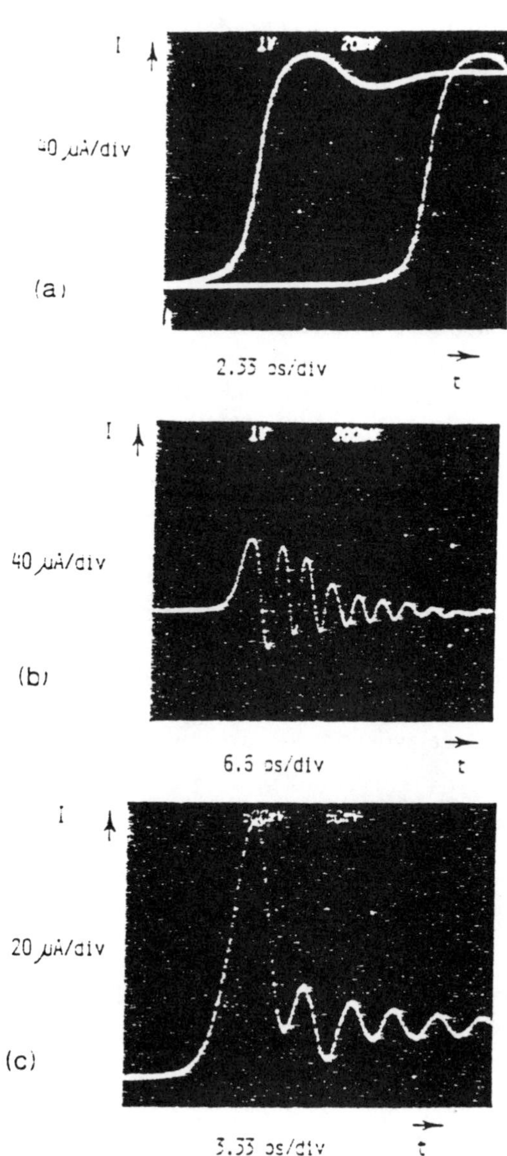

Fig. 10.30. Sampler-measured waveforms. a) Output pulse of an interferometer switching to the voltage state. Double exposure with one waveform shifted by 10 ps with the external delay line. The 10-90% rise time is 2.1 ps. b) Output pulse of an interferometer which switches to another vortex state. c) Output of a Faris pulser. (From P. Wolf, B. J. Van Zeghbroeck, and U. Deutch [48].) © 1985 — IEEE

would multiply the thermal noise power kT by roughly $10\times$. Since the threshold current fluctuations vary as $\Delta \mathrm{I} \sim kT$ ([13], Eq. III-14) this would result in greater time jitter. Consequently, if the time resolution is limited by internal rather than external factors better time resolution might not result. A significant advantage in testing room temperature circuitry would be gained by the ability to generate output pulses in the 100 mV range from a few high temperature superconductor Josephson junctions. Practically speaking, however, operation at temperatures of 10–27 K with refrigerators or liquid neon and the consequent enhanced supportability (and likely electronic performance) would greatly improve customer acceptance.

10.10. Transmission Lines

In all high-frequency, high-performance applications of superconductive devices, there is a requirement to transfer signals from one point to another. A unique important advantage of superconductivity is the prospect of doing this via transmission lines of very high bandwidth, which have exceptionally low loss and dispersion. In addition, this seems feasible for physically small and closely packed configurations which normally would result in unacceptable levels of cross-talk between the lines. Finally, because of these features one can even build long delay lines and deliberately coupled lines in order to perform some important signal processing functions.

A conventional normal-metal delay line is characterized by the usual inductance and capacitance per unit length, with dissipative terms due to the resistance of the line and its return path (ground) plane along with the dissipation of the dielectric medium. In the ideal case the impedance Z_o and delay/length are expressed thus:

$$Z_o = \frac{d_o}{W}\left[\frac{\mu_o}{\varepsilon_o \varepsilon}\right]^{1/2} \tag{10.19}$$

$$\tau = \left[\mu_o \varepsilon_o \varepsilon\right]^{1/2} \quad . \tag{10.20}$$

In a superconductor the London penetration depth allows the surface current to descend into the superconductor to a finite depth, on the order of 400 Å for Pb, 600 Å for Nb to 2000 Å for NbN. The new high temperature superconductors have a London penetration depth which is anisotropic, roughly 400 Å in one orientation and 3000 Å in the orthogonal direction. Within this region any electric field associated with the alternating currents will succeed in accelerating the normal electrons and hence result in losses which will increase with frequency. Hence the line will indeed not be lossless, but nevertheless will still have impressively low losses. Another aspect is that the superconducting electrons, lossless, will themselves be accelerated and decelerated by the driving source. Thus, in addition to the magnetic field, and hence stored energy, created by the current, there will be a kinetic energy stored by these electrons. As a consequence, the effective inductance/length will be increased which may significantly alter the impedance and

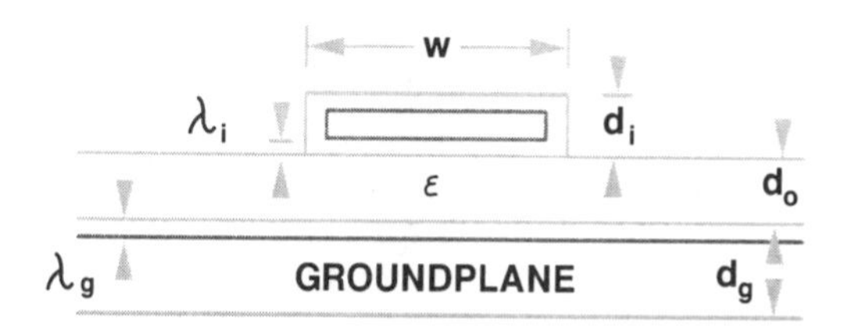

Fig 10.31. Superconducting transmission line. λ_i and λ_g are London penetration depths.

delay characteristics of the line. For lines where the London pene tration depth is not small compared with the dielectric spacing (Fig. 31). The governing equations for this are:

$$Z_0 = \frac{d_o}{W}\left[\frac{\mu_o}{\varepsilon_o \varepsilon}\left(1 + \frac{\lambda_g}{d_o}\coth\frac{d_g}{\lambda_g} + \frac{\lambda_i}{d_o}\coth\frac{d_i}{\lambda_i}\right)\right]^{1/2} \qquad (10.21)$$

$$\tau = \left[\mu_o \varepsilon_o \varepsilon\left(1 + \frac{\lambda_g}{d_o}\coth\frac{d_g}{\lambda_g} + \frac{\lambda_i}{d_o}\coth\frac{d_i}{\lambda_i}\right)\right]^{1/2} . \qquad (10.22)$$

The Meissner effect makes an important contribution to the transmission line behavior in "expelling" magnetic fields from the conductor. The magnetic field distribution between the transmission line and ground plane achieves a high value which is very localized (Fig. 32) and is practically frequency independent. The result is to produce a frequency independent phase velocity and a very low

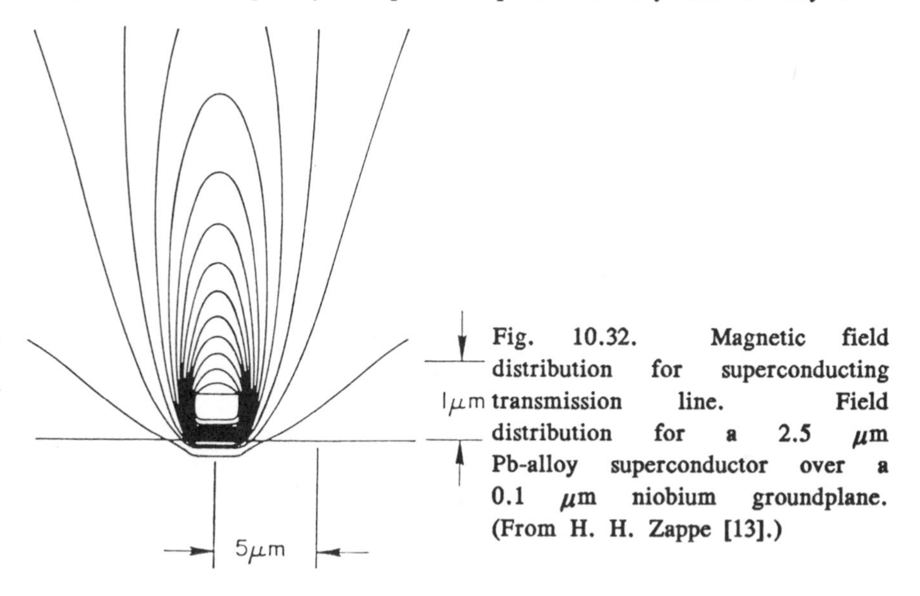

Fig. 10.32. Magnetic field distribution for superconducting transmission line. Field distribution for a 2.5 μm Pb-alloy superconductor over a 0.1 μm niobium groundplane. (From H. H. Zappe [13].)

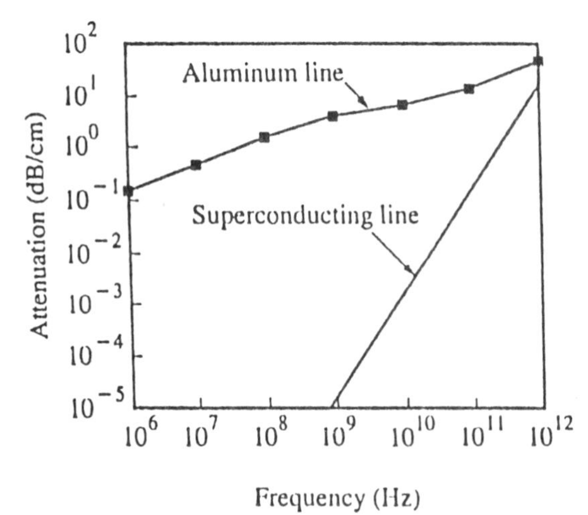

Fig. 10.33. Transmission line attenuation versus frequency. Calculated attenuation for a 2.0 μm linewidth, 0.5 μm thickness, and 1.0 μm dielectric at 77 K (from R. F. Pease and O. K. Kwon [49]).

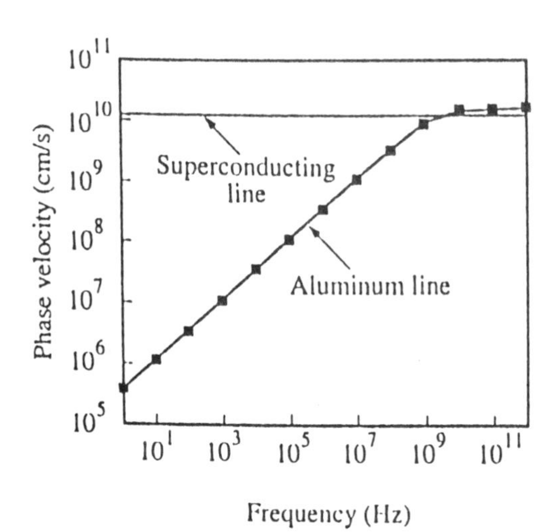

Fig. 10.34. Transmission line phase velocity versus frequency. Calculated phase velocity for a 2.0 μm line width, 0.5 μm thickness, and 1.0 μm dielectric at 77 K (from R. F. Pease and O. K. Kwon [49]).

inductive coupling between adjacent lines. The latter point permits very close line spacing with consequent small "cross-talk".

The current density requirement in such transmission lines depends upon both the driving voltage level and the desired line characteristic impedance. For low T_c Josephson junction circuits with voltage levels in the 3–10 mV range and line impedances of 1–10 Ω the current density is not stressed. For example, a 10 Ω impedance line driven at 10 mV calls for 1 ma of current (most circuits are in the 200 μa range) thus a 1-μ-wide, 2000-Å-thick line calls for a current density of $\sim 5 \times 10^5$ A/cm^2.

High T_c materials are being evaluated for possible use as interconnection lines between semiconductor chips which are already cooled to cryogenic temperatures for performance reasons: lower noise in the case of receivers, higher speed and higher reliability for digital systems. For digital circuits, line impedances of 50 Ω are common and voltage levels range from 1to 5 V; thus current levels are 20–100 mA. A line width of 2 μ and thickness of 0.5 μ results in a current density of from 2×10^6 to 10^7 A/cm^2. (Obviously widening the line greatly eases this condition, but these numbers are not out of sight.) It is always necessary to keep in mind the relative simplicity of using copper or aluminum at these temperatures. For short enough path lengths, wide enough lines, and frequencies below the gigaherz region considerable improvement is obtained by resistance reduction of a normal conductor. However, when large numbers of connection nets are involved (large circuit count and chip count) the increased line density and consequent reduction of wiring levels is very attractive (Figs. 33 and 34).

The most striking and useful exploitation to date of transmission lines has been the development of long dispersive delay lines carried out by Lincoln laboratories [45]. They construct two spaced transmission lines on the same substrate with a lateral separation which is carefully controlled by the lithography (Fig. 35). At

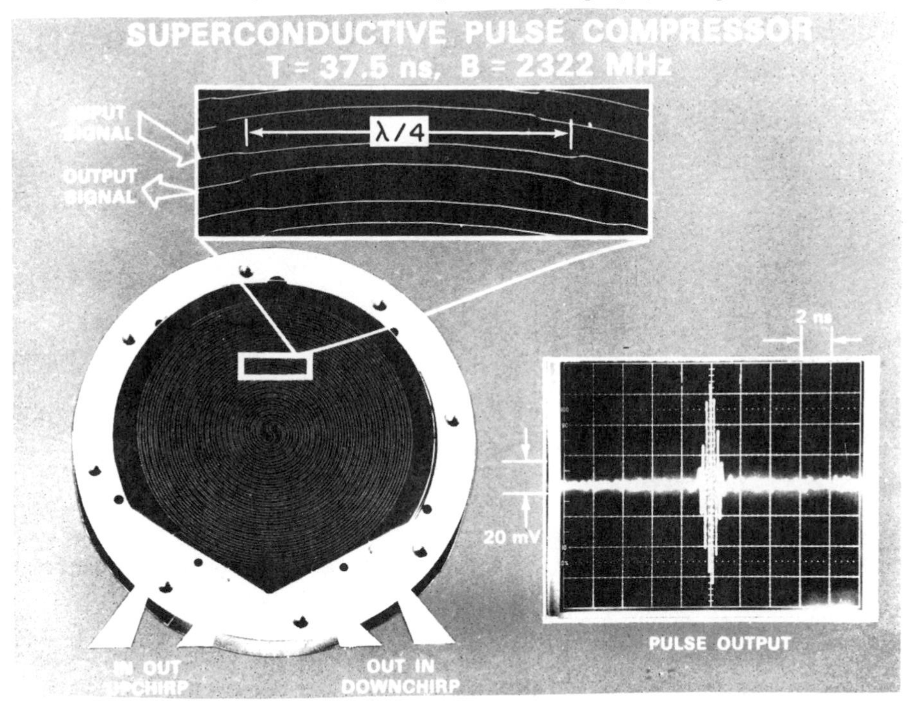

Fig. 10.35. Superconductive pulse compressor. Wafer containing two parallel superconductive transmission lines with $\lambda/4$ coupling sections between the lines. Result of compressing a "chirped" signal. (From R. W. Ralston [50].) © 1985 — IEEE

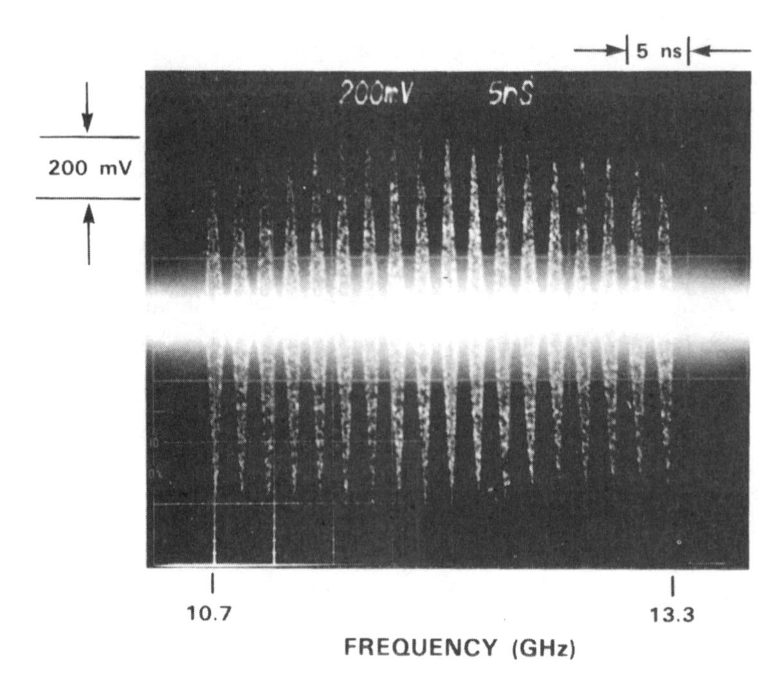

Fig. 10.36. Superconductive Chirp-Transform Spectrum Analyzer. Input signal: 10.7 to 13.3 GHz at -5dBm in 150 MHz increments (from [51]). © 1989 — IEEE

selected positions the lines are brought close enough together to cause the driven line to couple energy into the second line. Constructing this coupling section to be a quarter wavelength long results in the maximum transfer of voltage to occur at that frequency. These "taps" are usually chosen so that the lower frequency ones are positioned further down the line than those corresponding to higher frequencies. Thus, an input frequency spectrum can be linearly time delayed in its passage through the input–output transmission lines. If the input spectrum is itself linearly swept in time with the lowest frequency first such that the time spread matches the time dispersion of the line, an output compressed pulse will appear (Fig. 36). The line length and its quality determine the frequency resolution and bandwidth of the device. Lines 30 feet long (approximately 75 nanoseconds round trip delay) have been built and demonstrated using Nb metallurgy.

In considering applications for high temperature superconduc-

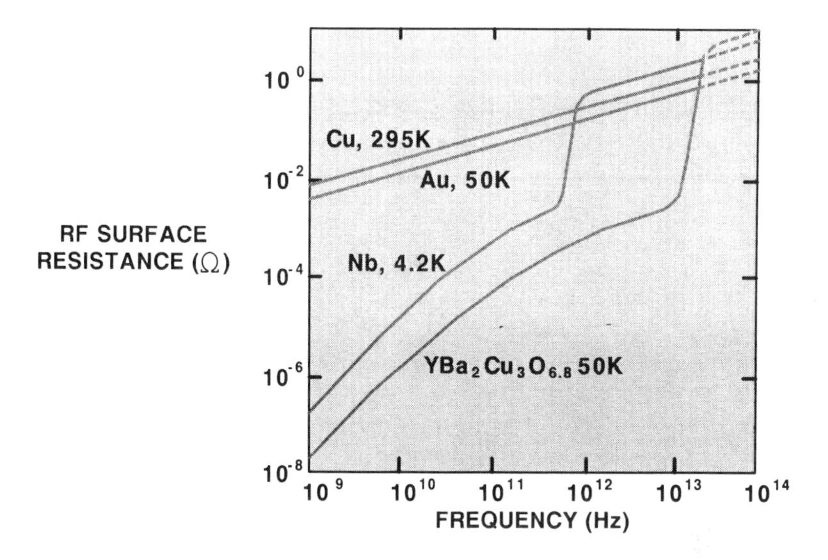

Fig. 10.37. RF surface resistance. Calculated surface resistance of YBCO based upon a scaling from low temperature superconductivity theory. Niobium curve is both calculated and observed; copper and gold are observed values. (From [51]).

tors such "chirp transform" filters are very attractive candidates. They require no active superconductive devices, only high quality lithographically determined films with a very low loss at GHz frequencies. In principle these goals are achievable. In practice, the substrates, insulators, and processing represent formidable "challenges". Preliminary estimates of the surface resistances versus frequency of these new materials project into a much superior performance over that of the present low temperature metals (if the physics scales and the usual "engineering details" are overcome). Figure 37 depicts the surface resistance of both superconductive and normal materials. The curve representing YBCO is obtained under the assumption of scaling from the well-established behavior of low temperature superconductive materials (achieved after much painstaking work).

10.11. Conclusion

The impact of high temperature superconductive materials will be felt in many important electronics systems. The active devices presently sought after are based upon our knowledge and understanding of tunneling phenomena, which background is very substantial. Reviewing the fundamentals of the phenomena involved is a mandatory exercise in order to not misuse or miss the transfer of that understanding to the new materials. It will also allow a determination of the expected and required characteristics, both experimentally and theoretically. In addition, there is always the opportunity to create a new device which can overcome a negative feature of the presently known ones, as well as the ability to invent new circuits.

The passive component applications are apparently more readily foreseen and more easily achievable. Here as well, successful use will depend upon a clear appreciation of the fundamental properties of superconductivity. And again, one should assess these needs and benefits by using the presently available materials.

Acknowledgments

This summary review would not have been possible with the persistent help of Mr. Carl Gardner, and the patience of Professor Jeffrey Lynn.

References

[1] D. A. Buck, *Proc. IRE* **44**, 482 (1956).
[2] B. D. Josephson, *Phys. Lett.* **1**, 251 (1962).
[3] I. Giaever, *Phys. Rev. Lett.* **5**, 147 (1960); *ibid.* pg. 464.
[4] J. G. Bednorz and K. A. Muller, *Z. fur Phys.* **64**, 189 (1986).
[5] H. Takogi, et al., *Jpn. J. Appl. Phys. Lett.* **26**, 189 (1986).
[6] C. W. Chu, P. H. Hor, R. L. Meng, L. Gao, Z. J. Huang, and Y. Q. Wang, *Phys. Rev. Lett.* **58**, 405 (1987).
[7] H. Maeda, et al., *Jpn. J. Appl. Phys. Lett.* **27**, Part II, L209 (1988).
[8] Z. Z. Sheng and A. M. Hermann, *Nature* **332**, 55 (1988); *ibid,* pg. 138.
[9] F. B. Silsbee, *J. Wash. Acad. Sci.* **6**, 597 (1916).
[10] J. W. Bremer, *Superconductive Devices*, (McGraw Hill, New York, 1962), pg. 17.
[11] W. C. Stewart, *Appl. Phys. Lett.* **12**, 277 (1968).
[12] D. C. McCumber, *J. Appl. Phys.* **39**, 3113 (1968).
[13] H. H. Zappe, *Advances in Superconductivity: Josephson Computer Technology*, edited by B. Deaver and J. Ruvalds, *NATO Adv. Sci. Inst.* **100**, pg. 51 (Plenum, New York, 1983).
[14] J. W. Bremer, *Superconductive Devices*, (McGraw Hill, New York, (1962), pg. 58.
[15] Reference [14], pg. 66.

[16] D. Shoenberg, *Superconductivity*, (Cambridge University Press, Cambridge, 1952), pg. 224.
[17] J. Clarke, *Phys. Rev. Lett.* **21**, 1566 (1968).
[18] S. Shapiro, et al., *Rev. Mod. Phys.* **36**, 223 (1964).
[19] M. Koyanagi, et al., *Future Trends in Superconductive Electronics*, edited by B. S. Deaver, C. M. Falco, J.H. Harris and S. A. Wolf (A. I. P., New York), pg. 187.
[20] M. T. Levinsen et al., *Appl. Phys. Lett.*, **32**, 776 (1977).
[21] R. L. Kautz, et al., *IEEE Trans. Mag.* **MAG-23**, 883 (1987).
[22] C. Hamilton, Private Communication.
[23] J. Matisoo, *IBM J. Res. & Dev.* **24**, 113 (1980).
[24] S. M. Faris, et al., *IBM J. Res. and Dev.* **24**, 2 (1980).
[25] S. Hasuo, *IEEE Trans. Mag.* **25**, 740 (1989).
[26] J. E. Zimmerman, et al., *J. Appl. Phys.* **41**, 1572 (1970).
[27] J. Clark and R. H. Koch, *Science* **242**, 217 (1988).
[28] J. E. Zimmerman, et al., *Jpn. J. Appl. Phys.* **26**, Suppl. 3, 2125 (1987).
[29] H. Nakane, et al., International Superconductivity Conf. (ISEC-87), Extended Abstracts, (Tokyo, Japan, 1987), pg. 411.
[30] J. P. Hurrell, et al., *IEEE Trans. Elect. Dev.* **ED-27**, 1887 (1980).
[31] W. Tsang and T. Van Duzer, *J. Appl. Phys.* **46**, 4573 (1975).
[32] H. H. Zappe, *IEEE Trans. Mag.* **MAG=13**, 41 (1977).
[33] T. Gheewala, *IBM J. Res. & Dev.* **24**, 130 (1980).
[34] S. Kotani, et al., *IEEE Trans. Elect. Dev.* **ED-33**, 379 (1986).
[35] R. C. Jaklevic, et al., *Phys. Rev. Lett.* **12**, 159 (1964).
[36] J. Clarke, et al., *J. Low Temp. Phys.* **25**, 99 (1976).
[37] J. Clarke and R. Koch, *Science* **242**, 217 (1988).
[38] J. P. Hurrell and A. H. Silver, *Future Trends in Superconducting Electronics*, *AIP Conf. Proc.* **44**, 437 (1978).
[39] See ref. [30].
[40] C. A. Hamilton, *IEEE Trans. Mag.* **MAG 19**, 1291 (1983).
[41] S. Faris, *Appl. Phys. Lett.* **36**, 1005 (1980).
[42] D. B. Tuckerman, *Appl. Phys. Lett.* **36**, 1008 (1980).
[43] P. Wolf, *IEEE Trans. Mag.* **21**, 226 (1985).
[44] TDR/Oscilloscope PSP-750, Hypres Inc., 500 Executive Blvd, Elmsford, N.Y., 10523.
[45] R. S. Withers, et al., *IEEE Trans. Mag.* **MAG 19**, 480 (1983).
[46] S. Kotani, et al., *IEEE IEDM Technical Digest*, **865** (1987).
[47] C. A. Hamilton and F. L. Loyd, *IEEE Elect. Dev. Lett* **EDL-3**, 335 (1982).
[48] P. Wolf, B. J. Van Zeghbroeck, and U. Deutch, *IEEE Trans. Mag.* **21**, 226 (1985).
[49] R. F. Pease and O. K. Kwon, *IBM J. Res. and Dev.* **32**, 643 (1988).

[50] R. W. Ralston, *IEEE Trans. Mag.* 21, 181 (1985).
[51] Proc. IEEE, Aug., 1989 (to be published).

General References

John W. Bremer, *Superconducting Devices*, (McGraw Hill, New York, 1962).

Josephson Computer Technology, NATO Advanced Study Institute Series, Vol. 100, edited by B. Deaver and J. Ruvalds (Plenum, New York, 1983).

Superconductor Applications: SQUIDS and Machines. NATO Advanced Study Institute Series, Vol 21, edited by Brain B. Schwartz and Simon Foner (Plenum, New York, 1977).

T. Van Duzer and C.W. Turner, *Principles of Superconductive Devices and Circuits* (North Holland, New York, 1981).

Chemical Formula Index

Subject Index

T

Permissions

Springer-Verlag would like to thank the original publishers and authors who granted permission for the use of their figures and tables in this book. The following contains the credit lines for those figures and tables.

[5.1] Reprinted from *Physics Review Letters* **58,** © 1987 by L.F. Mattheis.

[5.2] Reprinted from *Physics Reviews B* **35,** © 1987 by W.E. Pickett et al.

[5.3] Reprinted from *Physics Letters A* **120,** © 1987 by J.J. Xu et al.

[5.5] Reprinted from *Novel Mechanisms of Superconductivity,* © 1987 by Plenum Press.

[5.6] Reprinted from *Novel Mechanisms of Superconductivity,* © 1987 by Plenum Press.

[5.7] Reprinted from *Physical Review Letters* **60,** © 1988 by J. Zaanen et al.

[5.9] Reprinted from *Physical Review B* **37,** © 1988 by T.C. Leung et al.

[5.11] Reprinted from *Physical Review B* **37,** © 1988 by J.C. Fuggle et al.

[6.2] Reprinted from *Physical Review B* **35,** © 1987 by T.P. Orlando et al.

[6.3] Reprinted from *Physical Review Letters* **59,** © 1987 by M. Gurvitch.

[6.4] Reprinted from *Novel Superconductivity*, © 1987 by Plenum Press.

[6.5] Reprinted from *Novel Superconductivity*, © 1987 by Plenum Press.

[6.7] Reprinted from *International Journal of Modern Physics* **1,** © 1987 by Workd Scientific Publishing Company.

[6.8] Reprinted from *Physical Review B* **35,** © 1987 by T.P. Orlando et al.

[7.2] Reprinted from *Physical Review Letters* **61,** © 1988 by J.B. Torrance et al.

[7.5] Reprinted from *Physical Review Letters* **59,** © 1987 by M. Gurvitch and A.T. Fiory.

[7.15] Reprinted from *Phonons*, © 1972 by Flammarion Sciences, Paris.

Table 7.2 Reprinted from *Superconductivity*, © 1969 by Marcel Dekker Inc.

[8.3] Reprinted from *Journal of Applied Physics* **64,** © 1988 by K.B. Lyons and P.A. Fleury.

[8.10] Reprinted from *Physical Review Letters* **59,** © 1987 by G. Shirane et al.

[9.1] Reprinted from *Physical Review Letters* **60,** © 1988 by A. Aharony et al.

[10.3] Reprinted from *Superconductivity*, 2nd edition, © 1952 by Cambridge University Press.

[10.5] Reprinted from *Superconductive Devices* by J. W. Bremer, © 1962, by McGraw-Hill Book Company.

[10.6] Reprinted from *Superconductive Devices* by J. W. Bremer, © 1962, by McGraw-Hill Book Company.

[10.7] Reprinted from *Superconductive Devices* by J. W. Bremer, © 1962, by McGraw-Hill Book Company.

[10.9] Reprinted from *IEEE IEDM Technical Digest* December 1986, © 1986 by IEEE.

[10.10] Reprinted from *Journal of Applied Physics* **39,** © 1968 by S. McCumber.

[10.11] Reprinted from *Advances in Superconductivity*, © 1983 by Plenum Press.

[10.13] Reprinted from *Applied Physics Letters* **31,** © 1977 by Levinson et al.

[10.14] Reprinted from *IEEE Trans. Mag.* **23,** © 1987 by IEEE.

[10.18] Reprinted from *IBM Journal of Research and Development,* © 1980 by International Business Machines Corporation.

[10.19] Reprinted from *Advances in Superconductivity,* © 1983 by Plenum Press.

[10.24] Reprinted from *IBM Journal of Research and Development,* © 1980 by International Business Machines Corporation.

[10.25] Reprinted from *IBM Journal of Research and Development,* © 1980 by International Business Machines Corporation.

[10.26] Reprinted from *IEEE IEDM Technical Digest* **865,** © 1987 by IEEE.

[10.27] Reprinted from *IEEE Elec. Dev. Lett.* **3,** © 1982 by IEEE.

[10.29] Reprinted from *IEEE Trans. Mag.* **21,** © 1985 by IEEE.

[10.30] Reprinted from *IEEE Trans. Mag.* **21,** © 1985 by IEEE.

[10.32] Reprinted from *Advances in Superconductivity,* © 1983 by Plenum Press.

[10.33] Reprinted from *IBM Journal of Research and Development,* © 1988 by International Business Machines Corporation.

[10.34] Reprinted from *IBM Journal of Research and Development,* © 1988 by International Business Machines Corporation.

[10.35] Reprinted from *IEEE Trans. Mag.* **21,** © 1985 by IEEE.

[10.36] Reprinted from *Proceedings of IEEE* **77** (to be published), © 1990 by IEEE.

[10.37] Reprinted from *Proceedings of IEEE* **77** (to be published), © 1990 by IEEE.

Table 10. Reprinted from *IEEE Trans. Mag.* **25,** © 1989 by IEEE.